About Island Press

Island Press is the only nonprofit organization in the United States whose principal purpose is the publication of books on environmental issues and natural resource management. We provide solutions-oriented information to professionals, public officials, business and community leaders, and concerned citizens who are shaping responses to environmental problems.

Since 1984, Island Press has been the leading provider of timely and practical books that take a multidisciplinary approach to critical environmental concerns. Our growing list of titles reflects our commitment to bringing the best of an expanding body of literature to the environmental community throughout North America and the world.

Support for Island Press is provided by the Agua Fund, The Geraldine R. Dodge Foundation, Doris Duke Charitable Foundation, The Ford Foundation, The William and Flora Hewlett Foundation, The Joyce Foundation, Kendeda Sustainability Fund of the Tides Foundation, The Forrest & Frances Lattner Foundation, The Henry Luce Foundation, The John D. and Catherine T. MacArthur Foundation, The Marisla Foundation, The Andrew W. Mellon Foundation, Gordon and Betty Moore Foundation, The Curtis and Edith Munson Foundation, Oak Foundation, The Overbrook Foundation, The David and Lucile Packard Foundation, Wallace Global Fund, The Winslow Foundation, and other generous donors.

The opinions expressed in this book are those of the author(s) and do not necessarily reflect the views of these foundations.

THE
LAW AND
POLICY
OF
ECOSYSTEM
SERVICES

THE
LAW AND
POLICY
OF
ECOSYSTEM
SERVICES

J. B. Ruhl · Steven E. Kraft
Christopher L. Lant

ISLANDPRESS

Washington · Covelo · London

Library of Congress Cataloging-in-Publication Data

Ruhl, J. B.
The law and policy of ecosystem services / by J. B. Ruhl, Steven E. Kraft, and Christopher L. Lant.
    p. cm.
Includes bibliographical references.
  ISBN-13: 978-1-55963-094-8 (cloth : alk. paper)
  ISBN-10: 1-55963-094-9 (cloth : alk. paper)
  ISBN-13: 978-1-55963-095-5 (pbk. : alk. paper)
  ISBN-10: 1-55963-095-7 (pbk. : alk. paper)
   1. Environmental economics. 2. Ecosystem management. 3. Environmental protection—Decision making.   4. Biodiversity conservation—Law and legislation.   I. Kraft, Steven E. II. Lant, C. L. III. Title.

 HC79.E5R84 2007
 333.7--dc22
                                        2006036603

British Cataloguing-in-Publication Data available.

Book design by Brighid Willson

Printed on recycled, acid-free paper

Manufactured in the United States of America
09 08 07 06 05 04 03 02   8 7 6 5 4 3 2 1

Contents

Preface

This project has its origins in a conference on ecosystem services that Gretchen Daily of the Stanford Biology Department, Geoff Heal of the Columbia Business School, and Peter Raven, Director of the Missouri Botanical Gardens, organized at the Gardens in 1998. Having become intrigued by the concept of ecosystem services, which at the time was still relatively new even within ecological economics, the three of us eagerly attended and immediately noticed that, besides J. B., only one other lawyer was present in the audience of more than a hundred. The room was filled with ecologists, economists, and representatives from other social and physical sciences, but the contingent from law was conspicuously and troublingly thin. Law, after all, eventually has to enter the picture for ecosystem services to be put into operation as a meaningful policy driver. We left the conference thinking that a top-to-bottom exploration of the law and policy of ecosystem services was in order.

We hammered out an initial outline of this book around a sandwich shop table in Carbondale, Illinois, not long after the conference, but its scope and structure have gone through many iterations since then. The other lawyer present at the Missouri Botanical Gardens conference, Jim Salzman of Duke University School of Law, was of like mind about the importance of getting law on board, and toward that end had received a grant from the U.S. Environmental Protection Agency to examine opportunities for using ecosystem service values in decision making under then existing laws and regulations. Jim invited J. B. to join the grant team. Jim also spent a sabbatical year at Stanford in 2001–02, as did Geoff Heal, making Stanford the center of gravity at the time for interdisciplinary work on ecosystem services. Gretchen and Jim, along with Buzz Thompson of Stanford Law School, he organized a meeting of all involved at Stanford in 2000, the papers from which were published in 2001 in the *Stanford Environmental Law Journal.*

Although a milestone, the Stanford conference was of necessity exploratory, with all participants agreeing that much work lay ahead. In particular, it would be important for law to work closely with economics, ecology, geography, and other relevant disciplines, not only to understand where they had taken the theoretical research and practical applications, but also to ensure that those disciplines in turn appreciate the nature and limits of legal institutions and instruments. We designed our book to build the basic framework for that interdisciplinary conversation.

Many have helped along the way. Much of this book constituted J. B. Ruhl's dissertation in geography at Southern Illinois University Carbondale under the advisement of Chris Lant. The dissertation was also read and critiqued by Leslie Duram and Ben Dziegielewski of that department, Dan Tarlock of the Chicago–Kent School of Law, and Steven Kraft. Jim Salzman is owed special thanks, not only for what has already been mentioned, but also for his frequent collaboration with J. B. on the topic of ecosystem services in law and policy. In particular, he helped immensely in organizing the second conference on law and ecosystem services, held at Florida State University in the spring of 2006. Others who have played an instrumental role in shaping our thoughts on the topic include Buzz Thompson, Rob Fischman, Dan Tarlock, Robert Costanza, and Rudolf de Groot. We also benefited greatly from comments on early drafts by Federico Cheever and Robin Kundis Craig, and from the able research talents of Adam Schwartz, Ali Stevens, Bruce Hall, and Sethuram Soman.

In the personal support department, thanks of course go to our respective families, who have endured many years of talk about ecosystem services and "the book." Also, special thanks go to Annette House for serving as Chris's reader following eye surgery; her kind help kept this project on track. Our respective institutions, Florida State University and Southern Illinois University Carbondale, provided extensive support for our research. Finally, we thank our publisher, Island Press, for leading the way on the topic of ecosystem services starting with Gretchen Daily's groundbreaking book *Nature's Services*, and later Geoff Heal's *Nature and the Marketplace*. We are proud to follow in that lineage.

J. B. Ruhl
Steven Kraft
Christopher Lant

Introduction

Deep in the north Georgia hills, just a few hundred feet off the southernmost reaches of the Appalachian Trail, a small mountain brook marks the headwaters of the Chattahoochee River. The river meanders its way out of the Chattahoochee National Forest, through the quaint Bavarian-style town of Helen. From there the water soon empties into Lake Lanier, a huge reservoir north of Atlanta impounded in the 1940s by the U.S. Army Corps of Engineers' Buford Dam. Cool water spills out below the dam and works its way toward Atlanta, brushing by just north of that major southeastern city and then drifting westward toward Alabama. At West Point Lake Dam, the river veers more sharply southward and becomes the boundary between Alabama and Georgia. It passes by Columbus, Georgia, on its east bank, then later the Alabama plantation town of Eufaula. At Sneads, Alabama, where Lake Seminole is impounded, it joins the Flint River, which has its origins near the south side of Atlanta, and crosses into Florida. There it becomes the Apalachicola River, a ribbon of water slicing across a sparsely populated stretch of the Florida Panhandle and emptying into the Gulf of Mexico at the city of Apalachicola. This collection of rivers, over 750 river miles in all, makes up the Apalachicola–Chattahoochee–Flint River system, or "the ACF."[1]

The ACF drainage basin covers almost 20,000 square miles, within which one can find starkly different cultures and communities. At its northern reaches, for example, lie the modern boom city of Atlanta and its water playground, Lake Lanier. In addition to supplying residential and industrial water to urban Atlanta, Lake Lanier's 38,500 tree-rimmed surface acres are a boater's and retiree's heaven. Its shores are dotted with marinas, million-dollar homes, resort hotels, and golf courses. Houseboats as long as 120 feet are not uncommon. Its recreational economy generates billions of dollars in revenue annually. All of this depends, of course, on there being water in Lake Lanier.

Water is important, as well, to the agricultural communities that dot southwestern Georgia. The twenty-six southwest Georgia counties are dominated by agricultural economies, generating $1.6 billion in agricultural product revenue annually. These agricultural operations also use 325 million gallons of water per day, mostly for crop irrigation, and are projected to use 570 million gallons per day by 2050. Most of the irrigation water is drawn not from lakes or rivers but from the Floridan aquifer, a huge, highly productive limestone aquifer stretching from southern Georgia well into Florida. The relation between withdrawals from the aquifer and the chief surface-water resource in the area, the Flint River, is not fully understood.

At the opposite end of the ACF watershed from Lake Lanier, 544 miles from the headwaters of the Chattahoochee, lies the Apalachicola Bay, home to the most productive oyster beds in the nation and the center of a highly productive estuary. Life is so good in the bay that its oysters grow faster than anywhere on Earth (the bay supplies 10 percent of the nation's oysters) and many species of fish found in the Gulf of Mexico spend part of their lives there. The Apalachicola River itself, plus its floodplain of over 2,400 square miles, is home to one of the highest diversities of freshwater fish, amphibians, and crayfish in the nation. The Nature Conservancy lists the bay as one of the hottest biodiversity hotspots in the world. But life there is a far cry from the estates of Lake Lanier. A small but sustainable oyster and fishing industry has been based in Apalachicola for decades, but most oyster harvesters and fishermen live week to week in fairly hard-scrabble circumstances. Their very livelihoods depend on one thing above all else—water flowing out of the mouth of the Apalachicola River. But not just any flow. It has to be the right amount at the right time— the "natural flow regime" upon which the life cycles of many species in the bay depend. By and large, that's all the city of Apalachicola wants from the ACF system—water at the end of the pipe the way nature intended it to be delivered. The people there have no aspirations of withdrawing water to launch another Atlanta. There is but one traffic light in the entire county!

Alas, although Lake Lanier party boats, southern Georgia farm tractors, and Apalachicola Bay oysters are unlikely ever to cross paths, they are intricately connected players in battle over the fate of the water that courses through the landscape within which their respective domains are found. The chains that link these three worlds began forming in the 1940s, when Congress charged the U.S. Army Corps of Engineers with "taming" the Chattahoochee by erecting a series of major dams designed to impound water to meet a variety of human needs, mainly navigation. The ready supply of water proved irresistible to residential and industrial development throughout the region. Population growth in the ACF basin boomed, concentrated in Atlanta. The area became one of the

Figure o.1. Map of ACF basin and adjacent Alabama–Coosa–Tullapoosa (ACT) Basin showing major urban areas and reservoirs. (From Tri-State Water Commission.)

hottest regional economies in the nation. A hotspot of both biological diversity and economic vitality—the ACF had it all.

But trouble was on the horizon. A series of record droughts in the 1980s illustrated the limits of ACF water. In 1989, Georgia proposed diverting more water from the Corps' impoundments to quench Atlanta's thirst. Georgia then applied to the Corps to add yet another major impoundment in the state—this one on the Tallapoosa River just 5 miles from where it crosses into Alabama. Alabama, fearing that less water flowing into the state and along its boundary with Georgia would mean less potential for its own economic growth, immediately initiated litigation to halt both plans under a variety of federal laws, most prominently the National Environmental Policy Act. Florida, fearing that less water emptying into Apalachicola Bay could damage the bay ecosystem and the oyster and recreational fishing industries it supports, soon joined the fray.[2]

Western states have resolved their many interstate water allocation battles in three ways: (1) litigation before the U.S. Supreme Court under its original jurisdiction to resolve disputes between the states; (2) congressionally mandated allocation based on federal authority over interstate commerce; and (3) agreement between the states authorized through an interstate compact approved by Congress.[3] Because water disputes of any substantial magnitude have been rare in the East, these methods have been seriously field-tested in eastern settings only a few times, and not at all in recent history. But the ACF dispute was sizing up to be a western-style water war, with serious potential to head to the Supreme Court if the states could not agree. To avoid that high-stakes proposition, in 1992 the three states entered into negotiations that led in 1997 to an interstate compact to negotiate some more. The negotiations were protracted, focusing on each state's model of river flow conditions experienced under an array of climate and population projections. Unable to reach quick consensus, the states extended their self-imposed deadlines numerous times, hired respected mediators, and employed the best legal and technical experts money could buy, but to no avail. Negotiations broke down in 2003, and the states threatened to return to the courts.

When making its case, not surprisingly, Georgia has pointed to Atlanta—its population of 3 million, its booming economy, and the likelihood that both will continue to grow—as justification for it demanding a secure and increasing supply of ACF water (Thornley 2005). Accordingly, Georgia's primary negotiating position has been that it can guarantee delivery of minimum flows across the border to Florida, but no more. Florida, by contrast, points to the biological needs of oysters and other species in the Apalachicola River and Bay to press its case that Georgia should control its water appetite and guarantee sufficient flows to keep the downstream ecosystems healthy.

It is less than clear how the Supreme Court's existing interstate water dispute jurisprudence would balance these concerns. The basic theme of the Court's approach is to divide the interstate water so as to balance benefits and injuries with a sense of fairness to all states involved in the dispute. This doctrine of "equitable apportionment" takes into account a mix of factors, including state water law, economic impacts, climate conditions, available water use conservation measures, and the overall impact of diversions on existing uses (Tarlock 1985).[4] The doctrine has long been employed in the West but has only occasionally been used to resolve disputes between the eastern states (Abrams 2002). In the East or the West, however, no case has presented issues quite like those the ACF case would pose (Moore 1999). Usually the Court is called upon to decree an annual amount or minimum flow to which each state is entitled.

In the ACF case, however, Florida presumably would argue to the Court that, primarily for ecological reasons (albeit with incidental economic impacts), upstream states must deliver a particular "natural" flow regime that fluctuates throughout the year.[5]

Although the Court's equitable apportionment jurisprudence certainly leaves room for incorporating ecological factors into the analysis, the case precedents do not suggest how the Court would do so. The Court has suggested that "evidence of environmental injury" could play a role in balancing the water allocation equities between states,[6] and has even ruled that the doctrine applies not only to water but to allocation of resources that run within interstate waters, such as salmon and other anadromous fish.[7] And the Court has held that the doctrine imposes on states an affirmative duty to take reasonable steps to conserve and even to augment natural resources within their borders for the benefit of other states.[8] Yet, when downstream states claim injury from upstream diversions, the Court generally requires the downstream state to prove by clear and convincing evidence some real and substantial injury or damage.[9] The Court has yet to explain in applied terms what form and magnitude of environmental injury would satisfy that standard.

Florida and Georgia thus would pose a straightforward question, the answer to which is exceedingly complex: What is the injury to Florida that the Court should measure? No one disputes Georgia's claim that Lake Lanier and Atlanta form an economic engine of considerable magnitude, or that they make Apalachicola and its oyster industry look puny by comparison. On the other hand, no one disputes that oysters in Apalachicola Bay find natural regime flows valuable—indeed, indispensable for their survival. These are the conventional currencies of environmental policy, the way we have framed issues for decades. On the one side of the ledger are human economies expressed in the hard cash terms of prices; on the other side are ecological features expressed in the language of science. We count beer sales and oyster landings, water levels and wetland acres. Which side prevails may depend on political power, financial clout, or a judge's pen.

Yet, in whatever forum we find these interests in dispute—a congressional committee room, corporate office, or judge's chambers—seldom do we find the legal context counting all that matters. To be sure, resource commodities such as oysters matter, as do commercial products such as boats and human-supplied services such as fixing a farm tractor. These are the stuff of human economies. But there is more that is of value to humans than these, more that should be factored in the marketplace and respected in the law, but which is not. Watersheds like the one resting above Lake Lanier, for example, capture sediment and other pollutants that may foul the lake waters if not removed by this natural process. Riparian habitat like that found along the Apalachicola River regulates water temperature to the benefit of downstream species. And wetlands in the ACF floodplain protect adjacent areas from the hazards of flooding.

Humans would miss these benefits of "ecosystem services" if they were suddenly to disappear. Indeed, often we find it cost efficient to "produce" ecosystem services by replicating natural ecosystem structures, as in the case of "constructed wetlands," which have long been built and employed to remove nutrients and sediments from polluted water sources such as municipal wastewater and agricultural runoff (Kadlec and Knight 2004; Olson 1992; Steer et al. 2003; U.S. Environmental Protection Agency 2004). Yet, ecosystem service values derived directly from nature show up practically nowhere in our economy as it is structured, and much less so in the law supporting that structure. For example, wetlands, it turns out, also provide protection against the heat-radiation effect—heat radiating away from the ground on dry winter nights rapidly lowers soil temperatures and freezes the moist root zone—which is of value for preventing crop freezes, but one searches in vain for any recognition of this value in financial or policy marketplaces (Marshall et al. 2003). That ecosystem services have value is indisputable; however, what that value is and how to account for it in our day-to-day economic and legal decisions are far less clear.

The concept of ecosystem services is not new, but it is sufficiently recent that it is yet to be fully developed into coherent policy terms, and surely not yet into hard law to be applied. Mooney and Ehrlich (1997) trace references to "services" in connection with ecosystems as far back as 1970, but Walter Westman (1977) was the first to attempt to assign numbers to the values of what he called "nature's services," relying on the postulated technology costs of replacing or repairing impaired ecosystem functions. Soon thereafter, in a little-noticed article, Edward Farnworth and colleagues (1981) outlined one of the earliest comprehensive frameworks for considering the value of services provided by natural ecosystems. Edward O. Wilson later made ecosystem services a centerpiece in his epic study of biodiversity, *The Diversity of Life (1992),* and by the mid-1990s the discipline of ecological economics was well under way, with the journal by that name starting in 1989 and a full-length book on the topic (Costanza 1991) breaking the path for more to follow. A research team led by Robert Costanza grabbed national media headlines in 1997 with their estimate that global ecosystem service values were over $30 trillion (Costanza et al. 1997), and later that year the highly influential book *Nature's Services* (Daily 1997) established the ecological basis for ecosystem service theory in many different ecosystem settings. And with their publication of *Ecological Economics,* Herman Daly and Joshua Farley (2004) have firmly planted the discipline, including its focus on ecosystem service values, on the university curriculum landscape.

Nevertheless, despite a few prominent examples reported in the literature

(Daily and Ellison 2002; Heal 2000; Thompson 2000), practical applications of ecosystem service valuation theory remain few and far between. Like any estuary, for example, the vast commercial and recreational fishing economy in the eastern Gulf of Mexico depends on the integrity of the ACF flow regime, and the flood control and other benefits of intact riparian habitat along the river depend on that habitat remaining there (Mattson 2002). Indeed, recent work suggests that ecosystem service values provided in just the floodplain and estuary of the Apalachicola River in Florida could exceed $5 billion annually (Garrett 2003). In other words, immense economic benefits accrue *to humans* by maintaining the ACF under its natural flow regime conditions. Yet there is no mention of these ecosystem service values in any Corps study of the ACF, or in any report of ACF Compact negotiations, or, certainly, in any Lake Lanier Chamber of Commerce publication. Nor, for that matter, do we find reference to ecosystem services on any page of the Supreme Court's water allocation jurisprudence that may come to bear on the fate of the ACF.

The ACF is not alone in this respect. It is just one of many cases revealing the systematic failure of the legal framework of natural resource decision making to account for ecosystem service benefits. Other prominent examples include the following:

- When conducting a cost–benefit analysis of the U.S. Forest Service's 2001 proposed National Forest Management Act rule to limit future uses of large roadless areas of national forests—a total area of 60 million acres of public forestland and grasslands—the Office of Management and Budget (OMB) concluded that quantifiable costs in the form of lost jobs and forgone commodity extraction would exceed $180 million annually, but that the only quantifiable benefits would be $219,000 annually in the form of the saved costs of reduced road maintenance. OMB simply observed that "a variety of other nonquantifiable benefits" may accrue from the rule, "such as maintenance of air and water quality, recreational opportunities, wildlife habitat, and livestock grazing" (Office of Management and Budget 2002, 110). In other words, neither the Forest Service nor OMB considered the *value* of ecosystem services associated with 60 million acres of undisturbed forested lands, instead dismissing them as "nonquantifiable" and thus not counting toward the cost–benefit analysis (Ackerman and Heinzerling 2004).
- The National Environmental Policy Act (NEPA), one of the nation's premier expressions of environmental goals, requires each federal agency to study the effects of any major actions it carries out, funds, or authorizes and to provide the public an opportunity to review and comment on its published report of the study, known as an environmental impact statement (EIS). The Council on Environmental Quality (CEQ) has promulgated general regulations other

federal action agencies must follow in fulfilling their NEPA duties, including the scope and content of an EIS. However, nowhere in those regulations, or in the more particularized regulations each agency has adopted to implement the CEQ guidelines, are impacts to ecosystem service values required to be evaluated (Fischman 2001).

- Under Section 404 of the Clean Water Act, the U.S. Army Corps of Engineers administers a regulatory program that protects waters of the United States, including wetlands. Under this program the Corps has issued "wetland mitigation banking" guidelines that allow a developer intending to eliminate wetlands to compensate for that resource loss by purchasing "credits" from landowners who have created, enhanced, or restored wetland resources in large contiguous blocks. Yet nothing in the guidelines requires the Corps or the parties engaged in the "trade" of wetlands to consider the impact of the transaction on the delivery, location, and possible redistribution of ecosystem service values (Ruhl and Gregg 2001; Ruhl and Salzman 2006).

- The U.S. Fish and Wildlife Service (USFWS) must conduct a cost–benefit analysis of the impacts of designating the "critical habitat" of species listed as endangered or threatened under the Endangered Species Act. Although it has acknowledged that preservation of ecosystem service values is one benefit of protecting critical habitat, the agency has routinely refused to attempt to quantify those values in specific cases where it has proposed critical habitat designations (Millen and Burdett 2005; National Wildlife Federation 2004; U.S. Fish and Wildlife Service 2003).

There are many reasons why ecosystem services fail to be fully accounted for in decision-making settings as varied as these, but there are also many reasons why this should not be so. This book explores both sets of reasons. The primary objective of this project is to develop a framework for thinking about ecosystem services across their ecological, geographic, economic, social, and legal dimensions, and to evaluate the prospects of crafting a legal infrastructure that will help us build an ecosystem service economy as robust as the nation's economies for natural resource commodities, commercially manufactured products, and human-supplied services. To be sure, this will not be the first proposal to integrate ecosystem services into market economies. Geoffrey Heal is noteworthy among economists for making such a case in his book *Nature and the Marketplace* (2000), and a quickly growing body of journal articles does the same (see chapter 3). The United Nations Millennium Ecosystem Assessment project (2005) has moved the dialogue beyond academic discourse to concerted policy analysis. Yet proposals to date are largely conceptual in scope. It is one thing, for example, to postulate ecosystem service management districts with taxing and spending power (Heal et al. 2001), but quite another to sort out exactly how and where they would be established and invested with legal authority to act with respect to ecosystem services (Lant 2003; Ruhl et al. 2003).

In other words, the component that is least developed in the literature on ecosystem services is *the law,* particularly as it relates to property rights and governance institutions. While several authors have urged the need for foundational work in this field (Kysar 2001; Ruhl 1998; Salzman 1997), the ecological, geographic, economic, and social complexities of ecosystem services complicate any effort to forge such a body of law and policy. As Oliver Houck, one of the first lawyers to think about this problem, suggested in his early 1980s study of development in coastal Louisiana, law and policy have found it all too easy to ignore ecosystem services as much as economics had until then:

> The benefits from those uses that are damaging the area are measurable by the dollar. The values of the system in its natural state seem largely to defy measurement by this or any other standard and have therefore remained largely unmeasured and unaccounted for in individual decisions to build new canal systems, pipelines, and other developments. It is an easy frame of mind for developers and regulators to adopt. The more unmeasured the costs, the less one has to be concerned about them. (1983, 92)

Hence, that is the challenge this project undertakes—to take the discussion of ecosystem services out of the "easy frame of mind" and push it to the next level, at which serious and detailed law and policy implementation frameworks can be designed, tested, and implemented.

Part I starts by examining the *context* of ecosystem services through the lenses of three relevant disciplines: ecology (chapter 1), geography (chapter 2), and economics (chapter 3). Tremendous advancement has been made in the past decade toward improving our understanding of the ecological dynamics of ecosystem services, their geographic distribution across landscapes, and their economic value to human communities. But that improved understanding has pointed in most cases to the fact that ecosystem services are, by their very character, exceedingly complex in all three respects. Ecosystem services are not like other goods or services that move through our economy. They cannot be easily separated from their ecosystem bases, or moved around and delivered the way other raw materials or services are physically distributed. In short, ecosystem services, while clearly of tremendous value, are ecologically, geographically, and economically more complex than any other kind of commodity or service, which has made tapping into their value a challenge that has yet to be met.

The social and legal consequences of the complexity of ecosystem services are the subject of the chapters in part II, which provides a baseline for future work by examining the current *status* of ecosystem services in the law and society. First and foremost is the absence of any supportive system of property rights governing the production and use of ecosystem services (chapter 4), which renders them in many applications as public good resources subject to underprovision and overdepletion in the absence of some moderating influence. When property rights are as poorly designed as they are for ecosystem services, prescriptive regulations (chapter 5) and social norms (chapter 6) are often held out as the solutions to resource management problems. But here again the application of these institutional devices to ecosystem services has proven elusive. Although a consensus is building that ecosystem services hold tremendous values that we should seek to understand and incorporate into decision making about the environment, regulatory frameworks and social norms for efficiently managing ecosystem services have not materialized. The status of ecosystem services in law and society, in other words, is that they have none.

Part III introduces a series of nine empirical case studies that explore the causes and consequences of the lack of attention property rights, regulation, and social norms have given to natural capital and ecosystem service values. The case studies focus first on the application of ecosystem services to individual parcels of land (chapter 7) and to the hydrologic cycle (chapter 8). They then explore the realm of agricultural land use and watershed management through case studies of the Conservation Reserve Program (chapter 9) and the National Conservation Buffer Initiative (chapter 10) as important existing ecosystem service subsidy programs, the shift from crop-based (amber) subsidies to ecosystem service–based (green) subsidies in the United States and the European Union (chapter 11), and how these policies affect the economy and ecosystem service provision of a typical agricultural watershed (chapter 12). Part III then investigates the successes, failures, and potential of market-based instruments for encouraging investment in natural capital and the consequent delivery of ecosystem services in the realm of wetland mitigation banking (chapter 13) and tradable pollution permits (chapters 14 and 15).

Based on the foundational chapters in parts I and II and the lessons learned from the case studies in part III, part IV then forges an approach for the *design* of new law and policy for ecosystem services, working from the current baseline and taking into account the inherent limitations their ecological, geographic, and economic contexts present. The progression of the topics follows the choices that law and policy will have to make to put such an approach into action. First, it is essential to identify the important drivers of the existing status of natural capital and ecosystem services and to develop models of how they can be moved and the likely consequences of doing so (chapter 16). Policy choices then must confront the reality that taking more account of natural capital and ecosystem service values in natural resource decision making will not necessarily be a "win–win" for all stakeholders. Trade-offs are inevitable, and some people will be "winners" and others "losers" in the transition (chapter 17). Once policy is set, the appropriate instruments and institutions must be identified for policy implementation (chapter 18). In this sense, ecosystem services are likely to encounter the same tensions that environmental law in general has experienced as federal, state, and local governments, the courts, and interest groups jockey for position and authority. Only when all these choices are made in a cohesive, cogent institutional framework will the law and policy of ecosystem services have "arrived" and begun to fuse ecosystem services with resource commodities, manufactured products, and human-supplied services into a fully integrated decision-making framework for natural resources, one in which everything that matters is counted.

Ecosystem services are easy to take for granted until they are gone. As in the famous paradox of value that long puzzled economists, they have been more like water—essential for life, but so widely available they are easily obtained for free—than like diamonds, which are scarce and thus valuable despite having little practical use. But water in many parts of our nation is no longer so plentiful or so cheap. Similarly, as Gretchen Daily and Katherine Ellison (2002) put it, "ecosystem assets have the importance of water and are gradually acquiring the scarcity of diamonds as the human population and its aspirations grow" (11). One can only hope that long before the day comes when ecosystem services are as dear as diamonds, we will have formulated a law and policy of ecosystem services that allows us to manage them sustainably. To that end, we devote this work.

Part I The Context of Ecosystem Services

Law and policy depend on other disciplines to inform effective decisions about the appropriate institutions and instruments to use; hence a legal study of ecosystem services should not launch into analysis of policy failures and potential reforms without first building a foundation of the context of ecosystem services. To provide this context, part I examines ecosystem services through the lenses of three particularly relevant disciplines: ecology (chapter 1), geography (chapter 2), and economics (chapter 3). Tremendous advancement has been made in the past decade toward improving our understanding of the ecological dynamics of ecosystem services, their geographic distribution across landscapes, and their economic value to human communities. But that improved understanding has pointed in most cases to the fact that ecosystem services are, by their very character, exceedingly complex in all three respects. Ecosystem services are not like other goods or services that move through our economy. They cannot be easily separated from their ecosystem bases, or moved around and delivered the way other raw materials or services are physically distributed. In short, ecosystem services, while clearly of tremendous value, are ecologically, geographically, and economically more complex than any other kind of commodity or service, which has made tapping into their value a challenge that has yet to be met.

I   Ecology

It is tempting to overstate the case for ecosystem services, to try to find them everywhere simply because anywhere is in one or another ecosystem. But it is important not to confuse ecosystem *functions,* which are ubiquitous, with ecosystem *services,* which are the consequence of only some ecosystem functions. The critical difference between the two, and which makes the development of ecosystem services policy both complicated and controversial, is that ecosystem services have relevance only to the extent *human* populations benefit from them. They are purely anthropocentric. The ecology of ecosystem services, therefore, must be carefully defined in order to begin considering how to formulate a policy foundation for their management.

Ecosystems and Ecosystem Processes

Since Tansley's (1935) early description of the ecosystem as part of a continuum of physical systems in nature, decades of research and literature have been devoted to forging the concept into a scientific discipline (Brooks et al. 2002; Golley 1993). Modern ecologists describe ecosystems as the complex of organisms that appear together in a given area and their associated abiotic environment, all interacting through the flow of energy to build biotic structure and materials cycles (Blair et al. 2000; Millennium Ecosystem Assessment 2005). Ecosystems thus move and transform energy and materials through basic processes such as those listed by Virginia and Wall (2001):

Photosynthesis

Plant nutrient uptake

Microbial respiration

Nitrification and denitrification

Plant transpiration

Root activity

Mineral weathering

Vegetation succession

Predator–prey interactions

Decomposition

These and other ecosystem processes operate according to fundamental internal rules and constraints of physical and biotic systems. Energy transformation processes are essentially one-way flows, preventing reuse or recycling of the energy units. But nutrients can circulate through different components of an ecosystem, leading to what ecologists call nutrient cycles and nutrient pools. At its most fundamental level, ecology as a discipline is interested in describing and quantifying the factors that regulate energy transformation and nutrient cycling within an ecosystem as defined. And because these processes operate at many scales, ecological studies also take place at many scales. For example, photosynthesis can be measured and studied at scales ranging from the individual cell to the canopy of a forest ecosystem as defined. Often, therefore, it is as much a question of how to define an ecosystem as it is to understand how these processes work within it.

Ecosystem Functions

The process-based description of ecosystems has led to improved understanding of the functions ecosystems perform in natural settings. The transformation of energy and materials into vegetation structure, for example, provides habitat for other organisms. The decomposition of materials in the ecosystem builds soil structure. Each process under way in an ecosystem thus contributes to one or more of a set of functions associated with the ecosystem and with its relation to other ecosystems (Virginia and Wall 2001).

The same basic biological and chemical processes occur in all ecosystems, but different conditions yield different functional representations (Blair et al. 2000). It is like electronic circuitry—the same principles of electromagnetism apply in all cases, but different combinations of circuitry and voltage produce different functional applications. An inventory of just some of the functions typically associated with different ecosystem processes, and which we should expect to observe in different forms and magnitudes across ecosystems is provided in Table 1.1.

As this representation suggests, there is no one-to-one correspondence between ecosystem processes and ecosystem functions. In reality, many processes are needed to produce any of the defined functions. For example, a farm, which can be thought of as a highly modified and highly managed ecosystem, relies on biotic production, energy flow, decomposition, and nutri-

TABLE 1.1. Ecological processes and functions.

Biotic Production Processes	*Energy Flow Processes*	*Decomposition Processes*	*Nutrient Cycling Processes*
Providing prey	Enabling chemical reactions	Transforming and releasing gases and nutrients	Maintaining nutrient balance
Building habitat structure	Providing thermal habitat needs	Reducing debris buildup	Enabling energy transfer
Consuming nutrients	Regulating biological production	Building soil composition	Purifying water and soil

ent cycling to make possible its basic function of producing, say, corn. It is no different in the remote undisturbed depths of a rain forest. Hence, another key study theme of ecology is to improve our understanding of how the basic ecosystem processes work together to generate the functions vital to sustaining the ecosystem within its environment.

Ecosystem Structure and Natural Capital

Ecosystem functions contribute to the building of the ecosystem's physical structure, such as biomass (e.g., vegetation and wildlife) and abiotic resources (e.g., soil and water), which in turn supports the sustainability of the functions (Christensen et al. 1996; Daly and Farley 2003). Events that degrade ecosystem structure (e.g., overfishing in coral reef ecosystems) consequently disrupt the integrity of the associated ecosystem functions (Roberts 1995). These effects are important not only to the sustainability of the ecosystem but also to the sustainability of humans, given the importance of ecosystems to human well-being (Millennium Ecosystem Assessment 2003, 2005). This property—that ecosystem structure and functions provide for human needs and wants—is what makes ecology inevitably relevant to economics.

Ecologists thus analogize ecosystem structure to *capital* as that term is used in economic theory—the stock that possesses the capacity of giving rise to the flow of goods and services (Costanza et al. 1997; Ekins et al. 2003). Ecological capital, or "natural capital" as many ecological economists call it, consists of the ecosystem structure and functions that support the creation and flow of goods and services valuable to humans (Clark 1995; Costanza and Daly 1992; Daily and Dasgupta 2001). Other than in a totally artificial environment, such as a space station, "zero natural capital implies zero human welfare because it is not feasible to substitute, in total, purely 'non-natural' capital for natural capital"

(Costanza et al. 1997, 255). Yet, as with economic capital, we need not reach zero before we feel the effects of depreciating stock. As Daly and Farley summarize,

> [T]he structural elements of an ecosystem are stocks of biotic and abiotic resources (minerals, water, trees, other plants and animals), which when combined together generate ecosystem functions, or services. The use of a biological stock at a nonsustainable level in general also depletes a corresponding fund and the services it provides. Hence, when we harvest trees from a forest, we are not merely changing the capacity of the forest to create more trees, but are also changing the capacity of the forest to create more ecosystem services, many of which are vital to our survival. (2003, 106–107)

Another theme of ecology, therefore, is focused on understanding the impact of natural and anthropogenic events on the investment in and depreciation of natural capital in the form of ecosystem structure, and the consequent impact on the delivery of goods and services from the ecosystem (Deutsch et al. 2003; Ekins 2003; Ekins et al. 2003). Like our conventional economy, however, understanding cause and effect in the ecological economy is a horribly complicated undertaking given the complexity of the subject matter.

Ecosystems as Complex Adaptive Systems

The dynamic interactions of ecosystem processes, functions, and structural components have led many ecologists to describe ecosystems through the terms used in complex adaptive systems theory (Limburg et al. 2002), which provides a useful way of thinking about the difficulty of managing ecosystem services. Complex adaptive systems theory explores the behavior and properties of diverse, interconnected, autonomous agents (Holland 1995; Kauffman 1995). Systems composed of such agents—from immune systems to economies—are seen in physical, biological, or social contexts to generate feedback and feedforward loops among agents, through which the action of any one agent could affect many others, including the original actor. The aggregation of these feedback and feedforward loops produces the emergent behavior of dissipative system structure, which will inevitably exhibit dynamic nonlinear properties not found in or predictable from observation of any single agent in the system. Indeed, complex adaptive systems research focuses on the ways in which this emergent system behavior provides sustainability for the system as a whole by facilitating adaptation to external disturbances. On the other hand, the price of this adaptive capacity is constant change— a form of stable disequilibrium balanced between order and chaos (Kauffman 1995). Costanza sums up the difficulties of studying such systems:

> "Complex systems" are characterized by: (1) strong (usually nonlinear) interactions among the parts; (2) complex feedback loops that make it

difficult to distinguish cause from effect; (3) significant time and space lags; discontinuities, thresholds and limits, all resulting in (4) the inability to simply "add up" or aggregate small-scale behavior to arrive at large-scale results. (1996, 981)

It is no surprise that ecology has embraced complex adaptive systems theory for, as Simon Levin (1998, 431) has claimed, ecosystems are "prototypical examples of complex adaptive systems." Certainly the basic quality of complex systems exists in ecosystem dynamics, in that "we cannot understand ecosystems only by considering their separate components" (Bailey 1996, 16). John Holland, one of the leading figures in complex systems research, has explained the reasons why:

> Ecosystems are continually in flux and exhibit a wondrous panoply of interactions such as mutualism, parasitism, biological arms races, and mimicry. . . . Matter, energy, and information are shunted around in complex cycles. Once again, the whole is more than the sum of its parts. Even when we have a catalogue of the activities of most of the participating species, we are far from understanding the effect of changes in the ecosystem. (1995, 3)

Indeed, though perhaps misunderstanding the adaptive energy supplied by ecosystem diversity and its emergent system behavior, Tansley (1935) claimed that "the gradual attainment of more complete dynamic equilibrium . . . is the fundamental characteristic" of ecosystems, and that "the order of stability of all of the chemical elements is of course immensely higher than that of an ecosystem, which consists of components that are themselves more or less unstable— climate, soil, and organisms." But Tansley, like his counterparts, believed that equilibrium was "perfect," and that "its degree of perfection is measured by its stability" (301). Over time, however, the diversity–stability dimensions of ecosystem properties became increasingly appreciated (Pimm 1984; Tilman 1999), focusing research on the properties that bring dynamic, nonlinear disequilibrium to the table for ecosystems, and improving our understanding that complexity and diversity in ecosystems are, in fact, the properties most important to sustainability, but also the most vulnerable to human interference (Abel and Stepp 2003; Hartvigsen et al. 1998; Holling et al. 2002; Levin 1999; Limburg et al. 2002; Milne 1998).

Thus, ecologists today engage in complex systems-based research into such matters as soil–microbe dynamics (Young and Crawford 2004), linkages between aboveground and belowground biota (Wardle et al. 2004), the effects of disturbance events on forest structure outcomes (Savage et al. 2000), plant–plant interactions in response to environmental stress (Brooker 2006), mutualistic relations between plants and their pollinators (Bascompte et al. 2006), and the causes of "flips" in coral reef species assemblages (Moberg and Folke 1999). As such research unfolds, ecologists routinely account for two properties as the central products of complex adaptive system dynamics in operation: *resistance*—the ability of an ecosystem to withstand external stress without loss of function; and *resilience*—the ability of the ecosystem to recover from disturbance (Allison and Hobbs 2004; Carpenter and Brock 2004; Christensen et al. 1996; Folke et al. 1996; Holling 1996; Holling and Gunderson 2002; Tilman 1999; Virginia and Wall 2001; Walker et al. 2006). As Limburg and her colleagues explain,

> Complex, interactive systems tend to converge on stable states, or dynamic equilibria, in which flows and processes are balanced. To that end, they evolve stabilizing mechanisms. In ecological systems this propensity toward stability is measured by two emergent properties, resistance and resilience. Resistance measures how unyielding a system is to a disturbance and resilience measures how quickly a disturbed system returns to its equilibrium. (2002, 410)

Nevertheless, the more that is learned about these properties, the more researchers such as Christensen et al. (1996) appreciate that "[w]ith complexity comes uncertainty. . . . [W]e must recognize that there will always be limits to the precision of our predictions set by the complex nature of ecosystem interactions" (669). Add to this the nature of political and social institutions involved in ecosystem management as themselves exceedingly complex systems (Janssen 2002; Walker et al. 2006), and the problem of devising ecosystem services policies becomes all the more daunting.

Ecosystem Boundaries

The model of ecosystems as complex adaptive systems brimming with dynamic properties and unpredictable outcomes thus complicates one of the most fundamental starting points for ecosystem research and management—where do these complex entities begin and end? Tansley borrowed the "system" in ecosystem from physics, and in the strictest physical sense a system has boundaries that delimit it from its surroundings. But no ecosystem is perfectly delimited, or closed, in this respect (Bailey 1996). Wherever we might draw the physical "boundary" of an ecosystem for political, research, or other purposes, inputs of energy (e.g., sunlight) and materials (e.g., water) from outside its bounds will affect internal processes, and outputs of energy (e.g., increased water temperature) and materials (e.g., decomposition waste) will be returned to the producing ecosystem or become inputs delivered for use in other ecosystems (Blair et al. 2000). Moberg and Folke (1999), for example, document the intricate linkages of energy and materials that exist between mangrove forests, sea grass beds, and coral reefs, three discrete ecosystems found in tropical seascape regimes, and Holmlund and Hammer (1999) do the same in their study of the contri-

TABLE I.2. Examples of external inputs, internal process uses, and external outputs enabled by ecosystem processes.

	External Inputs	*Internal Process Uses*	*External Outputs*
Energy	Solar	Photosynthesis	Oxygen
	Water	Thermal flow	Thermal
	Wind	Desiccation	Nutrients
Materials	Water	Adsorption	Stream sediments
	Nutrients	Nitrification	Seeds
	Predator	Predator–prey	Animal waste

bution of fish to ecosystem services in the interface between terrestrial, aerial, and aquatic ecosystems. Countless other examples abound. Ecosystem processes, in other words, receive at least some energy and materials from outside, use energy to transform and recycle materials internally, thereby building ecosystem structure, and then move at least some energy and materials back to the outside. An inventory of just some of the possible external inputs, internal uses, and external outputs that are enabled and supported by ecosystem processes might include those shown in Table 1.2.

The "open" nature of ecosystems under this process-based conception presents difficult questions of boundary definition for research and management purposes (Ruhl 1999). Indeed, some commentators have gone so far as to argue that any effort to forge ecosystem-based policies is premature because we do not know enough about the biological and physical boundaries of ecosystems and thus cannot possibly develop effective policy (Fitzsimmons 1999). But this position seems calculated to preclude us from ever developing an ecosystem protection policy, for it will never be scientifically accurate to speak of an exact ecosystem "boundary." On the other hand, some commentators suggest that ecosystem boundaries be defined by a highly fluid set of criteria that would in theory allow tailor-made ecosystems based on ecological, economic, social, spatial, and temporal factors (Keystone Center 1996). Under that approach anything would qualify as an ecosystem depending on who is asked; little consistency of definition over time and space could be expected.

The challenge, therefore, is to make ecosystem delineation sufficiently precise for policy purposes without violating scientific sensibilities. Any such effort faces four major impediments presented by the open nature of ecosystem processes: (1) several smaller ecosystems may exist within a larger one, (2) ecosystems are interlinked and often difficult to separate, (3) boundaries of ecosystems expand and contract over time in response to natural and anthropogenic influences, and (4) ecosystems are ecologically rather than legislatively or administratively established features (United States Government Accounting

Office 1994). Yet, we "know" that the Everglades are not the Rockies—that somewhere between Florida and Colorado the two ecosystems have boundaries outside of which it is no longer scientifically (or politically) useful to think of being "in" either of them.

Commonality and connectedness thus provide practical themes for delineating ecosystem boundaries (Millennium Ecosystem Assessment 2003). Commonality comes in the form of basic structural units, such as terrain, vegetation type, hydrologic characteristics, and species assembly, with a well-defined ecosystem exhibiting shared structural characteristics over time and scale. Connectedness comes in the form of the interactions between ecosystem components, with a well-defined ecosystem exhibiting strong interactions among internal components and weak interactions across its defined boundaries. Plainly, the Everglades and Rockies share little in common and have, at most, weak interactions, making it impractical to think of them as examples of the same ecosystem type, much less of being in the same ecosystem.

To implement this pragmatic approach, we can turn to a variety of methods and criteria to serve as bases for a more precise, uniform method to delineate ecosystem boundaries (Bailey 1996). Some methods rely heavily on an intuition-based judgment process, and thus are influenced largely by who is in charge of drawing the lines. Explanation to and verification by decision makers become problematic in those circumstances. At the other extreme, some more precise and objective methods, such as digital-image processing, are exceedingly complex in application and have the effect of separating information about spatial and other characteristics (geology, landform, soils types, vegetation types) from the underlying ecosystem processes.

The most promising ecosystem delineation method, known as "controlling factors," relies on identifying certain key factors that strongly influence ecological processes and using them to partition the landscape into ecological units. This method has the advantage of allowing simplification, standardization, and verification, thus being the most appropriate method for translating the science of ecosystem dynamics into the political arena of ecosystem management. Among the controlling factors most often mentioned are vegetation, fauna, soil, physiography, watersheds, and aquatic biota (Bailey 1996; United States Government Accounting Office 1994). Each controlling factor candidate has its advantages and disadvantages from the scientific perspective— none provides the perfect ecosystem boundary delineation metric. For the present purposes of describing the ecology of ecosystem services, however, it is sufficient to observe that there are a number of scientifically useful ways of describing ecosystem boundaries, albeit all have limitations, and save for later the question of which method may best serve ecosystem service management policy.

The Benefits to Humans of Sustainable Ecosystems

Regardless of which metric is used to envision ecosystem boundaries, when described as a base of natural capital structure supporting important process-based functions, ecosystems assume both a biocentric component and an anthropocentric component. A babbling brook full of trout also provides peaceful solitude for the busy city dweller; indeed, simply knowing the brook exists may provide psychic pleasure without the need to visit it in person. What is habitat for ducks can also be a weekend retreat for duck hunters. And what provides sustenance for hardwood trees can also provide profit for timber companies. These examples illustrate the three primary categories that have conventionally been used for describing anthropocentric perspectives on ecosystem functions: (1) nonuse and other indirect existence benefits, (2) direct aesthetic and recreational use benefits, and (3) direct commodity consumption benefits (Ehrlich and Ehrlich 1992). Anyone can appreciate that these direct and indirect benefits of ecosystems rely on ecosystem structure and functions to construct what is of immediate value to humans—an image in one's mind, a pretty photo scene, a fishing hole, the wood of a sturdy tree.

Indeed, in addition to recognizing the value of ecosystems to wildlife and the environment, natural resource management and conservation laws are replete with references to these ways in which humans benefit as well. For example, the Wild and Scenic Rivers Act seeks to protect the "remarkable scenic, recreational, geologic, fish and wildlife, historic, cultural, and other similar values" of free-flowing rivers.[1] The Endangered Species Act acknowledges that imperiled species "are of esthetic, ecological, educational, historical, recreational, and scientific value to the Nation and its people."[2] And the Federal Land Policy and Management Act directs federal agencies to manage public lands so as to "provide food and habitat for fish and wildlife and domestic animals . . . [and] for outdoor recreation and human occupancy."[3] People are willing to, and do, pay for these benefits. The law, therefore, frequently seeks to manage ecosystems with human benefit in mind.

Defining Ecosystem Services as a Distinct Category of Ecosystem Benefits

The movement to define and describe ecosystem services recognizes what is also obvious, but which until recently has been largely unmentioned in laws like these—that ecosystem structure and functions also provide *service* values to humans beyond the direct and indirect benefits that are already so ingrained in our culture, economy, and policy (Ehrlich and Ehrlich 1992). To be sure, implicit in most natural resource laws is the understanding that ecosystems provide a wide range of benefits to humans, including "serving" us life-sustaining benefits. As Geoffrey Heal puts it, almost anyone can appreciate that ecosystem services provide "the essential, low-level infrastructure upon which human activities and built systems rest" (2000, 2). But the expression of that kind of benefit has been left at best implicit in law because there has been no scientific and economic foundation on which to build explicit policy goals. The science of ecology has largely been devoted to exploring the importance of ecosystem processes in natural contexts, but has ignored exploration of human service values until recently. Similarly, economics as a discipline focuses on pricing in markets, but without information from ecologists about the delivery to humans of ecosystem services, the market necessarily will underrepresent those values in pricing and resource allocation decisions. Researchers in both fields, however, have begun to bridge the gap, to fill in the very large hole of knowledge surrounding how *ecologically* important ecosystem attributes are *economically* valuable services to humans.

With that mission in mind, the ecology literature is burgeoning with efforts to identify and assign value to the service component of ecosystems. Many entries in the field take a rather broad view of what fits into the ecosystem services category. Daily, for example, defines ecosystem services as "the conditions and processes through which natural ecosystems, and the species that make them up, sustain and fulfill human life" (1997, 3). Her now-classic list of what fits under this wide umbrella is shown on the left-hand column of Table 1.3.

Similarly, in their famous article on ecosystem service and natural capital values, Costanza et al. (1997, 254) define ecosystem services as "flows of materials, energy, and information from natural capital stocks which combine with manufactured and human capital services to produce human welfare." They compiled a list of seventeen major ecosystem services, shown in the right-hand column of Table 1.3.

Many other lists of ecosystem services have been compiled, but while holistic inventories such as these certainly capture the essence of the ecosystem services concept, some further typology of kinds of ecosystem services is useful to inform the discussion of how to manage the ecosystems that provide the benefits (de Groot et al. 2002). For example, in 2001 the United Nations launched the Millennium Ecosystem Assessment, a massive international work program designed to meet the needs of decision makers and the public for scientific information concerning the consequences of ecosystem change for human well-being and options for responding to those changes. With that mission, it is no surprise that the project focuses heavily on ecosystem services (Millennium Ecosystem Assessment 2003, 2005), which it groups into four categories: provisioning services (e.g., providing food and water); regulating services (e.g., dis-

TABLE 1.3. Ecosystem services identified by Daily 1997 and Costanza et al. 1997.

Ecosystem Services Identified by Daily 1997	*Ecosystem Services Identified by Costanza et al. 1997*
Purification of air and water	Gas regulation
Mitigation of floods and droughts	Climate regulation
Detoxification and decomposition of wastes	Disturbance regulation
Generation and renewal of soil and soil fertility	Water regulation
Pollination of crops and natural vegetation	Water supply
Control of the vast majority of potential agricultural pests	Erosion control and sediment retention
Dispersal of seeds and translocation of nutrients	Soil formation
Maintenance of biodiversity	Nutrient cycling
Protection from the sun's harmful ultraviolet rays	Waste treatment
Partial stabilization of climate	Pollination
Moderation of temperature extremes and the force of winds and waves	Biological control
Support of diverse human cultures	Refugia
Providing aesthetic beauty and intellectual stimulation that lift the human spirit	Food production
	Raw materials
	Genetic resources
	Recreation
	Cultural

ease regulation); cultural services (e.g., recreation opportunities); and supporting services (services necessary for the production of other service types). Daily and Dasgupta (2001) provide slightly more detail, dividing ecosystem services into five subcategories: production of goods (e.g., pharmaceuticals and timber); regeneration processes (e.g., purification of air and water); stabilizing processes (e.g., control of pests and mitigation of floods); life-fulfilling processes (e.g., aesthetic beauty and existence value); and preservation of options (services that maintain ecosystems).

Holmlund and Hammer (1999) present yet another theme, and more detail, by distinguishing between fundamental ecosystem services that are essential for ecosystem function and resilience, and which are thus essential for human survival, and demand-derived ecosystem services, such as recreation, that are formed by human demand and may not be essential for sustaining

TABLE 1.4. Taxonomy of ecosystem services proposed by Holmlund and Hammer 1999.

Fundamental Ecosystem Services (essential for survival of ecosystems and humans)		Demand-Derived Ecosystem Services (satisfying human desires)	
Regulating Services (regulate ecosystem structure and processes)	**Linking Services** (provide links between ecosystems)	**Information Services** (providing humans useful ecological information)	**Cultural Services** (providing humans desired cultural and material uses)
Regulation of food web and nutrient balance	Active links (e.g., migration)	Information for assessing ecosystem stress	Food, fiber, and mineral supply
Regulation of structural processes	Ecological memory links (e.g., anadromous fish)	Information for long-term environmental monitoring	Recreational and aesthetic uses
Regulation of gas flux	Passive links (e.g., prey for migrating predator)		Human health uses

ecosystems or human society. Through their example of freshwater fish populations (255), their work suggests the categorization presented in Table 1.4.

The Distinction between Production and Use of Ecosystem Services

These kinds of lists and categorizations, however useful for conceptualizing ecosystem services, emphasize primarily one side of ecosystem services ecology—the *production* of service benefits. The process–function–structure–service progression of topics typically portrayed in the literature on ecosystem services provides tremendous insight into how ecosystem services come into being and, therefore, the importance of maintaining the integrity of the underlying natural capital that supports the ecosystem processes and functions. But focusing on the importance of ecosystem processes and functions to the sustained output of ecosystem services does not fully capture what is necessary for a complete description and understanding of ecosystem services—that is, the *use* of ecosystem services.

Although its primary focus is economic, the use-side perspective of ecosystem services is by no means simply a matter for economists to describe. It has a distinctly ecological foundation that is attuned to describing how, as a matter of ecological processes and functions, different ecosystem services actually find their way to becoming human benefits. A telephone company, for example, may have a well-developed understanding of how to lay telephone lines, erect cell towers, and build signal switching stations to make service available at a multitude of points. What it really needs to know in addition, however, is where people are likely to want to make or receive telephone calls. Indeed, companies pour significant resources into understanding their respective markets— the patterns of human demographics and behavior that allow them to match production to use. In the same vein, knowing about ecosystem processes and functions does not tell one all that is necessary for thinking about the valuation and management of natural capital and ecosystem services. Or, to put it more strongly, ecosystem processes and functions don't yield ecosystem services until they are used *by people.* It behooves ecology as a scientific discipline, therefore, to study not only the ecology of ecosystem service use but also its economy.

The Distinction between Direct and Indirect Use of Ecosystem Services

Once the topic of ecosystem service use is opened up, further refinement reveals the complexity of the subject. For example, in one of the earliest works on the topic of ecosystem services, Walter Westman (1977) differentiated between ecosystem functions that lead to ecosystem structure, such as habitat and the species that occupy it, and ecosystem functions that lead to ecosystem dynamics, such as gas fixation and release. To be sure, there can be no ecosystem structure without ecosystem dynamics, and vice versa, and thus structure and dynamics of ecosystems are interdependent and mutually supporting. But the distinction between structure and dynamics, while ecologically important, also provides useful insight for purposes of understanding how humans use different ecosystem services.

As noted earlier, the conventional way of thinking about ecosystem benefits presents three categories: (1) indirect nonuse and existence benefits, (2) direct aesthetic and recreational use benefits, and (3) direct commodity consumption benefits. The reason people do not think about ecosystem functions as providing services when they think about these three kinds of benefits is because that is *not* what they are consuming. To be sure, ecosystem services make all these benefits possible, as reflected in the Millennium Ecosystem Assessment's choice of the term "provisioning services" to describe this mode of ecosystem service use (Millennium Ecosystem Assessment 2003, 56–57; 2005, 40). But the user of these ecosystem benefits cares about the end result—the image, the scene, the hiking trail, the timber—just as a homebuyer cares about the finished house, not the many service providers who built it or its parts. Moreover, it is easy for people to describe a value for the end result in each case of these three types of ecosystem benefits. Sitting in her office in a distant city dreaming about a raft trip down the Colorado River, a lawyer knows how important it is to her that the Colorado River is there even though she is not using it at the moment. Hikers and hunters exhibit the value they place on aesthetic and recreational benefits through entry and permit fees, travel costs, and equipment expenditures. And a timber company can quickly determine the value of a tree as a commodity consumption benefit through the market for timber.

Not coincidentally, these three categories of ecosystem benefits depend for their value principally on ecosystem *structure*. It is the structure of the Colorado River the lawyer envisions in her office (nonuse), the structure of a mountain trail hikers use (recreational use), and the structure of trees that timber companies harvest (commodity use). The structural dimension of these benefits makes their valuation more tangible, whether directly through market uses of the structure or indirectly through nonmarket uses such as daydreaming and hiking. Hence people use, and thus value, the ecosystem services that make these structural benefits possible only *indirectly* (Costanza et al. 1997). This is not to say that this kind of indirect use of ecosystem services should not be recognized, but rather to acknowledge that their value is bundled into, and thus dependent on, the value of the nonservice benefit that is being enjoyed. The demand for the services, in other words, is a *derived demand* that depends on the demand for the structural benefit (Boyd and Banzhaf 2006; Holmlund and Hammer 1999). Indeed, Westman considered this structure-based kind of ecosystem service less difficult to value, or to envision as valuable, for this very reason—that is, anyone who can appreciate the value of the benefit ought also to appreciate that the ecosystem services contributing to it are also valuable.

As Westman additionally pointed out, however, there are some ecosystem functions that we use *directly*, even if we don't know it. We use gas fixation and release directly. We use pest control directly. We use flood regulation directly. We use thermal regulation of the atmosphere directly. These dynamics-based ecosystem functions, which the Millennium Ecosystem Assessment project aptly refers to as "regulating services" (Millennium Ecosystem Assessment 2003, 57–58; 2005, 40), are used as *direct service benefits*. When ecosystems do not provide these services at levels sufficient for our needs and desires, we either pay to find another way to provide them—that is, supply an alternative natural or technological mechanism to duplicate the service benefit—or suffer the consequences of going without.

Hence, it is useful to distinguish between two discrete categories of ecosys-

TABLE 1.5. Indirect benefits of structure-based ecosystem services and direct benefits of dynamics-based ecosystem services.

Structure-Based Benefits of Indirectly Used Ecosystem Services	Dynamics-Based Benefits of Directly Used Ecosystem Services
Water supply	Gas regulation
Soil formation	Climate regulation
Refugia	Disturbance regulation
Food production	Erosion control
Raw materials	Nutrient cycling
Recreation	Pollination
Cultural	Biological control
	Genetic resources

tem services in any model of how ecosystems deliver benefits to humans—service benefits used indirectly through structural components derived from ecosystems, and service benefits used directly through the dynamic processes of ecosystem functions. Bolund and Hunhammar (1999) drive this point home in their description of ecosystem services used directly in urban ecosystem settings, which include such regulating services as air filtering, microclimate regulation, noise reduction, rainwater damage, and sewage treatment. Taking the holistic list Costanza et al. (1997) present of ecosystem services as a broader example, the services can be divided into indirect and direct service use patterns as shown in Table 1.5.

From the perspective of formulating economic and regulatory policies for managing ecosystem services, this distinction between direct and indirect use will be of utmost importance, because it reflects the human perception of the service use values. In other words, it tells ecologists how to trace the ecosystem service from its point of production (its origin) to its point of human use. For indirectly used services, therefore, ecologists must reveal the link between a service and its contribution to the ecosystem structure components that people value in use and nonuse capacities. Conversely, for directly used ecosystem services, ecologists must reveal the manner in which ecosystem structure—the natural capital of the ecosystem—supports the processes and functions that provide the services for humans to use. While these appear to be flip sides of the same coin, further development in later chapters shows that fundamentally different economic and regulatory considerations apply, and that the most difficult problems for envisioning ecosystem service law and policy arise in cases of directly used service benefits.

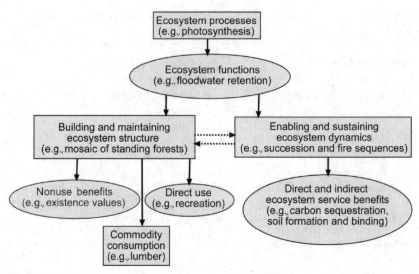

Figure 1.1. Relationships among ecosystem processes, functions, and services. The interactions between ecosystem structure and dynamics generate direct and indirect ecosystem services.

Building on what has been covered thus far in this chapter, the use-based model of ecosystem services can be represented as shown in Figure 1.1. One important feature inherent in this model is that the ecology of ecosystem services necessitates accounting for humans—where we are, how many of us are there, and what is of value to us in that setting (Pearce 1998). Without the human component, in other words, there are no ecosystem services of which to speak—the right branch of the model would end at "ecosystem dynamics" and the left side at "ecosystem structure." A complete ecological description of ecosystem services, therefore, cannot leave the description of the human component to economists, claiming that ecosystem services are at work but leaving their presence in human contexts unsubstantiated. Rather, the ecology of ecosystem services requires that ecologists of all disciplines do the job of tracing ecosystem processes and functions directly to human population receptors and of describing the services being delivered to that population, at which point economists can assist in the quantification of service values.

Is This Really Ecology?

It is the need to trace ecosystem services to human populations that causes some observers to question the wisdom of engaging in this brand of ecology

(Costanza et al. 1997; Doremus 2000). They worry that the focus on tracing ecosystem services to human beneficiaries will commodify ecosystems, and they fear that efforts to assign price or price-like values to ecosystem functions will detract from other policy grounds for ecosystem protection, such as the conservation of habitat for wildlife benefits and the protection of natural systems simply for their intrinsic values (Bockstael et al. 1995; Chen 2004). They also worry that the imprecision inherent in ecosystem service valuation may lead to inaccurate and misguided policy decisions (Gatto and De Leo 2000; Sagoff 1988; Toman 1998).

But it is not as if ecosystem services would not exist but for the efforts of researchers like Daily and Costanza. Ecosystem services are real. They have measurable value to humans, and whether we know their precise economic value or not, the fact that society has to choose how to allocate natural resources necessarily requires valuation of ecosystem services in some form or another (Costanza 1996). Failure to refine our understanding of their value, and the consequent inability to account for those values in regulatory and market settings and, more important, in the public mind, is unlikely to promote their conservation. As Pearce (1998) put it, "[T]he playing field is not level; rather, it is tilted sharply in favor of economic development. Two things have to be done to correct this situation. First, one has to show that ecosystems have economic value—indeed, that all ecological services are economic services. Second, a way has to be found to 'capture' the nonmarket values of ecosystems and turn them into real benefits for those who practice conservation" (23). Costanza et al. (1997) make the point more succinctly in urging that "although ecosystem valuation is certainly difficult and fraught with uncertainties, one choice we do not have is whether or not to do it" (255). To put it bluntly, it can't possibly help the cause of sustainable ecosystems to have ecologists sit on the sidelines of this endeavor, unwilling to engage in research on ecosystem service values.

On the other hand, neither is unbridled enthusiasm about the development of ecosystem services research and policy warranted. As the foundations for the ecology of ecosystem services form, danger lurks in two respects. First, the new focus on identifying and tracing ecosystem service values may lead to oversimplified conceptions of the underlying ecosystem processes and functions that make them possible. At the same time, a strengthened interest in consuming valuable ecosystem services could also lead to management of ecosystem processes and functions for their service values rather than for the sustainability of ecosystems generally. The ecology of ecosystem services must take precautions that both effects are understood and controlled.

The Problem of Oversimplified Portrayal of Complex Ecosystem Processes

As refinement of the ecology of ecosystem services leads to greater public appreciation of their value, demand for their continued delivery is likely to increase as well. But the public must understand—indeed, ecologists must help the users of ecosystem services understand—that ecosystem services are not services in the conventional economic sense. Services in the form of human labor can be used far more flexibly than can ecosystem services. If a developer wishes to obtain consulting on a building project, for example, engineers, planners, architects, lawyers, and other consulting service providers can be assembled into a consulting team. The developer can negotiate their consulting fees, obtain their services as needed, and replace those that do not perform adequately. By contrast, the use of ecosystem services presents far less flexibility. We know that within any given ecosystem the benefits we can confidently identify as ecosystem services are the result of a complex of ecosystem processes associated not only with that ecosystem but with all linked ecosystems. The services we use, therefore, cannot easily be selected for rate, location, combination, and other qualities as we can do for consultants. They are where they are and what they are, unless we alter the underlying ecosystem processes. Altering the processes is, of course, a complicated proposition for two reasons. First, the way in which ecosystem processes produce services may not be fully understood, so that we cannot know with confidence what the effects on service yield may be of changing one or more of the processes (Holland 1995). Second, we know that ecosystems are open dynamic systems, meaning that to alter ecosystem services provided from one defined ecosystem, we may need to alter processes of another ecosystem, but that doing so may affect process flows in yet another ecosystem (Bailey 1996).

The importance of complex adaptive systems research to the ecology of ecosystem services is inevitable from this trend in learning and appears widespread in the current literature (Bockstael et al. 1995; Ekins 2003; Ekins et al. 2003; Holling et al. 2002). Holland (1995) points out the general principle that "the coherence and persistence of each system depends on extensive interactions, the aggregation of diverse elements, and adaptation or learning" (4). Taking that principle to the ecosystem services context, Costanza et al. (1997) observe that "ecosystem services and functions do not necessarily show a one-to-one correspondence. In some cases a single ecosystem service is the product of two or more ecosystem functions whereas in other cases a single ecosystem function contributes to two or more ecosystem services" (254). There are, in

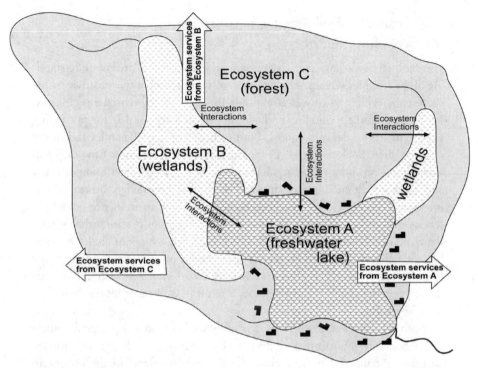

Figure 1.2. Spatial and functional relationships among ecosystem types and the services they produce. Lakefront homeowners primarily utilize services provided by the lake, but these are dependent upon services provided by the upstream wetland and forest ecosystems.

other words, extensive interactions and an aggregation of diverse elements in play within an ecosystem to produce ecosystem services of value to humans. Limburg et al. (2002) thus suggest that "an important function of understanding complex systems should be to inform decision-makers about when, or under what circumstances, an undesirable substantive state change is likely to occur, one that will diminish or enhance the value of ecosystem services" (410).

And there lies the rub. To the extent ecology is successful in demonstrating to humans the value of ecosystem services, people are apt to view the continued delivery of the services as the reason for managing the ecosystem. But, as noted earlier, ecosystem boundaries are in large part arbitrary. People may come to think of "that wetland" as providing "that flood control," whereas in fact "that wetland" is a complex system which, wherever we draw its boundary for management purposes, is open to interaction with other systems, as shown in Figure 1.2.

The Problem of Managing Ecosystems for Ecosystem Services

Greater interest in ecosystem services will inevitably lead to greater understanding of the complex relationship between ecosystems, and of the sensitivity some ecosystems exhibit to processes and conditions in other ecosystems. Increasingly, for example, ecologists and economists develop models for portraying ecosystem service delivery, models both of the ecosystem dynamics and of the service values (Bockstael et al. 1995; Costanza 1996), and over time they will surely build yet more robust models capable of improved description of cause and effect of different policy options. With that knowledge, however, may eventually come an improved understanding of how to influence ecosystem service delivery at one point by manipulating ecosystem processes throughout the chain of related ecosystems. But this prospect of a superior ability to regulate the delivery of ecosystem services presents its own latent potential in the form of unanticipated feedback and feedforward effects that could cascade throughout the chain of ecosystems as a result of ecosystem service "management" decisions.

Such cascade effects, which may amplify some services and degrade others throughout the interconnected chain of ecosystems, are inherent in complex adaptive systems and present especially difficult problems in the ecosystem management setting to handle strategically or deliberately (Christensen et al. 1996; Deutsch et al. 2003; Holling et al. 2002). There is evidence, for example, that managing biomass and productivity of tree plantations to maximize carbon sequestration also increases transpiration and rainfall interception, which in turn leads to diminished stream flows, increased soil salinization, and increased soil acidification (Jackson et al. 2005). Managing for one ecosystem service, in other words, has inevitable trade-off impacts for other ecosystem services. Hence it is one thing to observe that "modelers of systems usually look for boundaries that minimize interaction between the system under study and the rest of the universe in order to make the job easier" (Costanza 1996, 981), but a far more difficult challenge is actually to find those boundaries in ecological settings.

Understanding Ecosystem Service Trade-Offs

In summary, the ecology of ecosystem services must focus on trade-offs and synergies of many kinds and scales, and much research remains ahead (Daily and Dasgupta 2001). At the most fundamental level, the extraction of commodities from an ecosystem or the use of ecosystem resources for recreation necessarily affects the natural capital available not only for producing other commodities or for maintaining recreation, but also for supporting directly used ecosystem services. But the potential trade-offs and synergies go much deeper than this obvious level. Even the deliberate management of ecosystem services, rather than simply the commodities or uses they support, leads inevitably to the question of which service to favor when enhancing one diminishes or enhances another (Alcamo et al. 2006; Carpenter et al. 2006; Rodriguez et al. 2006). Indeed, the Millennium Ecosystem Assessment (2004a; 2004b) suggests that in many cases substantial changes (usually degrading) in the regulating services that ecosystems provide are the direct result of managing ecosystems primarily for their provisioning services (usually to enhance). Ecologists thus must improve the ability to describe the trade-offs and synergies between the indirectly used ecosystem provisioning services that support ecosystem structure and the directly used ecosystem regulating service benefits that are supported by ecosystem structure. And yet these effects may exist even between the directly used regulating benefits, because managing any ecosystem—any complex adaptive system—to promote one property may have the effect of altering another property.

There is, in short, much to be learned about the ecology of ecosystem services, both their production and their use. As Palmer et al. recently summarized,

> Maintenance of ecosystem services will require a considerably better understanding of the natural patterns and processes that sustain them. Innovative research must be initiated to answer crucial questions. Which ecological services are irreplaceable or too expensive to replace with emerging technology? What habitats must be protected to ensure that key services are provided? Which agents impoverish ecological services and how can their impacts be mitigated or avoided? How do individual, corporate, and government decisions sustain or degrade ecosystem services? (2004, 1251)

The importance of this line of research goes well beyond the development of ecology as a discipline, but rather has profound policy dimensions. The trade-offs between different services will necessarily affect different human populations differently, and questions of equity and efficiency will inevitably arise. Ecologists must accept that good ecological analysis in the conventional sense will not suffice to reveal this full dimension of ecosystem services to human populations (Boyd and Wainger 2002a). Alas, it seems that ecologists will be joined at the hips with economists to do so. But this much is certain, and should be of solace to ecologists: knowing how ecosystem services operate ecologically will not guarantee sound economic and policy decisions about the environment, but not knowing how ecosystem services operate ecologically will guarantee unsound economic and policy decisions. So economists have something to learn as well.

2   Geography

Although terms have changed and concepts have become more precise over the decades, geographers have been keenly interested in natural capital and ecosystem services, and in human impacts on these, throughout modern times. George Perkins Marsh's 1864 classic, *Man and Nature: Or, Physical Geography as Modified by Human Action,* a book written for many of the same purposes as this volume, illustrates this point well. Marsh begins, in nineteenth-century prose in the preface:

> The object of the present volume is: to indicate the character and, approximately, the extent of the changes produced by human action in the physical conditions of the globe we inhabit; to point out the dangers of imprudence and the necessity of caution in all operations which, on a large scale, interfere with the spontaneous arrangements of the organic or the inorganic world; to suggest the possibility and the importance of the restoration of disturbed harmonies and the material improvement of wasted and exhausted regions; and, incidentally, to illustrate the doctrine, that man is, in both kind and degree, a power of a higher order than any of the other forms of animated life, which, like him, are nourished at the table of bounteous nature. (3)

As early as the American Civil War, Marsh spoke of ecosystem services that would have served the Gulf Coast well in August 2005: "on many coasts, sand hills both protect the shores from erosion by the waves and currents, and shelter valuable grounds from blasting sea winds" (4). Marsh also attributed to abundant natural capital a substantial part of the success of Rome, although the phrase itself is not used:

> The Roman Empire, at the period of its greatest expansion, comprised the regions of the earth most distinguished by a happy combination of physical advantages. The provinces bordering on the principal and the secondary basins of the Mediterranean enjoyed a healthfulness and an equitability of climate, a fertility of soil . . . which have not been pos-

sessed in an equal degree by any territory of like extent in the Old World or the New. . . . Of these manifold blessings the temperature of the air, the distribution of the rains, the disposition of land and water, the plenty of the sea, the composition of the soil, and the raw material of some of the arts, were wholly gratuitous gifts. (7–8)

However, the slow, and occasionally rapid, depreciation of natural capital and subsequent failure of ecosystem services is similarly given an important role in Rome's decline:

> If we compare the present physical condition of the countries of which I am speaking, with the descriptions that ancient historians and geographers have given of their fertility and general capability of ministering to human uses, we shall find that more than one half of their whole extent—including the provinces most celebrated for the profusion and variety of their spontaneous and their cultivated products, and for the wealth and social advancement of their inhabitants—is either deserted by civilized man and surrendered to hopeless desolation, or at least greatly reduced in both productiveness and population. Vast forests have disappeared from mountain spurs and ridges, the vegetable earth accumulated beneath the trees by the decay of leaves and fallen trunks, the soil of the alpine pastures which skirted and indented the woods, and the mould of the upland fields, are washed away; meadows, once fertilized by irrigation, are waste and unproductive, because the cisterns and reservoirs that supplied the ancient canals are broken, or the springs that fed them dried up; rivers famous in history and song have shrunk to humble brooklets. . . . (9)

Given the critical importance of land use patterns in generating ecosystem services, geography, as a social science, has examined in considerable depth the social processes that generate landscapes where human influence is substantial. Carl Sauer, in his 1925 "The Morphology of the Landscape," conceptualized the landscape as a product of what today we would call "coevolution" between nature and society. His work contrasted strongly with the environmental determinist school dominant at the time, but which was resoundingly rejected in the 1940s due to its association with racism and imperialism. Nevertheless, Jared Diamond, in his Pulitzer Prize–winning book, *Guns, Germs and Steel* (1999), has reopened the issue of the powerful influence of geographical relationships and ecosystem characteristics on the unfolding of human history in a manner that rejects racism while embracing the archeological record, modern genetic analysis, and history as a natural laboratory. Diamond theorizes, for example, that the existence of domesticable wild plants such as wheat and barley and wild animals such as goats, sheep, and cattle launched the Fertile Crescent, the northern part of what is now termed the Middle East, onto a trajectory of social development that includes urbanization, political hierarchy, metallurgy, writing, and, unfortunately, pandemic diseases derived from livestock. The longitudinal diffusion of these social innovations east and west, and to similar latitudes in the Western and Southern Hemispheres, explains, according to Diamond, why Europeans, and to a lesser extent East Asians, have been able to dominate indigenous peoples of the Americas, Australia, and southern Africa.

Though he does not use these terms, Diamond's theory places natural capital and ecosystem services in the driver's seat of human history. Similarly, in *Collapse: How Societies Choose to Succeed or Fail* (2005), Diamond further traces the social processes and ecological conditions that have led historic societies such as those of Easter Island and the Maya to fall apart when natural capital was depleted and ecosystem services failed, and the threat of collapse of modern societies, such as Haiti and China, that are similarly depleting, through population and economic growth and poor resource management, natural capital that underlies necessary ecosystem service systems.

In bringing these ideas to bear here, however, two difficulties must be pointed out. First, for over a century geographers have not come to agreement over the raison d'être of the discipline, whether it is spatial analysis or nature–society relationships, and this leaves geography lacking a core theory (Turner 2002) that can be applied to the subject of ecosystem services. Second, geography has not to date embraced the ecological economic concepts of natural capital and ecosystem services. For example, the 12th edition of Harm deBlij and Peter Muller's *Geography: Realms, Regions, and Concepts* (2006), the most widely used text in the most widely taught geography course in U.S. colleges and universities, does not include "ecosystem services" or "natural capital" in the book's extensive glossary. Only very recently, with the beginnings of "sustainability science" (Kates et al. 2001) has a dialogue emerged between ecological economics and geography.

Rather than turning to ecological economics, geographers have generally pursued, on the one hand, the regional science approach that applies neoclassical economics to issues of space, and, on the other, the political-ecology school that applies the principles of political economy to human–environment relationships, especially human access to and management of natural resources. Piers Blaikie's 1986 classic, *The Political Economy of Soil Erosion in Developing Countries,* focused on economic and political inequality, especially as manifested in access to water and agriculturally productive land, as a primary force driving the poor, in an effort to maintain their subsistence, to overexploit the soil and extend agricultural production into marginal areas, thus leading to soil erosion and degradation. Today we would describe this turn of events as a rapid depreciation of natural capital with resultant failure of ecosystem services.

Blaikie's work founded the political-ecology school that reached its greatest influence in the 1990s, with applications primarily directed at environmental struggles in developing countries and with a markedly structuralist approach. Environmental degradation, poverty, and population growth were theorized as a vicious cycle ultimately caused by the economic inequalities inherent in capitalism, historic colonialism, and modern globalization, as manifested at local and global scales. With no access to prosperous livelihoods or productive resources such as land, growing populations in Africa, Latin America, and Asia rely upon large families (intensifying population pressure) and have little choice but to overgraze, deforest, over-irrigate, farm hillsides, overhunt, overfish, and otherwise liquidate local natural capital in an unsustainable manner. Yet local social relations and ecological variation are also important, giving rise to a complex geography of what we would here term natural capital investment, maintenance, depreciation, or liquidation.

At the same time, the structuralist emphasis of political ecology has been countered by an approach that emphasizes human behavior and agency. This *structure–agency* debate characterizes current geographic approaches to how society influences, but does not govern, individuals in their behavior toward nature and its capital. Why do some people choose to live on floodplains or refuse to evacuate when a hurricane comes bearing down? Why do others choose to convert from conventional to organic farming? What information needs to be provided and what emotional strings need to be pulled to convince people to stop littering and start recycling, or to conserve energy, or to invest in natural capital?

Part of our purpose in this chapter, therefore, is to bring the concepts of natural capital and ecosystem services to bear on geography, as well as to bring to bear the concepts and methods of geography to the problem of ecosystem service provision and delivery.

The Geographic Expression of Natural and Other Forms of Capital

In the ecological economics literature (see for example Costanza 2001; Lant 2006; Tietenberg 2005), a considerable effort is made to define nature as a capital asset, one that is the "factory" that produces ecosystem services. But, of course the human enterprise is also dependent upon other forms of capital that people produce themselves—financial, manufactured, intellectual, human, and social. At this point in the discussion, it is important to consider these various forms of capital in terms of their mobility and their consequent geographical expression (Table 2.1).

TABLE 2.1. The definition, mobility, and geographical expression of various forms of capital.

Form of Capital	Definition	Degree of Mobility and Cost of Transport	Direct Geographical Expression
Financial	Forms of money	Very high, nearly costless and instantaneous	None
Intellectual	Encoded knowledge in written and electronic forms	Very high, nearly costless and instantaneous	None
Social	Constructive formal and informal relationships among people	Variable. High for written forms and electronic communications; low for face-to-face interactions and relationships.	Communities and political jurisdictions at various scales
Human	Skills and knowledge inherent in individual people	Moderate. Mobile in the form of migration and travel.	Population, migration and travel patterns
Manufactured	Built infrastructures, buildings, and goods	Very low to moderate. Once built in a specific location, most infrastructures can be moved only at great expense; most manufactured items, however, can be readily transported at moderate cost.	Primary component of the built environment and human geography of places
Natural	Natural resources and ecological characteristics capable of producing ecosystem services	Low to none. Only valuable natural resource products can be moved at reasonable expense. Ecosystems are absolutely fixed in space.	Primary component of the physical geography and human ecology of places

While mobile forms of capital flow through geographically defined places, the characteristics of places are defined primarily by the immobile forms of capital that are fixed in specific locations. Manufactured capital, therefore, plays a critical role in defining the human geography of places as its built environment and economic infrastructure, while natural capital, fixed in place except when specific resources are extracted and transported through market exchange, is the determining factor in defining the physical geography and human ecology of places. Geographical places are substantially characterized, therefore, by the variable ecosystem service packages they receive which are derived, in turn, from constellations of natural capital and the spatial relationships among them.

Tracing Ecosystem Services to People

Tracing ecosystem services to human populations is complicated, not only by the complexity of the ecological processes, functions, and structures behind the services, but also by the geographic distribution of natural capital that is the source of ecosystem services, of the channels of service delivery, and of the service users. In each of these respects, ecosystem services might exhibit patterns that range from being relatively discrete and predictable, making them easy to identify and map, to being highly variable in nature, which would complicate their geography. Of course, any one ecosystem may present a complex landscape of different but interrelated ecosystem services exhibiting a mixture of such properties, making it difficult to pinpoint the geographic context of a single identified service.

To further confound our ability to provide geographic representations of ecosystem service distribution, the patterns that are exhibited may vary depending on the spatial and temporal scales of ecosystem processes and of our observation of those processes, with the most difficult to define being those involving numerous sources from which service values are relevant only on a cumulative basis. For example, what looks like a discrete source of an ecosystem service on a local scale on an immediate basis, may be one of thousands of diffuse sources of a different service operating in unison at a regional level over longer periods of observation. Because ecosystem processes operate over a wide range of spatial and temporal scales (Christensen et al. 1996; Kremen 2005; Millennium Ecosystem Assessment 2005), knowing which set of scales is most appropriate for making particular economic and social policy decisions is no straightforward matter. Indeed, the valuation of ecosystem services is as profoundly influenced by their geography as by their ecology. As Konarska et al. (2002) suggest, "The distance of the ecosystem to a population center, the fragmented nature of many ecosystems, the purchasing power of people in various parts of the world, and the spatial scale at which the ecosystem extent is measured, all can influence the valuation of ecosystem services" (492). Hence, as this chapter explores, even in cases where the ecology of an ecosystem service is well understood, identifying and locating the service for practical uses in law and policy will be challenging in many settings.

One critical issue, for example, is how people's willingness to pay (WTP) for various types of ecosystem services, a subject we explore in more depth in chapter 3, is related to their distance from them. In a California study, Pate and Loomis (1996) found that WTP declined with distance for a wetland improvement program and a selenium contamination control program, such that WTP has a "spatial half-life" (i.e., declines by 50 percent) for every 295 miles for wetland improvement and for every 140 miles for selenium control. WTP for a river and salmon improvement program, however, did not decline significantly with distance. A related study using surveys distributed nationally found that WTP for the removal of two dams on the Elwha River in Olympic National Park in Washington to greatly improve salmon habitat did decline with distance, but by only 13 percent over a distance of 2,560 miles (Loomis 1996). These studies alone, however, illustrate how little is known about the effects of geography on use, nonuse, and indirect use values of ecosystem services, despite its critical importance. Salmon, like wildlife of the African savanna or the Arctic National Wildlife Refuge, represent "charismatic megafauna" carrying aesthetic and existence values at national to global scales, and recreational, ecotourism, and option values over long distances. But many ecosystem services, like nutrient and sediment retention in wetlands or crop pollinators, are more pragmatic and local in significance, while others, like carbon sequestration or reductions in emissions of carbon dioxide or methane, are globally cumulative. These complexities make it difficult to define regions for the provision of ecosystem services along the same lines as fire and school districts that deliver social services, where the geographical area benefiting from a service can be held responsible for financing its provision.

The Complex Landscape of Ecosystem Services

A hiker seeking a shade tree under which to rest on a hot summer day has little trouble identifying the physical space within which occur the source, delivery, and use of the tree's offer of respite from the sun's rays. It's quite simple: find the shady spot on the ground and you've found the ecosystem service, ready to be used. It would be more difficult, however, for a person walking down a path through a deep canopy forest on a hot summer day to make the same straightforward observations. No single tree provides all the shade along the path. Trees that don't shade the path—even trees quite distant from the path—contribute to microclimate temperature and humidity regulation within the forest, effects that may be enjoyed by our hiker while on the path. Other structural and dynamic features of the forest may also contribute to the conditions the hiker finds pleasurable while on the path. It's easy to know where, when, and how to reap the benefits of these services—just stay on the path. But it is far more difficult to identify their sources and their modes of delivery.

Any such exercise in tracing ecosystem services from provision source to service user can be unpacked into six components that must be identified and defined, as shown in Figure 2.1. The provision source and the service user are obviously necessary ingredients in any such description. In between these two

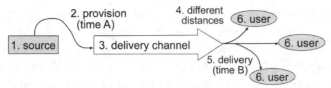

Figure 2.1. Tracing ecosystem services to human populations. This requires that we identify and define (1) the source(s), (2) the timing of the provision of the service entering the delivery channel, (3) the channel of delivery, (4) the delivery distance, (5) the timing of the delivery of the service for use, and (6) the user(s).

nodes we must also define the timing of the provision of a service ready to be put into the channel of delivery, the channels by which the service is delivered to users, the physical distance over which delivery occurs, and the timing of the delivery of the service ready to be used, which, depending on delivery distance, latency of the service benefits, and user decisions, may not correspond with the timing of the service provision.

Each of these components can assume different characteristics leading to profoundly different spatial and temporal patterns. As channels of delivery, for example, rivers and the atmosphere present strikingly different qualities. For purposes of constructing a more complete model, therefore, each of the components can be further refined using three descriptive categories: discrete, ambient, and variable.

Components with discrete characteristics are readily identifiable and predictable in behavior. Thus, for example, the shade tree offering respite to our weary hiker is a discrete point at which the service is provided. The timing of the provision of the service is predictable to the second, and the channel of delivery is well defined by the tree's shadow. The distance over which the service is provided is local to the source, and the timing of its delivery is also predictable to the second (and in this case contemporaneous with the time of provision). All that's left is the service user, our hiker, who is clearly a discrete user in this case.

By contrast, the hiker in the forest is enjoying the benefits of an ambient ecosystem service regime. There is no particular tree, patch of soil, or body of water that accounts for the hiker's experience. Rather, the area known as "the forest" is providing a suite of services provided and delivered in relatively predictable patterns through the diffuse network of the forest's regional microclimate. And as we add more hikers, swimmers, a few campgrounds, and a nearby town, we find that enjoyment of these services is dispersed among a regional population.

The difference between the shade tree and the forest in these examples is largely a matter of emergent system properties. Within the forest are numerous discrete shade trees, each of which, in a well-defined radius, offers relief from the heat. When taken as an accumulation of trees and their surrounding resources, however, the forest exhibits the emergent properties of its microclimate (Bailey 1996; Christensen et al. 1996; Gottfried et al. 1996; Holling et al. 2002; Millennium Ecosystem Assessment 2003; Savage et al. 2000). Although the forest, because of its ambient properties, may present a less predictable service regime than does a single tree, it follows patterns familiar to the regional user population and is predictable in that sense.

Of course, there is variability in both settings that defies prediction—a lightning strike that takes out the shade tree, or a major fire that alters the forest canopy and ground cover. Indeed, there may be settings in which ecosystem service regimes are or become highly variable over time horizons relevant to service users. During extended drought periods, for example, the amount of disturbance within the forest from increased fire, depleting water supplies, and death of wildlife and vegetation may make the provision and delivery of microclimate and other benefits quite spotty and unpredictable. It becomes increasingly difficult to predict when, where, and for how long different benefits are available for use. Indeed, for some ecosystems, such as deserts, variability may be the rule rather than the exception.

To summarize, the foregoing discussion leads to the taxonomy of possible ecosystem service source, delivery, and use patterns shown in Table 2.2. In terms of mapping ecosystem services from their natural capital provision source to the ultimate service users, service regimes that follow completely discrete or completely ambient patterns through the chain of delivery present the least challenge. Discrete regimes can be mapped linearly within a well-defined spatial dimension from point source to point user, following predictable provision and delivery timing patterns. Likewise, completely ambient regimes can be mapped as larger areas within which numerous sources and receptors are distributed and the service is provided and received predictably, albeit diffusely, on a relatively predictable basis throughout the user region. Completely variable service regimes, by contrast, present considerable tracing difficulties, as any attempt to capture them as linear features could be underinclusive at any par-

Table 2.2. Taxonomy of ecosystem services in terms of spatial and temporal relationships.

Type	Provision Source	Provision Timing	Delivery Channel	Delivery Distance	Delivery Timing	Service User
Discrete	Point	Predictable	Conduit	Local	Predictable	Point
Ambient	Diffuse	Predictable	Basin	Regional	Predictable	Diffuse
Variable	Spotty	Irregular	Unstable	Ebb and flow	Irregular	Spotty

Tracing Ecosystem Services to Human Populations

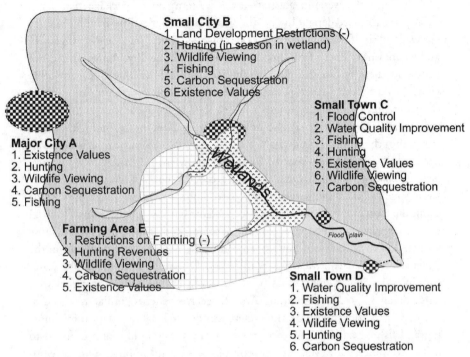

Small City B
1. Land Development Restrictions (-)
2. Hunting (in season in wetland)
3. Wildlife Viewing
4. Fishing
5. Carbon Sequestration
6 Existence Values

Small Town C
1. Flood Control
2. Water Quality Improvement
3. Fishing
4. Hunting
5. Existence Values
6. Wildlife Viewing
7. Carbon Sequestration

Major City A
1. Existence Values
2. Hunting
3. Wildlife Viewing
4. Carbon Sequestration
5. Fishing

Farming Area E
1. Restrictions on Farming (-)
2. Hunting Revenues
3. Wildlife Viewing
4. Carbon Sequestration
5. Existence Values

Flood plain

Small Town D
1. Water Quality Improvement
2. Fishing
3. Existence Values
4. Wildife Viewing
5. Hunting
6. Carbon Sequestration

Figure 2.2. Spatial relationships in ecosystem service utilization. This simple case of ecosystem service delivery from a riparian wetland area illustrates many of the geographic complexities in identifying ecosystem service beneficiaries. While all areas would benefit marginally from the carbon sequestration occurring in the wetland, other ecosystem services are more space-dependent, and some, such as flood control, occur only under extreme circumstances. Small City B and Farming Area E pay an opportunity cost for conserving the wetland area.

ticular time, and any attempt to encircle the area within which they might occur could be overinclusive at any particular time. To complicate matters, moreover, it may often be the case that in tracing a particular ecosystem service from provision source to service user, we find that it exhibits a combination of discrete, ambient, and variable patterns along the way, and that the pattern changes over time.

For example, Figure 2.2 depicts a riparian wetland area located along the junction of several tributaries to a river and occupying 5 to 7 percent of a watershed near its center. Wetlands are a particularly good example of multifunctional ecosystem service "factories" producing wildlife habitat, flood control, carbon sequestration and storage, and improving water quality through such processes as denitrification. But unlike the goods from a factory that, in this age of globalization, can be delivered and sold all over the world, the ecosystem services from the wetland generate an ecosystem service package that varies from place to place depending on the geographic relationship of the places to the wetland itself. Major City A, for example, lies largely outside the watershed, and so does not benefit from flood control or water quality improvement, but people in the city can still drive to the wetland itself for hunting or wildlife viewing, or to the river downstream of the wetland for fishing or perhaps canoeing, all of which vary in quality with the season of the year and even the daily weather. Residents may also hold existence values for the biological diversity within the wetland or its characteristics as a place. Nonuse or existence values could reasonably be expected to decay rapidly with distance from a typical wetland like the one illustrated. Even if this ecosystem service package is relatively small in value per capita, it may have quite a large total economic value given the large population receiving it in Major City A. Small Towns C and D receive larger per capita ecosystem service packages than Major City A because C, lying on the floodplain downstream from the wetland, is dependent on the wetland for flood control while both C and D use the river for public water supply, water whose quality has been improved by the nearby upstream wetland. Small City B and Farming Area E receive benefits similar to those of Major City A, depending on the preferences of the local population for hunting and fishing and the environmental attitudes they hold that are linked to existence values. However, preservation of the wetland imposes land development restrictions on these communities that have an economic cost and structure the political debate over preservation of the wetland. Were interests in Small City B or Farming Area E to convert part of the wetland to urban or agricultural uses, there would be an *ecological opportunity cost* imposed on all five areas as their ecosystem service packages were marginally diminished. As shown in chapter 13 where we examine wetland mitigation banking, these theoretical issues are central to policy reform. Clearly, *space matters* in considering ecosystem services.

Mitsch and Gosselink (2000) argue that the relationship between area and ecosystem services is not linear, but there is evidence that 3 to 7 percent of temperate-zone watersheds should be maintained as wetlands in order to facilitate a variety of ecosystem services. This implies that if the percentage is already higher, the marginal ecological opportunity costs of drainage are lower than if the percentage is in this range. However, at the lower end of the range, ecosystem service provision can collapse entirely and remaining wetland remnants may have little value. Conversely, wetland restoration and preservation are most valuable per acre in watersheds within this range because the marginal improvement in ecosystem services is greatest, and in heavily populated areas, because the number of beneficiaries is higher.

If we take this notion to a larger spatial scale, populous and prosperous regions, such as the northeastern United States, much of western Europe, or Japan, often preserve local ecosystems for the valuable ecosystem service packages they provide, even if they have to import a greater portion of their food and other raw materials—the production of which imposes ecological opportunity costs on the regions producing and exporting the commodities, not the often more powerful regions importing them. So just as places A through E in the small-scale example presented earlier can have markedly different interests with respect to the riparian wetland, so also, at larger geographic scales, regions can compete for ecosystem services through the geography of natural resource–based commodity production and patterns of trade. In particular, mining, large-scale timber harvesting, intensive agricultural production, and other natural resource–based, export-oriented industries that severely diminish local ecosystem service provision often take place in regions that are less populous and less prosperous than the regions that are importing the raw materials. To a considerable extent, then, environmental politics is a geographically based struggle over ecosystem service packages that are tied to specific places.

Accounting for Spatial Scales and Ecosystem Panarchies

One tool geographers use to organize this kind of spatial landscape complexity, and thus to make research and analysis more fruitful, is the concept of scale. *Scale* as used here refers to physical dimensions in space and time, and can apply to either the phenomenon under study or the study of the phenomenon (Millennium Ecosystem Assessment 2003). A particular ecosystem process, for example, may be characteristic of an associated spatial dimension, say, a forest microclimate, and time dimension, such as seasonal. The scale of an ecological property, therefore, is defined by the spatial and temporal dimensions at which the property has the most coherence; whereas at other spatial and temporal scales the property takes on little importance (Limburg et al. 2002). Studies of such phenomena also take place at different physical measurements, as in 1 acre versus 1,000 acres, and different time durations. Choice of scale matters for the very simple reason that an ecosystem service that might be visible at one level of phenomenon and study, such as the forest microclimate studied over a season, may be far less visible if a different scale is used, such as a multistate watershed examined over ten years (Millennium Ecosystem Assessment 2003). Moreover, some processes interact with others across scales, whether they are visible at all scales or not, thus making the study of cross-scale properties as important as studies conducted within scales (Christensen et al. 1996; Limburg et al. 2002).

Simply put, *scale matters.* We know that in order to intelligently formulate ecosystem service law and policy, we must be able to describe the chain of delivery from provision source to service user. This will require a deep understanding of not only the ecology of ecosystem services but also the spatial and temporal scales that we use to define the discrete, ambient, and variable characteristics of their geography. There is no superior scale to use for all such purposes, making it important that we recognize the consequences of scale selection in every setting. Accounting for spatial and temporal scales is thus a vital role for geographers in the development of a comprehensive body of ecosystem services law and policy.

The spatial scale associated with a particular ecosystem property is a function of numerous factors, including the home and migratory range of organisms, the area of influence of disturbances, and the distance over which materials are transported and remain ecologically influential (Millennium Ecosystem Assessment 2003). Some properties have very large spatial scales, such as the distribution of carbon dioxide in the atmosphere and its influence on global climate. Some properties have quite limited spatial scales, such as the effects of a shade tree. The mix of services provided by a particular piece of land or body of water, and their values, thus depends largely upon the scale of study one chooses.

Choosing scales of study implicates not only ecosystem scales but also social, political, and economic scales. For example, one segment of society, represented by a local government unit, may value a forest stand for one set of services vital to its local economy, but a larger segment of society may value the regional watershed within which the forest is placed for another set of values operative at regional economic scales (Gottfried et al. 1996). In a very significant way, therefore, the adoption of a particular scale for articulation of ecosystem service values dictates the type of problems likely to be identified and addressed, the range of policy options that can be considered, and the distribution of ecosystem services among the population (Millennium Ecosystem Assessment 2003). Thus it is important for social, political, and economic institutions to have available clear descriptions of characteristic ecological scales and an appreciation of the impacts that the selection of scale for policy assessment can have on ecosystem service valuation.

The dominant model of ecosystem spatial scales has been one built around a *hierarchical* depiction of how ecosystem properties at different scales relate (Holling et al. 2002). Because spatial scales by definition run from small to large, it is convenient to think of ecosystems as being consistently inserted, or nested, into each other from smaller scales to larger scales, with each level subsuming the environment of the ecosystems one level below it (Bailey 1996). The boundaries between each scale level represent the transition points at which interaction of ecosystem processes is more active "vertically" up and down the hierarchy than "horizontally" across a particular level (Costanza 1996). To be sure, this is frequently how ecosystem landscapes organize: the watershed of an ephemeral stream fits along with several others within that of a perennial stream, which in turn fits along with several others of its scale into the watershed of a small river, and so on up to a major river basin. Many ecosystem processes associated with riverine ecosystems can be studied at any of those scales, just as the effects of photosynthesis can be studied at the scale of a leaf, or a tree, or a copse, or a forest, or a region of the nation. It will frequently be the case, however, that the complex adaptive system properties of ecosystem processes lead to new processes emerging as we go up the "ladder" of scales, processes that were not present or evident at lower levels (Bailey 1996). Hence there is much appeal to the conceptualization of ecosystems as falling within a hierarchy of scales.

The hierarchy model, however, has had the unfortunate tendency to promote a top-down representation of scale in which "higher" levels exercise vertical control and dominance of ecosystem processes at "lower" levels (Holling et al. 2002). The common use of the term *hierarchy* would suggest that servient lower levels contribute materials and energy "up" to higher levels, whereas dominant higher levels regulate "down" to lower levels. To be sure, larger spatial scales by definition set physical limits on lower scales, but neither controls the other. In fact, although "nested," smaller within larger, the complex adaptive system properties of ecological processes lead to strong linkages within and between ecosystem scales. The emergent properties that appear at one level of a nested set of complex systems do not owe only that level for their appearance; rather, the system as a whole, including processes at lower levels, make emergent properties possible.

Holling et al. (2002, 73) thus suggest that each level in the so-called hierarchy is better understood as an "adaptive cycle," and that what in fact become nested are not simply hierarchical spatial units but interrelated adaptive cycles, with the cycles in each level potentially influenced by all or some of the cycles at other levels. Hence they abandon use of the term *hierarchy* to describe ecosystem scale and instead adopt the term *panarchy*, after the Greek god Pan, to represent the synthesis and change inherent in the nested adaptive systems (Holling 2004; Holling et al. 2002).

The panarchy model of ecosystem spatial scale, as shown in Figure 2.3, has important implications for ecosystem service policy. It dispels any assertion of scale superiority based on size and more vividly illustrates the point that there is no universally appropriate scale of study or management. This throws into question the arguments that are frequently made in the political sphere that local-, state-, or federal-level institutions are "better" for ecosystem planning and management purposes (Millennium Ecosystem Assessment 2003). Local

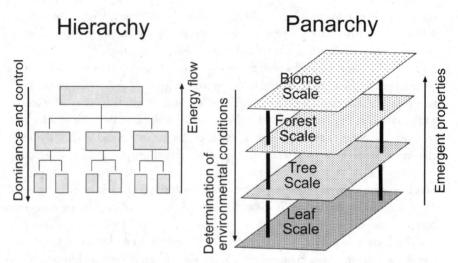

Figure 2.3. The panarchy concept as contrasted with hierarchy. In a hierarchy, control originates at the top and cascades down to lower levels in the hierarchy, while energy is transferred up the hierarchy. In a panarchy, larger-scale phenomena occur as emergent properties of small-scale phenomena but in turn provide structural conditions for and constraints upon smaller-scale phenomena.

social and political institutions are likely to value locally delivered ecosystem services relatively more than will the average member of a regional populous, and centralized state or federal institutions are likely to exhibit preferences for ecosystem services that emerge at larger scales or diminish less strongly with distance. For example, oil and gas drilling in the Alaskan National Wildlife Refuge (ANWR) has long been supported by Alaskans and their elected officials in Washington for the oil revenues and economic development it promises to Alaskans, but opposed until 2005 by senators and congressmen from the "Lower 48" because the arctic wilderness and its wildlife represent a national ecological treasure. Thus managing for ecosystem services through the lens of just local or just regional ecosystem scales fails to recognize cross-scale interactions and the importance that ecosystem processes outside the vision of a particular social, political, or economic scale have to the values that relevant interest group populations seek to gain. Many poor resource management decisions find their roots in this mistake (Christensen et al. 1996). In short, we cannot optimize local ecosystem services by managing only local ecosystem properties, and we cannot optimize regional ecosystem services by managing only regional ecosystem properties. As the Millennium Ecosystem Assessment summarizes,

> If cross-scale interactions in ecological and social systems affecting ecosystem services are common, then it should not be expected that

there is generally a single most appropriate level for response or policy. While responses at certain levels or scales can have disproportionately greater significance or impact, appropriate responses at different levels are in general needed in concert to achieve desired results. (2003, 124)

The upshot for geographers is that multiscale spatial assessment models of ecosystem service landscapes will be a vital component of ecosystem service policy formulation. Surely not every ecosystem service must be accounted for at every assessment scale, but the assessment must provide insight to the linkages that counsel which different scales must be considered in concert to make appropriate policy choices (Rodriguez et al. 2006). For example, the Millennium Ecosystem Assessment (2003, 125) recommends simultaneous use and integration of large- and small-scale assessments so as to identify the important dynamics of the overall system delivering ecosystem services. Tracing ecosystem services from source to user, in other words, is complex not only at any single scale but even more so as a consequence of the need to conduct assessments at multiple scales and then integrate the findings from all relevant scales into a final spatial model.

Accounting for Temporal Scales and Alternative Trajectories

The panarchy model of ecosystem scales, through its depiction of adaptive cycles operating at different, interacting spatial scales, also recognizes the temporal dimension of scale and its close relation to spatial context. Any multiscale spatial model of an ecosystem service, in other words, changes over time as adaptive cycles unfold and, indeed, as ecosystem boundaries move with time. Hence, time scales must be accounted for as well in order to build a complete geographic model of ecosystem services.

Temporal scales of ecosystem properties are defined by factors such as organism life span, material pool depletion rates, and periods between disturbances (Millennium Ecosystem Assessment 2003). Different ecosystem processes have different "speeds" based on characteristics such as the material and energy cycles involved and the response times to disturbance. Fast processes change more rapidly than the organisms or other ecosystem components they influence, whereas slow processes exhibit gradual change relative to the internal dynamics of their ecosystem. In a savanna ecosystem, for example, annual grasses are associated with fast processes, perennial grasses with slower processes, and shrubs and grazers with yet slower processes (Holling et al. 2002). As with spatial scales, which run from small to large, temporal scales thus run from slow to fast, which means that some ecosystem services may be more visible at one temporal scale than another. In the savanna ecosystem, for example, services associated with shrubs may be lost in an assessment conducted over one rainy season, whereas services associated with annual grasses may be overrepresented.

In general, faster ecological processes are associated with higher variability than slower processes (Millennium Ecosystem Assessment 2003). For this reason, slower processes, like the relationship between large and small spatial scales, provide the relatively stable boundaries within which fast processes operate (Holling et al. 2002). Nevertheless, slower processes, though generally more stable than fast processes over short time frames, do change, and over time may become more difficult to steer back to a previous course. A single fire in a grassland ecosystem has dramatic effects on annual grasses and less effect on shrubs. An extended period of drought, however, may introduce a prolonged and more geographically extensive fire regime that eradicates the shrubs. The annual grasses will return when the rains return, but reestablishing the shrubs will take far longer, and they may never reappear in the same assembly as before, if at all, depending on an intervening introduction of invasive species, depletion of soils, or shift in wildlife assembly.

Ecosystems, in other words, have trajectories that play out over time (Savage et al. 2000). Like any complex adaptive system, at any point in time there is an array of alternative future trajectories (Holland 1995; Kauffman 1995). Which path the ecosystem takes, and how far that path diverges from the previous trajectory, will depend largely on the degree of sensitivity the ecosystem exhibits to changes in conditions. Edward Lorenz, the founder of chaos theory, called this sensitivity to initial conditions the "butterfly effect." Resistance and resilience are not infinite. At some point, indeed, slow process change may reach a threshold of irreversibility, a "bifurcation point," setting the entire ecosystem, including the fast and slow processes, onto a course toward a completely different set of dynamic equilibria (Holling and Gunderson 2002). Desertification, for example, is often an irreversible consequence of poor grazing or irrigation practices that affect enormous tracts of land in the Middle East thousands of years after these regions emerged as the cradle of Western civilization. George Perkins Marsh's example of the Mediterranean lands serves as an equally powerful example. Soil has a long memory; once thresholds are crossed, it can take enormous spans of time to rebuild natural capital through ecological processes.

Temporal ecosystem service scales thus present the same modeling and management challenges as do spatial scales. Any particular ecosystem service may depend on fast and slow ecosystem processes for its provision, thus requiring multiscale models. And management for an ecosystem service based on any particular temporal scale risks missing the complete picture, thus raising the possibility that processes operating at different scales will interfere with the management objectives over time, and vice versa. If, for example, we were interested in securing the benefits associated most directly with fast processes operating in the savanna ecosystem, it would be a mistake to neglect the indirect effects slower processes have on their viability, and vice versa. Yet this is a mistake often made, as social, political, and economic systems often pursue higher returns over shorter time scales, thus neglecting the importance of longer time scales to those very pursuits (Christensen et al. 1996; Millennium Ecosystem Assessment 2003). Geographic models of ecosystem services thus must also account for temporal scales across multiscale dimensions in order to provide a complete tracing of ecosystem services from source to user.

Accounting for Cumulative Impacts across Nonlinear Scale Domains

Of course, space and time exist together and are often correlated into what is known as a "scale domain" unifying spatial and temporal models. There is a strong association in ecosystem dynamics between small and fast processes on the one hand, and large and slow processes on the other (Holling et al. 2002). So, for example, characteristic scales of different ecosystem processes may play out over a spectrum of scale domains from small–fast to large–slow, as in the in-seeding of annual plants (patch–annual), to tree replacement (forest–decade), to vegetative succession (region–century), to plant migration (biome–millennium), to species extinction (global–geologic period) (Millennium Ecosystem Assessment 2003). The geography of ecosystem services explores how these different scale domains work and interact to produce ecosystem services and make them available to human populations in spatial and temporal contexts.

People are generally concerned with the small and the fast, as in what happened today in the neighborhood. To the extent people increasingly appreciate the value of ecosystem services, therefore, we can expect them to favor services that are characteristic of small–fast ecosystem process scale domains capable of delivering tangible benefits on an immediate, local scale domain. As the preceding discussions of spatial and temporal scales have suggested, this often leads to the misperception that the services directly associated with small–fast processes can be sustained simply by managing small–fast processes, making the importance of large–slow scale domains harder to impress on social, political, and economic dialogue. Cairns and Niederlehner sum up the problem concisely:

> People most easily appreciate ecosystem services and environmental problems (threats to or failures in ecosystem services) that are intense, local, and immediate. . . . However, as an ecosystem service or environ-

mental problem becomes less intense, more widely dispersed, and occurs chronically, its perception directly or personally is more difficult. In addition, cause-and-effect relationships become less obvious, more uncertain, and, therefore, less likely to motivate action. Low-intensity stresses that cause subtle rather than obvious damage, damage that is spotty or thinly dispersed over a wider area, and damage that will likely occur only over the long term are all less obvious threats to human quality of life, harder to quantify, and less likely to motivate a management response. (1994, 936)

As this insight suggests, "the mismatch between the spatial and temporal scales at which humans make resource management decisions and the scales at which ecosystem processes operate presents the most significant challenge to ecosystem management" (Christensen et al. 1996, 678). The most difficult of this kind of phenomenon to model and manage involves two properties we know are commonly found in complex adaptive systems such as ecosystems: emergent properties arising from cumulative impacts, and nonlinear trajectories arising from sensitivity to initial conditions. The cumulative impact property of emergence involves the aggregation of numerous actions with minor effects at one scale domain into relatively significant effects at another scale domain. One familiar example in the environmental context is the massive loss of extensive wetland resources that has occurred over large areas and long time frames primarily as the result of countless discrete events of small-scale conversion of wetland areas to farming and, more recently, urban development (Ruhl 1999). No single conversion of a wetland to a farm or shopping mall seems very important locally this week, but the aggregate of many such events through time and space can produce massive environmental degradation at larger scales over longer time frames. The slow accumulation of carbon dioxide and methane due to fossil fuel combustion and certain agricultural activities is another obvious example where each automobile, each rice farm, each head of cattle, each electricity consumer has a trivial impact on climate, though aggregately over decades to centuries, the impact is profound.

Moreover, the aggregation effect in cumulative impact scenarios like these cannot be assumed to be, and seldom is, a linear relationship confined to the same spatial and temporal scale. As the Millennium Ecosystem Assessment (2005) observes, "nonlinear changes, including accelerating, abrupt, and potentially irreversible changes, have been commonly encountered in ecosystems and their services" (88). Examples of these agents of nonlinearity include disease, invasive species, algal blooms, and climate change. The effect of aggregation of events in small–fast scale domains, therefore, may not emerge in large–slow scale domains until after an extended period of time, but then may do so quite suddenly, with a bang so to speak. At that point it may be too late to restore the trajectory of the large–slow processes to its previous path, and a quite divergent new path may begin to affect even the small–fast processes—in other words, the threshold of irreversibility for the entire set of ecosystem processes, from small–fast to large–slow, may have been crossed (Carpenter et al. 2006).

Cumulative impacts with nonlinear, cross-scale emergent effects are the most difficult to model and predict. Cairns and Niederlehner explain the problems inherent in such an undertaking:

> Cumulative impact assessment recognizes that individually minor stresses can be significant when they are aggregated through time or space. As such, the scales on which various environmental problems are studied are not always sufficient to recognize cumulative environmental outcomes. . . . Only by expanding the scale of interest can cumulative effects of human actions on ecosystem services be addressed. However, expanding the scale of interest depends on ecological models whose accuracy often cannot be definitively established. (1994, 936)

In response to this challenge, Bailey (1996) is representative of geographers who call for the development of a more disciplined approach to ecosystem geography as "the study of the distribution pattern, structure, and processes of differentiation of ecosystems as interacting spatial units at various scales" (15). Landscape ecology, with its focus on the role of patches and corridors, fragmentation and connectivity, in regulating flows of water, nutrients, and species though space, represents a related school of thought that has important implications for the management of ecosystem services that emerge from landscape pattern (Forman and Godron 1986). For example, landscape ecology has taught us that riparian corridors are critical in filtering water as it flows from the land surface to stream channels and for facilitating the migration of both plant and animal species (Malanson 1993). The fragmentation of forests can reduce their ability to harbor forest interior species, such as spotted owls, while enhancing habitat for edge species, such as white-tailed deer (Forman and Godron 1986). Much as will be the case for ecologists, therefore, geographers have opened a new horizon for research and modeling around the unifying theme of multiscale assessments of ecosystem services.

The rapid emergence of geographic information systems (GIS), remote sensing satellites, and computing power in general has also made possible a new, very rigorous mode of geographical analysis and modeling that is relevant to the problem of ecosystem services. Specific land-ownership and land use units can now be represented in a GIS; their topography, vegetation cover, soil, and climate can be characterized; their spatial relationships to one another can be measured; and their natural capital can be assessed. Future landscapes can be modeled based on probabilities of land units converting, say, from crop production to suburban tracts, or from pasture to crop production, due to not only environmental characteristics and spatial relationships but also the economic and policy environment that the landowner responds to as an independent agent.

Moreover, environmental process models from the huge and global, like the global circulation models that use supercomputers to predict future climatic changes due to greenhouse gas emissions, to watershed-scale models that assess relationships between land use and water quality, are beginning to make possible a rigorous assessment of natural capital and ecosystem service provision. Nevertheless, due to the feedback mechanisms and nonlinearities inherent in both ecosystems and societies as complex adaptive systems, reliable predictions of the effect of deliberate management actions or of specific policy initiatives on ecosystem services remain elusive. Advancing these ecological and geographic foundations will be critical steps toward building the institutional capacity needed to trace ecosystem services from sources to users, which in turn will be essential in refining the third and final contextual component needed for policy analysis in ecosystem services—economic theory.

3   Economics

Tracing ecosystem services from source to user tells us who benefits but not by how much. Nor does an appreciation of the ecological value of ecosystem processes translate directly into a measure of their economic value to people. Something more than a description of the ecology and geography of ecosystem services is needed to complete an exploration of their overall policy context. In short, we require a method for their *economic* description—how allocative decisions are made regarding their uses and how relative value is assigned to them.

One economic metric—some would say the most relevant measure of relative economic value—is price assigned through market transactions. Yet, as ubiquitous as ecosystem services are, it is nevertheless intuitively obvious that generally there are few if any functioning markets in which they are traded. One does not have to purchase photosynthesis or the radiation screening effects of the ozone layer, and therefore no data on market price are available for them. Ecosystem services are, for the most part, free for the taking; however, this does not mean they are without value (Repetto 1992).

To be sure, it may be costly to produce more ecosystem services, such as through the construction of wetlands for their water treatment effects. And it may be costly in some cases to gain access to ecosystem services if, for example, someone else owns the land on which they are being delivered. But why and when would anyone make such an investment? In other words, how do we know that ecosystem services are economically valuable, so valuable as to motivate economic behavior such as investment, production, purchasing, maintenance, or protection?

The quick answer is we know that without ecosystem services, we all die. Or, more realistically, with widespread degradation of ecosystem services, eventually some people would die and many others would be substantially worse off than they are today. But the same could be said of medicine, timber, electricity, physicians, engineers, and a host of other goods and services. The real question, therefore, is not whether we know that ecosystems are economically valuable, but whether we know how valuable they are compared to other goods and services. Only then could we address resource allocation issues that will inevitably arise given the inherent trade-offs among different ecosystem services and, more generally, between ecosystem services and other benefits of technology and natural capital.

This chapter examines the question whether ecosystem services can be assigned some measure of meaningful monetary value notwithstanding the lack of markets. Economists generally employ three techniques for describing price-like values for goods and services that are not traded in markets—the avoided cost and replacement cost methods, revealed preference methods such as travel costs and hedonic pricing, and stated preference methods such as contingent valuation. Each technique has its own set of limitations, but each can also reveal the tremendous monetary values of ecosystem services that are currently "hidden" from the marketplace. Given the level of knowledge about ecosystem services that members of society have at any time, the results of these techniques are first approximations of the values assigned by individuals to ecosystem services and the functions they perform.

Of course, assuming perfectly competitive markets, market prices would tell us more about ecosystem service values than do the surrogate nonmarket methods. In particular, markets can provide relevant information about the economic value of goods and services across many scales. Also, markets, being complex adaptive systems in their own right, can adapt more seamlessly with the evolution of ecosystems and ecosystem services. And as the coevolutionary dynamic between markets and their subject matter unfolds, market prices provide a more continuous, reliable measurement of the value people attach to a good or service. Hence, as informative as nonmarket valuation techniques may be for understanding the monetary value of ecosystem services in general, they may fall short of market-based valuation for purposes of guiding reliable resource allocation decisions. Economists interested in ecosystem services— practitioners of ecological economics—thus make it their business to examine the prospect of whether, when, and how to establish markets or other market-based policy actions designed to motivate and allocate investment in a sustainable ecosystem services economy (Daly and Farley 2003).

Basic Principles of Market Economics

Shouldn't we be happy to leave it that ecosystem services are valuable and free for the taking? After all, getting something for nothing is a good thing, right? Not necessarily. If we care about the capacity of ecosystems to continue to provide ecosystem services, and if we care about who is capable of benefiting from them, then we ought to care about the potential problems that arise from their being "free" or the conditions under which the "assigned" market values are unreliable due to conditions known as market failure.

In a world of perfectly competitive markets[1] in which well-informed producers and consumers who are price takers participate in a multiplicity of factor and product markets, the prices resulting from the interplay of the forces of supply and demand created by the participants provide important information to the participants in the economy. The entrepreneur combines price information from the markets in which he is selling his product with price information from the markets in which he is purchasing his factors of production or inputs to determine how much product to produce and how much input to purchase. In the process of doing so, in theory at least, he compares the cost of producing one more unit of output to the additional revenue obtained from selling that one more unit. As long as the additional cost is less than the additional revenue, the entrepreneur expands production.[2] As this process is carried out across many firms producing the same output, the result is the supply curve or function in the market for the particular output. Those entrepreneurs who can produce less expensively than others will make more profit than those who have higher costs.

While the entrepreneur is producing output, he is also determining how much input he requires. In this case, he is interested in how much additional value he will obtain from adding one more unit of input to the production process. That is, what is the unit's marginal value product. As long as the unit's marginal value product is less than the market price of the input, it will be profitable for the entrepreneur to expand production by demanding more of the input. The price of the output is already included in the determination of the additional value obtained from one more unit of input. Across all of the firms using the input, similar analysis results in the demand for the input in the factor market. In short, our entrepreneur integrates information from two distinct types of markets, those for the products he produces and sells and those in which he procures the inputs and services required to produce the desired output. In terms of ecosystem services, both types of markets are relevant for our discussion. Owners of natural capital producing flows of ecosystem services could well be participating in both types of markets.

Consumers, based on their preferences, income, and a number of other factors, participate in a myriad of product markets where in combinations with others they create the forces that result in the demand side of the product markets.[3] Some of the goods or services the consumers demand might be ecosystem services traded in these markets. Finally, when we combine consumers selling their labor and intellectual services, renting out their land and real estate, lending their financial capital along with owners of natural and environmental resources selling units of their resources and other entrepreneurs producing goods and services used only by others as inputs, we have the source of the supply curves or functions in the factor markets.

Given this system of markets—and there are thousands of them—as the forces of supply and demand interact in each market, factors of production move to those uses in which they earn the highest return. This return is derived from the demand for the products in which the factors are used as inputs. In this way resources move to what is called their "highest and best use"—the use that results in the greatest return or level of satisfaction over and above their cost of utilization.

Through the interactions of literally millions of participants in thousands of product and factor markets, individual consumers seek to maximize utility and income while entrepreneurs seek to maximize profit. Individuals and the markets are continually adjusting to new information, new technology. If individual participants in the markets try to sell a product to consumers at a price higher than consumers are willing to pay or more than the competition believes consumers will pay while still covering their costs of production, those individuals will soon be without customers. Similarly, if a producer tries to undersell the market, unless her cost structure is significantly lower than her competition, she may sell her product in the short run but be out of business in the long run when she is unable to cover her operating and fixed costs. This equilibrating force of individuals operating through impersonal markets while at the same time allocating resources among competing uses to those ends in which they have the highest value is what Adam Smith meant by the invisible hand—the foundation principle of early classical market theory (Farnworth et al. 1981).

Given the assumptions that we started out with as we began to describe our simple economy, a key assumption was that our consumers and our entrepreneurs were price takers—no one is in a position to set prices or unduly influence the operation of the markets as participants interact. Another assumption implied but not stated was that there were markets for the relevant goods and services. The situation results in what is referred to as a Pareto[4] efficient outcome (Daly and Farley 2003; Katz and Rosen 1998): the economy is operating at some point on the Pareto frontier such that it is not possible to make someone better off without making someone else worse off. Once on this frontier defined by the utility functions of its members reflecting their preferences and given the other assumptions, the frontier can be thought of as defining aggregate social welfare. One of the questions we will return to is the role of ecosystem services in the preceding model of the economy: to what extent are they meaningfully allocated among competing uses on the basis of relative prices determined through the interaction of supply and demand?

Total Value, Marginal Value, and Capital Valuation

A price for a good or service derived from this market economy provides a rich amount of information to individuals and society about resource values that is used by the individuals in making allocative decisions. In the most straightforward sense, market price can be used to compute the total value of a good or service in an economy. One simply needs to know the price per unit times the number of units sold, and that is the total value of the good or service. But according to neoclassical economic theory, the most important piece of information the market price conveys is the *marginal value* of the good or service— the price one must pay to secure one more unit of the good or service. From the perspective of the entrepreneur, the marginal value tells her how much additional value she will obtain as she adds one more unit of input to the production process or how much value she will acquire if she sells one more unit of output. For the consumer, it indicates how much he will be willing to pay to consume one more unit of the good or service. Total value tells us how much of the good or service is flowing through the economy, which is certainly useful to know, but marginal value tells us how valuable participants in the market economy consider the good or service is relative to other goods and services—important information, if you are an entrepreneur, in determining how much of the product to produce; or if you are a resource owner, in determining how much of the resource to make available; or if you are a consumer, in determining how much of the good or service to consume.

In efficiently operating markets the interactions of many economic agents result in forces of supply and demand arriving at a price for the particular good or service that is being traded in the market of interest. However, if we step back and look at individual participants in the market, we are very likely to find individuals who demand the product—that is, buyers—who are willing to pay more for the good or service than the market price. Nevertheless, they get to participate in the market at the equilibrium price and thus enjoy a bonus— what economists call consumer surplus—in the form of being able to buy the good or service for a price less than they were willing to pay. On the other hand, there are just as surely some people who would have purchased the good had the price been just a little bit lower. And then there are buyers for whom the price is just right—that is, if it changed just a little bit upward they would drop out of the market. Similarly, there are some individual entrepreneurs who are able to provide the product or service to the market for a cost less than market price. These individuals enjoy a bonus referred to as producer surplus. The lower the entrepreneur's cost relative to the market price, the larger his producer surplus. On the other hand, there are other entrepreneurs who would have provided more of the good or service to the market had the price been just a little bit higher. A consumer is comparing the market price to the additional satisfaction obtained from consuming one more unit of the product, that is, her marginal gain; the entrepreneur is comparing the additional revenue or value from selling one more unit of product compared to its cost of production.

It is in this sense that economists refer to price as the indicator of marginal value, for it defines the line, or margin, between people willing to purchase the good or to supply the good in the market versus those not willing (Pearce 1998; Simpson 1998). The market prices of goods and services, therefore, tell us how much just a little bit more of something is worth relative to a little bit more of anything else. The discovery in classical economics of this informational content of prices led to the solution of the so-called paradox of value—the puzzle of why water, essential to life and of immense total value, is far less expensive in the market than are diamonds, which are of little practical value. So long as the supply of water far exceeds the demand, it doesn't matter how important to life it is—it will be cheap on the market (Opschoor 1998). However, in this paradox of value, a key bit of information is that both the diamonds and the water are actively traded in markets and as such their respective interacting forces of supply and demand arrive at prices that act as signals for all of the participants in the markets: in the case of diamonds—diamond miners, mine owners, jewelers, final consumers; in the case of water—spring owners, bottlers, distributors, consumers. Armed with this information, these participants can make decisions such as expanding or contracting production, adopting new technology, or making new purchases.

For the entrepreneur or the owner of natural capital that can be employed in a variety of productive activities, the economy and its market prices provide vital information useful in the assessment of the relative value or profitability of those alternative activities. For example, if we have a large tract of land that yields a flow of ecosystem services with little or no input on the part of the owner, one way to assign a value to the parcel providing the flow of services would be to capitalize the annual net value of the ecosystem services. This of course assumes that there are active markets in which our landowner is able to sell the ecosystem services. Alternatively, the same parcel of land may be suitable for a number of alternative uses after it is "developed" through its combination with additional amounts of capital and labor. For each of these alternative uses, the owner can develop a capital budget in which she projects a timeline for each alternative use and the associated costs of development, costs of operation, and income or revenue so that she can forecast the net cash flows over the life of each alternative use for the parcel. Once the projected net cash flows are determined, these can be discounted to the present by her discount rate and the net present value[5] of each alternative calculated. Assuming our decision maker is economically rational, she would select that alternative with the greatest net present value or capitalized value. Throughout the whole process, our decision maker has relied on the markets in the economy to provide her with the relevant information—that is, prices—so that she could conduct her analysis. This, in short, is why market prices provide important signals as to the most efficient allocation of capital resources like land or other forms of natural capital.

Ever since Adam Smith first described the market's operative mechanism, many people have criticized market theory for the normative implications of its ostensibly nonnormative premises. Many object to its assumption that people are inherently single-minded maximizers of utility (Ostrom et al. 1999; Simon 1957) and have the mental abilities necessary to analyze all of the relevant data to make maximizing decisions (Simon 1957). And market economics says nothing about who gets what—who will be the haves and who the have nots— on the way to maximizing total social welfare as defined by individual preferences; it also says nothing about societal welfare as a whole (Kelso 1977). Indeed, it says nothing about the particular form of the social welfare being maximized. In other words, it does not account for what might be perceived as distributional inequities and other social consequences one might find objectionable even about a perfectly competitive market outcome (Daly and Farley 2003). Neoclassical market theory also rests on a set of assumptions about nontrivial matters such as the information available to market participants, including the ability to arrive at representative discount rates, many of which may be unrealistic when applied to the complex nature of ecosystem services in our contemporary world.

Putting those concerns aside for the moment, there is no reason why, *in theory*, the basic economic model cannot be applied to ecosystem services to avoid inefficient resource allocations. Natural capital resources are capable of providing, among other things, economically valuable services to humans. The capitalized value or net present value of a natural capital resource could thus be computed for its variety of ecosystem services and compared to its value when put to other uses, such as production of commercial timber or conversion to a shopping mall. Whichever use comes out on top in that analysis is the one the owner, and society, should choose unless there are strong reasons otherwise.

In theory, this works just fine. In application, however, this model faces numerous obstacles to its implementation in the context of ecosystem services.

Public Goods and Positive Externalities

To understand why the basic model of market economics has so much trouble gaining traction in the context of ecosystem services, it is useful to start with the simple question of why ecosystem services are, or appear to be, freely available. Indeed, much of this attribute has to do with the ecology and geography of ecosystem services—they are the result of ecosystem processes that operate in open complex ecosystem settings and that deliver services to humans though a myriad of different landscape settings. Even when we know exactly how an ecosystem service is provided and precisely where its natural capital source is physically located, it can be quite difficult to allocate it through the mechanisms of the market's "invisible hand." How, for example, would anyone sell or buy photosynthesis? Moreover, even when someone can control the provision of an ecosystem service, such as the owner of land on which is located a wetland area that provides downstream flood control benefits, whom would the person charge for the service, and how?

The market-based resource allocation mechanism works best when the goods and services being traded are *private*[6] in nature—that is, when sellers can control their distribution and deny access to them unless the price demanded is paid. Private goods and services thus are what economists call "excludable" and "rival." They are excludable on the supply side in the sense that exclusive ownership is possible, as in food, a car, or a home, thus allowing the seller to ensure that only the buyers can use or consume the good or service (Daly and Farley 2003; Randall 1983). They are rival, also known as "subtractable," (Buck 1998; Ostrom et al. 1994) on the demand side in the sense that each person competes with all others for the benefits of use either in consumption or as an input, as in seats at a ball game, a glass of wine, or a ton of steel used in manufacturing (Daly and Farley 2003). These supply-and-demand qualities allow the supplier of the good to charge for its use and to take advantage of competition for access among those demanding the good or service in order to obtain the highest possible price given the dynamics of the market (Frischmann 2005).

Ecosystem services, by contrast, are often much closer in economic behavior to what economists characterize as *public goods,* the classic example of which is air,[7] in that they are either not completely excludable, or not completely rival, or, in the case of a pure public good such as air, neither excludable nor rival to any degree (Daly and Farley 2003). Some public goods may exhibit more or less rivalness depending on the scale of usership. A country road may appear to be wide open to a lone driver, but a traffic backup caused by a farm tractor may irritate that same driver, upset that he must share the road. This effect, known as "congestion," can lead to competition for access to a public good (Daly and Farley 2003). Although congestion is usually not a factor in the use of ecosystem services such as photosynthesis, it can be a problem in the use of other ecosystem services, such as the cleansing services of the body of water being strained by the dumping of waste in it thus interfering with other uses.

The challenge for an owner of a wetland area, for example, is that she cannot practicably charge a price to dictate which downstream landowners derive the flood control benefits of the wetland area, and the downstream owners cannot crowd each other out for access to the benefits. Hence, if one person does for some reason pay for the flood control services, her use of them does not deprive other downstream landowners of the benefit. So why would anyone pay for the flood control services? Consequently, while the wetland is privately owned and as such is a private good, the ecosystem service of flood control it provides is a public good and one from which the wetland owner can derive little direct benefit. Alternatively, if we were considering the ecosystem service of duck habitat provided by the wetland, the owner might be able to derive some economic benefit from the sale of exclusive hunting rights to duck hunters. As just described, the public good aspect of many ecosystem services derived from privately owned natural capital results in the owners of the natural capital undervaluing it, or at least in them not considering the possibility of value derived from natural capital's diverse streams of ecosystem services when placing a value on the natural capital. In essence, this is a problem of missing markets or what economists refer to as market failure[8]—there is no way for the demand for flood mitigation services from wetland parcels to become an effective demand in the market to which wetland owners can respond.

Of course, in many instances people generally don't pay for ecosystem services, a problem discussed in more detail below. But more important, because many ecosystem services behave like public goods or close to it, owners of natural capital resources seldom have any real incentive to provide the ecosystem services in the first place. As far as the owner of the wetlands is concerned, the flood control benefits are what an economist would call a *positive externality*. An externality is any cost or benefit from the production or consumption of a good or service that is not borne or enjoyed by the producer or consumer but is borne or enjoyed by a third party (Baumol and Oates 1988; Simpson 1998).[9] A chemical factory that freely pollutes the air and faces no consequences for the environmental or health damages of the pollution is creating a negative externality—through the act of polluting, the owners of the factory are using the air as a dump or sink for some of their waste products. Because the associated costs of waste disposal are not borne internally by the producer, the producer does not need to recover them in the market and thus will perceive the market clearing price as providing more returns over his production costs than would have been the case were the pollution control costs paid by the producer as part of his costs of production. The results include the pollution and the physical and monetary damages it creates for others in society, an oversupply of the chemical whose production results in the pollution, an overcommitment of resources to the production of the chemical, and a lack of incentive to find production practices that are less polluting per unit of production. From an economic perspective there is market failure with the resulting price signals for the commodities involved being unreliable indicators of their values and the invisible hand resulting in suboptimal results from the perspective of social welfare.

Conversely, the wetland owner who cannot charge for flood control benefits accruing downstream derives no gain from them and thus will not take them into account when determining the value of the land as a capital resource. Alternative uses that transfer the land into some other function, such as farming or shopping malls, may appear to yield a higher net present value, but might in fact be lower in value, or at least less attractive in relative value, were there a way for the wetland owner to capture revenue from the flood control benefits.

As Farnworth et al. (1981) identified decades ago, this reveals the central economic quality that challenges the formulation of ecosystem services law and policy—that is, because ecosystem services are often the positive externalities associated with ownership of natural capital resource, resource owners seeking to maximize gain do not take them into account when deciding how to use the resource (Balmford et al. 2002; Daily and Ellison 2002; Daly and Farley 2003; Guo et al. 2000; Heal 2000). If other uses such as farming or shopping malls yield a higher return than whatever the stream of uses provided by the ecosystem services from the same units of natural capital yields, the resource owner is likely to turn to those alternatives without taking the unrealized value of the ecosystem service into account. That would be the economically rational course of action. But of course, if all resource owners behave in that individually rational manner, the supply of ecosystem services will diminish, even though we all know they are vital for our enjoyment of life, and the supply of farms, shopping malls, and other land uses will be too high. Hence, a conundrum: if we assume we have a perfectly competitive, market-based economy, then the result just outlined is the socially optimum one arrived at through the operation of the invisible hand working through a multitude of markets and millions of individual decision makers acting to maximize their income/utility or profits. However, if the result is the loss of critical ecosystem services necessary to sustain life—a critical component of social welfare, one could argue—then we would have to question if the perfectly competitive, market-based economy outlined earlier is really accomplishing all it is supposed to do, at least from a societal perspective. Earlier in this section, we pointed out two areas in which there is market failure, where the market fails and as a consequence the results are suboptimal: externalities and public goods.[10] As Costanza and Farber (2002) summarize, "[a]ctions that treat resources, the environment, or ecosystems as if they were 'free' when they are not . . . can only lead to reductions in potential human welfare or increased real costs of maintaining flows of ecosystem services in the long run" (367–68).

There is more than economic theory to substantiate this concern. Balmford et al. (2002), for example, reviewed over three hundred case studies to compare estimates of marginal values of goods and services delivered by an ecosystem when relatively intact versus when converted to typical forms of human land uses. They describe their findings as

> highlight[ing] the fundamental role of market failures in driving habitat loss. In most of the cases we studied, the major benefits associated with retaining systems more or less intact are nonmarketed externalities, accruing to society at local and global scales. Conversion generally makes narrow economic sense, because such external benefits (or related external costs, as in the damage caused by shrimp farming) have very little impact on those standing to gain immediate private benefits from land use change. (952)

So, indeed, we should not be happy that ecosystem services are free for the taking.

The Power of Information

Many ecosystem services truly behave as public goods, in the sense that there is absolutely no way through a conventional market system for the person owning the natural capital resource that provides an ecosystem service to make it rival and excludable. Of course, not all natural capital resources owners are seeking to maximize utility in the form of economic gain. Public entities,[11] for example, own substantial resources and may have no intention of behaving according to the principles of rational economic gain. In other words, public entities may pursue policies, such as maximizing ecosystem service provision, that reflect social rather than economic agendas. Moreover, some private entities, such as land trusts, have goals in mind other than maximizing economic gain. But the reality is that a significant proportion of natural capital resources are in the hands of private landowners who *do* wish to maximize primarily economic gain.

It is not the case, however, that all ecosystem services will behave as pure public goods when the natural capital resource owner seeks to maximize economic gain. More realistically, the ability of the owner of land to control the provision of ecosystem services flowing from the land falls on a continuum from none at all to a great deal. Indeed, in the absence of proscriptive regulation or some other legal constraint on the use of property, many landowners have the practical option of converting land that provides ecosystem services, such as flood control from wetlands, to some use that does not, such as a shopping mall. In these cases, therefore, the owner at least has the ability to regulate the volume and timing of ecosystem services provisioning simply by manipulating the attributes of the land that supply the service. This equates to the functional power to veto use of the service by others, thereby forcing them to appreciate the economic value of the service as they suffer the consequences of its absence in the form of, say, flood damage or the costs of other flood control measures. With sufficient information about the economic value of the services to others, therefore, we should expect to observe owners with this level of control attempting to engage in negotiations with the service beneficiaries to charge a fee for continued supply of the service (Buchanan and Yoon 2000).

It turns out that market prices are not the only source of such information. Consider, for example, the findings of researchers studying the effects of proximity to undisturbed forest habitat on the productivity of coffee plantation trees (Kremen et al. 2004; Ricketts 2004). In Costa Rican coffee plantations, for example, Ricketts (2004) found that coffee trees close to forests were 20 percent more productive than trees farther away. The study attributes this incremental gain in productivity to the fact that the trees receive a greater number of visits from wild pollinators and also enjoy a greater diversity of wild pollinator species, thereby providing more stability in the available pollinators over the course of seasonal and natural disturbances. The effect of this "free" ecosystem service translated into $62,000 in additional annual revenue for a typical plantation compared to one not enjoying the pollination service benefit.[12] The value of the forest habitat were it transformed into grazing land (which is considered a "high-value" use of the land under current thinking) would be about $24,000 per year for a parcel of the size needed to generate the observed pollination benefits. Clearly, then, coffee plantations nearer to these forest habitats are reaping an ecosystem service–based use of the forested land that is capable of being valued in hard dollar terms and compared, in this case quite favorably, to an alternative use of the land.

Before the owner of the forested tract learned of this research, the pollination services the tract provides would have behaved for that owner's purposes as positive externalities. Indeed, the owner quite possibly would not even have known about them as an ecological phenomenon, much less as an economic benefit. Assuming the forest owner does not have a legal duty to continue providing the pollination services to the coffee plantation and faces no general legal restriction against removing the forest resources, conversion of the forest into grazing land would have been an economically rational choice.

Once the productivity-enhancing value of the pollination for the coffee plantation is known, however, the forest tract owner is not resigned simply to bemoaning the fact that this newfound economic benefit is literally flying away. Rather, because the forest tract owner has the capacity to eliminate or reduce the opportunity for pollination by managing levels of forest cover and density, it is possible to open the door to economic bargaining. In short—and still assuming no legal duty to provide the pollination services or general restriction against manipulating the forest resources—the forest tract owner could give the coffee plantation owner a choice: provide compensation for keeping the forest, and thus the pollination services, intact or else the forest will be converted to grazing. Naturally, to be of interest to the forest owner, the compensation would need to be an amount that exceeds the value to the forest owner of putting the land into grazing. But given the values the research has revealed are associated with the pollination services, the coffee plantation owner may very well conclude it makes good economic sense to provide such compensation. The two parties, in other words, now have the opportunity, knowledge, and motivation to strike a bargain.

To be sure, this opportunity for selling the ecosystem service will not produce as seamless a valuation system as a fully operational, multiagent market could supply. In this bilateral "market" of one seller and one buyer, where the parties wind up as the final compensation point will be more up to relative knowledge asymmetries, bargaining skills, and bravado than any sort of market clearing price. But the negotiation of compensation between the two parties does provide a very tangible measure of ecosystem service value for them in the defined circumstances. In short, research providing direct knowledge of the economic benefit of an ecosystem service is a powerful tool that may prompt natural capital resource owners to initiate negotiated transactions that bring about more rational use of the resources. Nevertheless, a number of barriers exist to replication of this experience more broadly into a market system of resource allocation that takes ecosystem service values into account.

Information Costs and Alternative Nonmarket Valuation Techniques

One obstacle to making the kind of ecosystem service transaction described in the coffee plantation scenario more prevalent is the cost of information (Balmford et al. 2002). The kind of research that led to a direct understanding of the value to coffee plantations of wild pollinators can be expensive and take considerable time to conduct. Such research is also unlikely to provide generalizable information about the value of wild pollinators that would be useful in other settings, such as the value of wild pollinators to apple orchards in Maine. As a practical matter, in other words, it would require a tremendous investment of time and resources to generate this kind of direct information about economic value for all ecosystem services in all their delivery settings. Indeed, the research may cost more to conduct for particular benefited parcels than it is worth from an economic perspective—it would make little economic sense for the parties in the coffee plantation scenario to spend $400,000 to discover that the pollination services are worth, say, $10,000 annually to the particular coffee plantation.

Fortunately, economists can employ other, less costly and time-consuming methods of value estimation in the absence of operating markets. These approaches, which have been used increasingly in the context of ecosystem services, fall into three general categories—the avoided cost and replacement cost methods, revealed preference (also known as inferential valuation) methods such as travel costs and hedonic pricing, and stated preference methods such as contingent valuation (Augustyniak 1993; Champ et al. 2003; Colby 1989; Van Wilgen et al. 1996; Goulder and Kennedy 1997; Simpson 1998; Wilson and Carpenter 1999; Heal 2000; Loomis et al. 2000; de Groot et al. 2002; Farber et al. 2002; National Research Council 2004a). There may also be noneconomic indicators, such as certain ecological attributes, that could act as surrogates for economic value (Boyd 2004; Boyd and Wainger 2002a; Gustavson et al. 2002; Salzman 1997).

The avoided cost and replacement cost methods focus on the economic costs associated with the consequences of and alternatives to the particular ecosystem service. In the coffee plantation pollination scenario, for example, costs that are avoided by maintaining the wild pollination include not only the lost coffee bean production but also perhaps higher costs of financing and insurance as a result of lower productivity performance and any other costs caused by the lower yield per tree. The avoided costs, therefore, could well exceed just the forgone coffee bean revenue of $62,000 annually. On the other hand, there may be no need to suffer all those avoided costs if there is a cost-effective substitute for the wild pollinators. Domestic pollination services by professional beekeepers, for example, are used throughout many crop-producing regions. If the coffee plantation could secure such services for, say, $30,000 annually and recover the coffee bean yield to the same levels as experienced when relying on wild pollinators from the forest tract, doing so would make economic sense.

The revealed preference methods rely on imputing some of the value of one good or service that is traded in a market to the presence of another attribute that is not traded in a market. In hedonic pricing, for example, the land values of different coffee plantations can be inferred to reflect coffee tree productivity, which is affected by, among other things, access to wild pollinators. If the pattern of coffee plantation land values is shown to rise with closer proximity to forested tracts, therefore, it may be reasonable to assign that incremental value to access to wild pollination. Similarly, the travel cost method assumes that if people spend more time and money to travel to one location versus another, it must be because of some superior attribute in the former, to which the value of the incremental expenditures can be assigned.

The stated preference method of contingent valuation involves direct surveys of people to ask how much they value a particular attribute of a good or service. In the context of ecosystem services, for example, a survey could ask coffee plantation owners how much they would be willing to pay to prevent loss of nearby forest habitat. While this method is the only one that can reveal useful information about nonuse values (Baker 1995; Gatto and De Leo 2000), it is unnecessary to employ it when one of the other methods is reliably available.

Using any of these methods to examine ecosystem service values may also reveal correlations between economic value and ecological attributes, thus allowing the ecological attribute indicators to take over as the generalized valuation method. For example, hedonic pricing may show a highly significant correlation between the land values of coffee plantations and the density of the forest canopy within 1 mile of the plantations. Once that correlation is established, hedonic pricing would no longer be necessary if a forest canopy analysis could be performed quickly and inexpensively, say, from a standard aerial photograph or satellite image.

Any or all of these methods may prove useful in assigning generalized estimates of economic value to ecosystem services at costs that are not prohibitive to negotiated transactions. For example, it may take just one phone call to find out the cost of domestic pollination services, which the coffee plantation owner then could place as the upper limit on paying compensation to the forest owner. On the other hand, each of the methods has serious limitations, particularly in the context of trying to value ecosystem services (Daily et al. 2000; Wilson and Carpenter 1999).[13]

As is explained in chapter 1, managing ecosystems for specific ecosystem services is difficult to begin with, and where successful is likely to alter other ecosystem service flows from the same ecosystem as well as from other connected ecosystems. To provide a complete value estimate, therefore, avoided cost methods must account for the positive and negative economic effects felt throughout the chain of connected ecosystems (Southwick and Southwick 1992),[14] and replacement cost methods must ensure that all the lost service values in all the connected ecosystems, not just the one under consideration, are replaced to full value. Because they focus on total value, moreover, neither the avoided cost method nor the replacement cost method can be used to assign marginal value to the ecosystem service. That is, they do not yield a value that is reflective of the decision making of the participants in the market economy outlined earlier in the chapter.

The complexity of ecosystems also complicates any effort from revealed preferences to assign value inferentially to any single attribute, such as pollination services, through hedonic and travel costs methods. Similarly, contingent valuation methods have been criticized generally as unreliable given the propensity people may have to answer survey questions without a complete appreciation of the context or without a sober assessment of their own behavior when money really is on the line and opportunity costs[15] become more real and apparent.

Finally, using ecological indicators to draw conclusions about economic value requires that at least one of the valuation methods was employed at some point in the process of correlating ecological attributes with economic consequences, so is only as reliable, at best, as was that method in the context in which it was used. Hence, while the nonmarket valuation techniques may alleviate the problem of information costs for some cases, they do not take the place of market-based prices as a metric of economic value.

The Problem of Transaction Costs

Assuming the coffee plantation owner and the forest tract owner have equal, low-cost access to reliable information about the value of pollination to the coffee plantation, the costs of substitutes for wild pollinators, and the value to the forest owner of conversion to grazing, basic economic theory suggests that they will come to an agreement that efficiently allocates all the resources involved. In this scenario, given the relative values, the efficient outcome clearly will be for the coffee plantation owner to pay the forest owner somewhere between $24,000 (the grazing value) and $30,000 (the cost of domesticated pollination services) annually to maintain the forest. They are both individually better off under this option than under their respective alternatives, and society as a whole enjoys a more efficient use of resources.

So why does this not happen regularly where information is readily available about the values of ecosystem services? One important obstacle is what economists call *transaction costs*—these include the costs of acquiring sufficient information for contracting, bargaining, and consummating the transactions as well as attendant costs of monitoring, policing, and enforcing transactions and agreements after the fact (Bromley 1978; Coase 1937; Williamson 1979). Basically, the costs of consummating and monitoring transactions such as the purchase of wild pollination may be so high as to offset the efficiency gains so much as to make the transaction not worthwhile to the interested parties. In the case of pollution, this hurdle was explored by Ronald Coase (1960) in his famous article "The Problem of Social Cost." Coase explored the kind of two-party transaction described in the pollination scenario to show how the parties, assuming each had sufficient wealth to "buy out" the other's interest, *could* negotiate an efficient level of pollution regardless of the initial allocation of property rights as long as transaction costs were assumed to be zero (Baumol and Oates 1988; Cole 2002; Randall 1983). More important, however, Coase's primary purpose was to show how the transaction costs associated with such bargaining interfere with this possibility in more realistic settings (Daly and Farley 2003; Edwards 2003; Simpson 1998; Zerbe and McCurdy 1999).

Consider, for example, what would happen were the forest area supplying the wild pollination owned not by a single person but by several dozen people in separate parcels. This would complicate the coffee plantation owner's ability to negotiate as each owner would have to be located and separate negotiations may need to be held. Each parcel owner may have particular special demands or limitations, meaning that no standard form of transaction can be used. Moreover, perhaps not all the different forest parcels provide the same pollination benefit per acre, or, to make matters even more complex, perhaps the pollination benefit is not proportional to forest area, meaning that there is no point to the coffee plantation owner of buying any forest parcels unless some minimum aggregate area can be secured. All of these complicating factors add cost to the negotiation process itself, making the value of securing the wild pollination less attractive than it may have been when compared to using domestic pollination services.

As the discussion of ecosystem service geography in chapter 2 revealed, the physical landscape of ecosystem services is often likely to present just this kind of set of barriers to negotiation. Far from resembling the "small-number" scenario of the one-on-one negotiation between a single forest owner and a single coffee plantation, the ecology and geography of ecosystem services will more often produce what Baumol and Oates (1988, 10–11) call the "large-numbers case." For example, flood control benefits of wetlands may often be experienced for long stretches of river systems. There may be numerous owners of different wetland parcels responsible for the supply of the benefits, as well as numerous owners of different benefited parcels. Even to initiate the kind of transaction in mind, a sufficient number of wetland parcel owners would need to identify each other, agree to threaten to eliminate the wetlands (assuming no legal restriction), evaluate the flood control benefit values, identify the flood control beneficiaries, develop a strategy for negotiating with them, and devise a method for allocating any payments received among the group. Or, each wetland parcel owner may say, why not just sell out to a shopping mall developer? Clearly, transaction costs like these must be overcome in order to make the relative value of ecosystem services take on real value in any market sense. If the cost of engaging in the market outweighs the gain to be had by doing so, there will be no market, no motivation for the transaction.

Open-Access Resources and Free Riders

As is the habit of economists, the additional assumption of no transaction costs can be made in order to advance the economic description of ecosystem services. Now, however, our forest owner faces the problem that control over the pollination ends at the property line under our private property system (discussed in more detail in chapter 4). In other words, once the pollinator crosses the property line, the pollination service is nonexcludable—the forest owner can manipulate the pollinator habitat to reduce or eliminate the pollination benefit but cannot control what happens once the pollinators leave the forest. The pollinators might visit the coffee plantation with which the forest owner has entered negotiations, but they might not, or if they do, they might visit other coffee plantations as well. As such, pollination, like many other ecosystem services, behaves as what economists call an *open-access resource*.[16] All the forest owner can do is prevent *everyone* from enjoying the benefit, but once the benefit is made available to anyone, it is available for all to capture.

Of course, there are some limitations on how up for grabs pollination is in our scenario. To begin with, anyone seeking it must operate a coffee plantation within range of the pollinators. Presumably, moreover, pollinators do not have boundless energy, so closer proximity to the forest may be an advantage. From the perspective of the coffee plantations, therefore, pollination may exhibit some degree of rivalness, as is true for many open-access resources (Daly and Farley 2003). But within these conditions any coffee plantation in the relevant area can expect to enjoy at least *some* pollination benefit so long as the forest owner maintains *any* pollinator habitat.

Undaunted by this concern, and (we assume) enjoying zero information costs and zero transactions costs, it should nonetheless be easy for the forest owner to visit each coffee plantation with a proposed compensation figure. Yet why would any coffee plantation owner pay the price asked? Rather, an economically rational owner will hold out to see if any other owner pays the price, because once pollination is secured for one coffee plantation, it is secured for all!

Economists refer to this prospect as the *free rider* problem, and it plagues the management of open-access resources (Daly and Farley 2003; Farnworth et al. 1981; Simpson 1998). As Ostrom et al. (1999) observe, "when resource users interact without the benefit of effective rules limiting access and defining rights and duties, substantial free-riding in two forms is likely: overuse without concern for the negative effects on others, and a lack of contributed resources for maintaining and improving the [resource] itself" (279). Generally, in the absence of the ability to control the distribution of the resource itself—a given with ecosystem services such as pollination, flood control, and photosynthesis—these free rider problems can be overcome only through collective action by the users of the resource, whether voluntary or coerced. For example, the coffee plantation owners could all agree to contribute to a pot of funds with which to negotiate with the forest owner as a united front. This approach, however, suggests transaction costs including monitoring and enforcement. There may be informal enforcement mechanisms for such agreements—from a cold shoulder at the local market to exclusion from a coffee seller's cooperative—but these have both practical and economic costs as well. Alternatively, a public authority could impose taxes on the coffee plantations to raise the funds to secure the forest habitat, yet here again there will be administrative and enforcement costs. But the merits of these and other options of collective action are getting ahead of our story—more on them later in part II—for the main point to be made here is that free rider problems are a background concern that, if not addressed, will inevitably complicate efforts to turn knowledge of ecosystem service values into actual economic returns through multiparty transactions.

Economic Valuation and Ecosystem Service Scales

Let's say we have achieved the unlikely state for a particular ecosystem service of having overcome information costs, transactions costs, and free rider problems. In other words, we have identified a relatively inexpensive and widely available nonmarket method of estimating the value of the service, the owners of the natural capital resources that provide the service can easily identify and negotiate with the beneficiaries of the service, and the service can be provided in a way that prevents free-riding. Overlaying this set of assumptions on the coffee pollination scenario may open the door to frequent negotiation between coffee plantation owners and forest owners, leading, over time, to an efficient allocation of coffee plantations, forested land, and grazing in the region. Another problem for a more general economic description of ecosystem services, however, stems from the spatial and temporal scales at which the value of the service is known versus the scales at which it is being delivered in a particular setting. In other words, transporting the coffee pollination experience elsewhere, or even long into the future for where it transpires, is an unreliable exercise.

There is always some risk in moving between scales with limited economic data. Knowing the gross national value of, say, annual tomato production does not provide anyone the slightest insight about the price of tomatoes tomorrow in Pittsburgh. And of course the reverse is also true—the price of a single tomato at a local market does not alone provide information about the overall tomato economy for a region or nation. Just as with geography, therefore, spatial and temporal scales complicate the description of economic phenomena, with ecosystem services being no exception (Balmford et al. 2002; Bockstael et al. 1995; Opschoor 1998).

At least with tomatoes, however, the product is relatively fungible and market prices are available, making movement between spatial scales and time periods less incoherent than is the case for ecosystem services. A person can walk into the produce section of a market in San Francisco and, with little effort, draw comparisons to tomatoes he or she purchased a day or a year before in Miami. But the flood control values of a wetlands area depends tremendously on location and timing.[17] An analysis of value of avoided cost in one month of the flood control benefits of a wetlands area in an urbanized Florida county may be wildly divergent of an analysis conducted for the same wetlands during a dry season a month later, and even more so for a wetlands of equal size in, say, a remote area of Wyoming. Thus, when Costanza et al. (1997) computed annual global value of different ecosystem services, their figures provided relatively little utility for value estimation of the same ecosystem services at local levels (Pearce 1998; Salzman 1997).

And as much as location matters, so does size. The replacement cost analysis for an area of wetlands cannot be assumed to provide a unitized value, such as so many dollars per acre. A wetlands area half the size of another may provide only one-tenth the flood control value, making generalized per-acre estimates from either baseline useless without a full understanding of the continuum of proportionate service benefit effects experienced when total area varies. Once again, as an example, the global ecosystem service value figures Constanza et al. (1997) compiled cannot reasonably be used to estimate per-acre values for the different ecosystem types associated with each service.

Ultimately, moreover, time will also matter for valuation of ecosystem services not only because of variable ecological conditions but also because of variable social and technological conditions. As population in an area increases, and along with it residential and commercial development expands, the same wetlands area may produce flood control benefits of the same or less ecological value but of increasing economic value. On the other hand, the importance of flood control to rapidly populating urban areas may lead to technological advances that lower the replacement cost value of wetlands, making them less valuable relative to other land uses over time as the replacement technology becomes less costly and more widely available (Brown and Lant 1999).

These kinds of spatial and temporal discontinuities have led many to question the reliability of even simple habitat trading programs designed to provide compensatory ecological value when habitat is degraded as a result of land development. These so-called banking programs allow a land user to develop habitat at one location and compensate for the ecological losses by purchasing "credits" in areas that have been enhanced and managed to deliver the same ecological values. Without careful design and implementation, such programs, even in the best of circumstances, run the risk of not producing truly compensatory ecological values (Salzman and Ruhl 2000). As is discussed in more detail in chapter 13, these concerns are compounded when ecosystem service values are taken into account, for even if ecological functions are fully compensated on a regional scale through creation or restoration of new service-providing natural capital, presumably a different local human population is enjoying the ecosystem service values (Brown and Lant 1999; Ruhl and Gregg 2001; Ruhl and Salzman 2006).

In short, because ecological functions are counted as ecosystem services only where and when humans are benefited, the ecosystem service value of any ecosystem function depends not only on the ecological and geographic variability of ecosystems across space and time but also on where people are, when they are there, and their respective levels of demand for different services. These demographic variables are difficult enough to track even in robust market conditions—consider the price variation of similar housing stock across the nation. Doing so for ecosystem services, where virtually no market data are available and value estimates rely for now on indirect measures and surrogate indicators, is only that much more complex an undertaking.

To some extent, moreover, the problem of scale is what separates ecological economics from neoclassical economic theory. Even if neoclassical economic theory had all the information at its disposal to model answers to the scale problems mentioned above, the one limit it cannot avoid is that the earth's resources are finite. Yet, as Kysar (2003) points out, neoclassical economics treats resource scarcity as merely a factor that will be reflected in price, as if the pool of resources is essentially infinite. Similarly, McMichael et al. (2003) contend that neoclassical economics "implicitly assume[s] that the world is an open, steady-state system" and that the discipline therefore has "a limited ability to appreciate that the fate of human populations depends on the biosphere's capacity to provide a continued flow of goods and services" (1919). We are well past being able to enjoy that naïveté, however. It is no longer foolish to ask whether it is possible that "the market," even in its most perfect applications, might lead us to exhaust fisheries or alter the climate. It is not sufficient to respond, as neoclassical economics might suggest, that as our global resources approach unsustainable levels, the profit motive will yield entrepreneurs of space colonization. For the foreseeable future, Earth is a closed system.

In short, economic theory may provide powerful explanations for why people do not invest in or conserve natural capital resources for ecosystem services provisioning, but it has no answers for what we will do if the services run dry. Ecological economics, forged as a discipline by Boulding (1966) and practiced today by such leading figures as Daly (Daly and Farley 2003), and Costanza (1991), to whose work we return later, uses economic theory to develop solutions to avoid reaching that decidedly nonoptimal position.

Economic Models and Complex Adaptive Ecosystem Properties

In some cases there may be good reason for confidence that comparison across spatial and temporal scales is reliable for a particular ecosystem service given well-known proportionate cross-scale relations, and that the economics of the service, distorted as they may be by all the factors discussed above, will nonetheless avoid complete exhaustion of its natural capital base. Hence the final concern with the estimates of nonmarket ecosystem service value is perhaps the most vexing in the long run in terms of policy development, namely, that nonmarket value estimates are essentially *models* of economic value rather than the direct measure that market prices provide. In the case of ecosystem services, the subject matter of these models—ecosystems—behaves inherently as a complex adaptive system subject to emergent properties, feedback and feedforward loops, and nonlinear patterns of evolution. Unless economists can comprehensively capture and account for these properties—a challenge that even ecologists have yet to tackle successfully—the nonmarket valuation methodologies will provide at best a snapshot to use as a value estimate (Bockstael et al. 1995; Daily et al. 2000; Brock et al. 2002; de Groot et al. 2002; Gustavson et al. 2002; Opschoor 1998; Westman 1977).

Indeed, even market prices face this problem when attempting to make long-run predictions about the economic value of goods and services produced in such complex systems. Because price reflects the marginal values for the good or service of interest held by the participants in markets, price works best as an indicator of performance when working with incremental, proportional systems, and thus is not a good indicator of the future if very small changes in the goods or services being valued can lead to very large, nonlinear changes in the ecosystems underlying them that could shift their availability or quality dramatically (McMichael et al. 2003; Simpson 1998). Because ecosystems are often subject to such discontinuities and disproportionate cause–effect relationships, the market value of an ecosystem service at any one moment, and of the natural capital from which it is provided, may not be stable very long into the future. Hence it is not entirely clear—indeed, far from it—that the individual incentives that play such a prominent role in neoclassical economic theory, when inserted into the reality of highly complex ecological systems, will necessarily align with social goals of a sustainable base of natural capital resources (Wilson 2001).

Gowdy and McDaniel (1999) emphasize this point in their study of the environmental destruction of the Pacific island of Nauru. After colonization in the late 1880s, phosphate strip mining began in the 1900s and was carried out under what neoclassical economics would describe as perfectly rational, welfare maximizing behavior. After the phosphate was depleted, however, the island of Nauru was depleted in all other senses—devoid of natural capital and located 3,000 kilometers northeast of Australia, it could not support a functional society as it once had.

Market theory and reality departed on Nauru. As Gowdy and McDaniel observe, "basic to neoclassical economics is the sanctity of individual choice, the fungibility of economic goods and inputs, and faith in the market system to bring forth substitutes as relative prices change" (333). The concern of many ecological economists, however, is when those assumptions include the "weak sustainability" premise that technological capital can provide perfect substitutes for natural capital and, hence, all that matters is that an economy save more than it depreciates in *combined* capital. But that premise cannot be applied reliably without a complete understanding of the system attributes of natural capital over time. Even if a technology and a natural capital resource appear fungible at the moment in the relevant spatial scale, they may depart substantially over time, and as Nauru illustrates, time matters. Just as important, in the case of Nauru, as the Nauruians worked to maximize their individual utility/income, aggregate social welfare was not necessarily maximized (Kelso 1977). As Bulte et al. (2004) summarize, "although some economists have attempted to focus on . . . models that incorporate temporal variability and spatial scale, there remains a gap between stylized economic models and recent ecological thinking. There is a strong need to bridge the gap between ecological theory and the economics of natural resources by incorporating variability, complexity, scale, and uncertainty into current economic models" (421).

The Tragedy of Ecosystem Services

So far this chapter has identified a number of general obstacles to the formation of markets, and even more simple multiparty transactions, to put the economic value of ecosystem services into operation:

• Because many ecosystem services behave like public goods or close to it, owners of natural capital resources from which ecosystem services flow perceive them as positive externalities for which the owner derives no gain.

- Even when the resource owner can regulate the volume and timing of ecosystem services flowing to users and has some sense that they are economically beneficial to the users, information costs, transaction costs, and free rider problems more often than not will impede the ability of the owner to engage in transactions with users from which some economic gain could be derived.
- Even when there is an opportunity to engage in such transactions, the price the owner is able to command may prove to be an unreliable indicator of the value of ecosystem services for different spatial and temporal scales.
- And even when we have reason to be confident that spatial and temporal conditions are sufficiently proportional in the near term to allow cross-scale use of price evaluations, those cross-scale relations may be unstable over time given the propensity of ecosystems to behave like complex adaptive systems.

The combined effect of these features is twofold compared to what might be expected in a competitive, market-based economy. First, owners of natural capital resources from which ecosystem services could flow have less incentive to employ the resources for that purpose. The economically rational landowner will not perceive the offsite benefits of ecosystem services as a potential source of significant economic gain. And on the user side of the equation, there is less incentive for the beneficiaries of ecosystem services flowing from other people's resources to invest in their conservation or, for that matter, do anything other than simply reap as much of the benefits as possible as fast as possible.

Of course, any time the owners of the capital have little incentive to use it to produce a particular good or service *and* consumers of the particular good or service have little incentive other than to grab it while they can, eventually the good or service will dwindle in supply and/or quality. This predicament has led to analogies (Heal 2000) between ecosystem services and the problem explored in Garrett Hardin's famous article "The Tragedy of the Commons" (1968), in which he charted the resource depletion problem of open-access property regimes. Hardin used an example of seasonal grazing land open for use by any member of a particular community. To be accurate, Hardin's scenario describes what is referred to as a common-pool resource within which access is open (nonexcludable) only to members of the defined community, rather than a fully open-access resource in which anyone at all can use the resource (Cole 2002; Ostrom et al. 1999). Nevertheless, because common-pool resources are, in the absence of rules governing the group, an open-access resource within the group membership, a resource degradation problem can arise in either case (Cole 2002). If the grazing land can support, say, 100 cattle indefinitely without becoming degraded, then it would make economic sense for no more than 100 cattle to be grazed over the season. However, each individual grazer in the community might not see it that way. An economically rational grazer could perceive some advantage to loading as many cattle as possible onto the land, for at the end of the grazing season that grazer recovers all those cattle in full but shares the cost of any resource degradation with all the other grazers. If many grazers think in this economically rational manner, before too long far more than 100 head of cattle will be on the land, leading to ultimate ruin of the grazing land. As Hardin summed up the problem,

> The rational herdsman concludes that the only sensible course of action for him to pursue is to add another animal to his herd. And another, and another . . . But this is the conclusion reached by each and every rational herdsman sharing a commons. Therein is the tragedy. Each man is locked into a system that compels him to increase his herd without limit—in a world that is limited. Ruin is the destination toward which all men rush, each pursuing his own best interest in a society that believes in the freedom of the commons. Freedom of the commons brings ruin to all. (1968, 1244)

Naturally, this kind of behavior is unlikely to promote conservation of ecosystem services, which is the reason why, as discussed in part II of this book, the group in charge of the common-pool resource frequently turns to the formation of property rights, regulations, and social norms to manage against ruin. These institutional frameworks provide the means for enforcing excludability *within* the common-pool resource group.

For some ecosystem services, however, the natural capital that supplies them is already securely in the hands of private individuals, not communal groups. Yet even in those cases the incentive structure can lead to degradation of the natural capital resource, not through overexploitation, but rather, in what Heller (1998) and others (Buchanan and Yoon 2000) call the "anti-commons," through *under*exploitation. As in the pollination scenario, in many cases the only way for knowledgeable resource owners to leverage value from ecosystem services produced on their land is to strategically veto their use by others, principally by threatening to destroy the natural capital that supplies them. As resource owners "make good" on that threat, for a while at least the service beneficiaries can simply rely on the free services provided by other lands. Over time, however, as more resource owners veto the use of services by converting natural capital to other uses, or by destroying it altogether, the supply of ecosystem services dwindles. By the time service beneficiaries appreciate that the scarcity of natural capital has turned ecosystems from water into diamonds, it may be too late to restore the stock of natural capital in time to turn the services spigot back on.

Part II examines the extensive literature examining solutions to commons and anti-commons resource problems. The dilemma, however, is an appropriate way to close this part of the book in its examination of the ecological, geographic, and economic context of ecosystem services. In short, if ecosystem services are seen as economically valuable at all, it is generally in their derived demand state of contributing to the provision of a commodity, or of a recreational use, or of the physical structure essential for nonuse values. As for supplying ecosystem services available for direct use, who cares? Economically rational individuals won't care so long as there are more lucrative uses of the natural capital resource and the ecosystem services it produces are free for the taking. To borrow a phrase, this is the Tragedy of Ecosystem Services.

Appreciating Marginal Ecological Opportunity Costs

Obviously, however, at some point people will have to care about the Tragedy of Ecosystem Services. Unless society is confident that it will indefinitely be technologically and economically feasible to replace all diminishing ecosystem services with alternatives, at some point there will be no choice but to find a way to alter what is economically rational behavior for the individual resource owner. A skeptic might observe that as ecosystem services become ever scarcer, eventually a market in them will emerge and the problem will self-correct, so what's the worry? This is the end of the paradox of value story mentioned above: water, essential for life, is worth less than diamond baubles because it is far more plentiful relative to demand, but as it becomes less so, its marginal value will rise and more will be done to conserve it.

The same is likely to be true of ecosystem services, but there are a couple of hitches. By the time scarcity alone focuses economic investment on ecosystem services, we may not have sufficient natural capital resources available to provide the services in the quantities and at the quality demanded, and we may not be able to create enough either. Ecosystems are not like machines in mothball, ready to gear up when demand justifies. The reason for the decline of ecosystem services will be the continued degradation of ecosystem functions that support natural capital resources. The degradation that ecosystem functions suffer is not likely to be fully reversible. Indeed, it is more likely that the decline of ecosystems by that point will in many cases have become *irreversible,* or at least very hard to steer toward a new direction more conducive to facilitate the provision of ecosystem services. Furthermore, for markets and prices to emerge for ecosystem services, the impediments to those markets discussed in this chapter will have to be overcome. This is no small challenge. In the case of diamonds and water, the markets and the institutions surrounding them are quite well established, which is not true for ecosystem services.

So, is there no hope? Not at all, provided we act before ecosystem services become as scarce as diamonds. Resource owners can easily determine the marginal economic opportunity cost of *not* converting wetlands into shopping malls. Our economy is brimming with readily available information to remind them of that "folly." What they need to know is the marginal *ecological* opportunity cost of making the "smart" move by not converting the wetlands, and what that means to them economically. Ecologists can provide substantial insight as to the former, but so far we lack much reliable information about the latter. One won't find it in the financial pages. It is essential, therefore, that economists team with ecologists to inform society of the marginal value of ecosystem services by illustrating the opportunity cost of transferring natural capital resources out of ecosystem service production and into some other use (Costanza and Daly 1992).

Doing so will ultimately require focusing on ways of correcting the incentive structure that leads to the Tragedy of Ecosystem Services. For Hardin, only a fool in open-access property regimes would voluntarily cede from the economically rational but resource-depleting behavior, given the risk that others will not follow, so Hardin and many others before and after him have explored ways to alter the incentive structure that leads to this problem. Generally three avenues have been proposed for doing so: better-defined property rights, prescriptive state regulation, and social norms. Part II examines these three institutional domains to determine the extent to which *existing* social and legal frameworks provide a foundation for the necessary reorientation of incentives regarding the provision and use of ecosystem services.

Part II   The Status of Ecosystem
Services in Law and Policy

The study in part I of the ecological, geographic, and economic complexities of ecosystem services suggested significant consequences for the formulation of law and policy. Knowing where to go, however, requires knowing where things stand today. Part II thus provides a baseline for future work by examining the current *status* of ecosystem services in the law and society. The chapters in this part make several significant findings about that status. First and foremost is the absence of any supportive system of property rights governing the production and use of ecosystem services (chapter 4), which renders them in many applications as public good resources subject to underprovision and overdepletion in the absence of some moderating influence. When property rights are as poorly designed as they are for ecosystem services, prescriptive regulations (chapter 5) and social norms (chapter 6) are often held out as the solutions to resource management problems. But here again the application of these institutional devices to ecosystem services has proven elusive. Although a consensus is building that ecosystem services hold tremendous values that we should seek to understand and incorporate into decision making about the environment, regulatory frameworks and social norms for efficiently managing ecosystem services have not materialized. The status of ecosystem services in law and society, in other words, is that they have none.

4 Property Rights

Although there is much yet to be learned about ecosystem services, what is already known demands attention from the discipline of law. Notwithstanding all the complications dwelt upon in the previous three chapters, in many cases the underlying ecosystem processes are well understood, the service can be traced from source to user, and we know the service is valuable. So the obvious question is, what can be done to solve the Tragedy of Ecosystem Services? As a starting point, any answer should consider how different configurations of property rights respond to the economic incentive problems discussed in chapter 3—externalities, information costs, transactions costs, and free riders.

This chapter explores the existing law of property as it relates to natural capital and the flow of ecosystem services. It is designed to provide a grounding in the basic foundations of property rights as a legal regime—what they are, how they are expressed and enforced through law, and why we might turn to them as a mechanism for designing the law and policy of ecosystem services. The chapter then turns to a detailed examination of how the modern law of property treats ecosystem services in three different property ownership contexts: individual private ownership, group-owned common property, and state-owned public property. As an example of how adaptable property rights can be (or not be) to complex resource allocation questions, the chapter closes with a discussion of rights in water.

Property Rights and Private Property

Americans have a well-known love affair with "property rights." Nevertheless, if one were to ask any number of Americans what property rights entail, at least as many different answers would likely be offered. Most people associate property rights with some conception of *private* property capable of *individual* ownership, but that does not answer the question about the scope of the rights an individual owner of private property holds. Everyone knows, for example, that

87

the Fifth Amendment to the United States Constitution commands that "private property [shall not] be taken for public use, without just compensation." But the Constitution contains no elaboration on the contours of private property, what it means to have it taken, how to establish whether it is taken for public use, and what fixes just compensation for such a taking. To this day much remains unclear about those fundamental aspects of rights in private individual property. In short, defining what individual property rights are is not nearly as easy as saying we have them.

At the most basic level, defining property rights goes to the difference between saying something is "mine" and that it is "my property." The bare assertion by one person that a tract of land belongs to him or her might simply invite the same assertion by another person. A property system is just one way of resolving their competing claims. Other means include a duel, a game of poker, or a footrace. While some disputants may be willing to resolve their competing claims by a cut of the deck, there would be no guarantee that a third party would not then come along to make the same claim against the winner and suggest physical combat as the means of resolving any dispute. Defining a system of rights in property is a way out of this unsavory possibility.

A system of property rights, in other words, avoids us having to rely on "might makes right" by establishing three general conditions with respect to the person who, in the eyes of the property rights system, legitimately lays claim to being the owner of land or other goods and services. The first is a well-defined set of interests in the property that the owner enjoys relative to others, such as to make use of the property and to exclude others from doing so. The second is a system for enforcing those interests against others who might improperly contest them. And the third is a means for divesting the interests, so that the owner can sell or otherwise transfer all or some of the property to others, who then in turn will enjoy the ability to have those interests enforced and to divest them. If these conditions are met, the foundations of a private property system are firmly in place (Ostrom et al. 1999).

It is quite evident, however, that these foundational conditions will not materialize without cooperation (Ostrom 2000a; Rose 1990a). If everyone who might contest the right to a particular good or service in question does not agree to abide by the property system, those who refuse might try to exert force to take the bounty. On the other hand, if a large enough number of people agrees to abide by the system, the few who refuse and attempt to exercise force are simply prosecuted as trespassers and thieves. Hence the level of cooperation generally associated with what passes for a system of private property in the classic sense, as opposed to cultural norms that may provide looser and potentially less reliable versions of rights, enforcement, and divestiture, is some form of sovereign state that has the power to promulgate and enforce laws its citizens must follow. The essence of private property, therefore, is the *state* and its ability to enforce *law* (Cole and Grossman 2002). As Jeremy Bentham long ago famously observed, "property and law are born together and die together" (1882, 113). In the modern context, Eric Freyfogle, who has established himself as a leading thinker in property law and ethics, has put it more bluntly in observing that "private property is a form of power over *people,* not *land*" (2006, 12).

Of course, there is a bit of irony to the fact that a system of private individual property rights, to function as intended, requires the public to coerce its government to coerce it back (Krier 1992). Only a deep and sophisticated level of cooperation between the members of the relevant group could be expected to lead to such a system. But as some property law scholars have pointed out (Posner 1979; Rose 1990a), if the group is capable of such a level of cooperation, why does it need a system of private individual property with coerced rights and state enforcement? Why go so far as to resort to coercion? That question deserves and receives further exploration later in this chapter, but first it is important to establish the central features of the private property system as so described and to explore what virtues it is supposed to offer in matters such as the Tragedy of Ecosystem Services.

Property Rules and Liabilities

Once armed with the three central features of private property, the business of law is putting the system into operation through the formulation and enforcement of *rules* and *liabilities.* To begin with, of course, property law must prescribe the doctrine for knowing who owns a particular tract of land. But that ownership question is not what presents the policy challenges for natural capital and ecosystem services. The concern is not who owns the land where the natural capital produces the services or the land where the service benefits are enjoyed—that is usually perfectly clear—but rather what the respective owners can do with their property and can reasonably expect of each other in return. Defining the *relationships* between property owners is the far more complex aspect of any property system. And in the domain of law, any question of relationship between people is ripe for developing a set of rules and liabilities designed to lead to desired behavioral outcomes.

Consider the case explored in chapter 3 involving pollination of a coffee plantation by bees and other wild pollinators calling a nearby forest their home habitat. The owner of the forest tract considers it reasonable to convert the land to grazing in order to increase economic gain, but the owner of the coffee plantation thinks that the forest owner's decision is unreasonable given the economic losses the plantation suffers from the reduction in pollination. To work as intended, the private property system must have a means of deciding who wins—that is, which landowner must yield to the other's interests.

One means of resolving the matter is to devise a rule that allows one property owner to force the other to change behavior. For example, suppose that the property system specifies that landowners have the right to continued flow of pollination services. If we were to enforce this pollination right through a rule, the coffee plantation owner in our example could employ the state to enforce the right against the forest tract owner and prevent the conversion of the forest to grazing uses. The essence of a rule-based system is this ability of a property owner whose right is being injured—in this case the coffee plantation's right to pollination—to enforce the right through an *injunction* remedy.

An alternative means of resolving the competing interests is to establish liability principles that define which property owner must compensate the other if the two wish to engage in incompatible land uses. Thus, instead of being able to force the forest tract owner to maintain the flow of pollination to the coffee plantation, under a liability-based system the coffee plantation owner could employ the state to force the forest owner to pay for the economic losses that result when pollination services dwindle. The essence of a liability-based system is this ability to enforce a *compensation* remedy.

It is readily apparent that these two approaches, rules versus liabilities, can present vastly different relational dynamics between the two parties. Under the rules-based approach, it is whichever landowner the property system recognizes as the injured party who is empowered to make the decision whether and how much of the injurious activity to allow. Once the state has demonstrated its willingness and power to enforce the rule through injunctive relief, the injured party can negotiate to sell all or a portion of the right to enforce the injunction to the enjoined party. If the value of continuing the injurious activity is high enough, the price the injured party could command, if bargaining is approached strategically, could exceed the cost of the injury, meaning that the injured party could experience an economic windfall.

By contrast, under the liability-based system it is the party causing the injury that gets to decide how much injury to cause and thus how much compensation to pay. The only right the injured party has is to the damages compensation, and if the value of continuing the injurious activity exceeds the amount of the damages, the injured party can simply pay the compensation and thereby force the injured party to continue to suffer the physical conditions that lead to the injury. So long as the payment fully compensates the injured party for the present value of all past and future injuries, the injured party is made "whole" in the eyes of the law.

Where the value of the injurious activity exceeds the cost of the injury, therefore, the choice between enforcing property rights through rules versus liabilities makes a difference. Under a rule-based system, the injured party could bargain with the other to sell the right of enforcement at a price that exceeds the cost of the injury but is less than the value of the injurious activity. That would make economic sense for both parties. Similarly, under a liability-based system, the party causing the injury would have to pay just the cost of the injury as compensation to be entitled to continue the activity, which also would make economic sense for both parties. In both situations, true to the Coase theorem as outlined in chapter 3, the parties should negotiate to an efficient solution that "internalizes" the relevant externalities in the party causing them and leaves both parcels in their most productive uses (Medema and Zerbe 2000).

On the other hand, where the cost of the injury exceeds the value of the injurious activity, both approaches are likely to result in the activity not occurring. Under a rule-based approach, the injured party would not agree to sell off the right to enforce the rule at a price that would make economic sense to the party causing the injury—there's no point in paying a price to do something that yields less gain than the price. Similarly, under a liability-based approach, the party causing the injury would find no profit in paying the necessary compensatory damages to be able to continue the injurious activity. In short, the gain to the party causing the injury of paying to continue the activity must exceed the cost of the injury to the other party in order for bargaining to occur (Medema and Zerbe 2000).

Of course, these postulated outcomes depend on the unrealistic assumptions that parties in a rule-based system are perfectly rational (i.e., not strategic or spiteful) and face no substantial costs of bargaining (i.e., zero transaction costs), and that the state in a liability-based system faces no substantial costs of determining the amount of compensation to require (i.e., zero information costs). Plainly, these assumptions are seldom valid. Even in simple bilateral negotiation contexts, or even particularly so, the lack of a competitive market combined with strategic bargaining and, in many cases, a mutual lack of trust, may lead each party to attempt to derive so much gain from the deal that an agreement is never attained (Posner 1998).

A voluminous body of literature debates the advantages and disadvantages of rules versus liabilities under alternative assumptions about transaction costs and information costs (Calabresi and Melamed 1972; Kaplow and Shavell 1996; Krier and Schwab 1995; Kysar 2003; Smith 2004). As Swanson and Kontoleon (2000) thoroughly summarize, the classic battleground for this debate focused on the law of nuisance, the age-old, judge-made common law doctrine that provides no landowner may unreasonably interfere with another person's use and enjoyment of his or her property. Courts develop common law doctrines such as nuisance when they resolve disputes between resource owners affected by their decisions (Frazier 1998). As courts in one jurisdiction decide similar disputes, often with the aid of decisions from other jurisdictions, courts and lawyers distill common law doctrines and the policies behind them. American property law has this common law tradition at its core.

A major goal of American property law, and of the common law doctrine of nuisance in particular, is to promote efficient use of land. Appropriately, therefore, Coase used nuisance law as the context for his famous theorem, thus prompting many economists and legal scholars since then to continue in the tradition as they explored the relative merits of rule-based and liability-based remedies. The Coase theorem in its "strong" form postulates that regardless of how rights are initially assigned, negotiations among the relevant parties will result in an efficient and invariant allocation of resources (Medema and Zerbe 2000). As Coase observed, however, this outcome depends on zero transactions costs, zero information costs, and perfectly competitive markets. The law of nuisance, by contrast, must deal with the real world, where the costs of bargaining and of information are anything but zero and the initial allocation of rights can matter depending on the relative economic positions of the parties.

Indeed, the history of nuisance law reflects these conditions. When cases involved pigsties stinking up neighborhoods, it was a relatively easy matter for courts to devise and enforce nuisance doctrine as a rule-based system of injunctive relief. The economic value of the pigsty was usually far less than the cost of the injury to surrounding properties, and its generally noxious character did not cry out for the more flexible liability-based remedy. A rule-based approach became the norm for nuisance law, meaning that if a court found a particular land use a nuisance, it was enjoined (Ellickson 1973; Lewin 1986).

As industrialization pitted pollution and noise from factories against surrounding land uses, however, the value of the injurious activity frequently became appreciably higher than the cost of the injury, at least as represented by the parties seeking the injunction. Facing the prospect of shutting down a factory that employs half a town just to spare the senses of a few residential or agricultural landowners, some courts strained to find no nuisance liability even though the activity was surely unreasonably injurious.[1] Eventually, in order to avoid playing fast and loose with the substantive doctrine of what constitutes a nuisance, courts abandoned the use of injunctive relief as the exclusive remedy and allowed the party committing the nuisance to pay compensation.[2] This liability-based approach, of course, worked handily for socially valuable industrial land uses, allowing them to continue operating while internalizing the costs of their offsite effects, at least with respect to the persons bringing the claim in court, by paying compensation to injured parties (Ellickson 1973; Lewin 1986).

The objective here is not to resolve the debate over rules versus liabilities in property law, but rather to illustrate that if natural capital and ecosystem services are to have their space under the umbrella of private property rights, the distinctions between rules and liabilities may prove important. Before delving into that thicket, however, it is important to ask why anyone should bother. In other words, what is it about private property rights, and a system of rules and liabilities in particular, that would lead us to think that they may help resolve the Tragedy of Ecosystem Services?

Why Private Property?

Given the foregoing discourse on the nature of private property and how enforcement of different rules and liabilities puts into action the desired configuration of property rights, what implications does private property have for ecosystem services? What magic does the private property system have at its disposal for changing the way people think about and act toward natural capital and ecosystem services?

A good starting point is to consider how resource users would behave in the complete absence of any system of rights for a particular resource—that is, with no defined interests, no enforcement, and no means of divestiture. In this kind of absolute *open-access* regime, as was shown in chapter 3, we would expect to find substantial free-riding in the form of individuals being willing to overuse the resources without concern for effects on others and being unwilling to invest in the maintenance of the resource (Ostrom et al. 1999). Two conditions are necessary for a property rights system to overcome these behavioral incentives. First, rights that provide some degree of exclusivity of access to a resource must be established and enforced in order to prevent open access and its invitation to free-riding. That much is a direct result of the private property system conditions—if the state is willing and able to enforce exclusivity of access, open access is no longer possible unless the owner of the resource agrees. Second, the property system must also provide incentives to the owner of the resource to invest optimally in the resource rather than overexploit it. But how this effect is accomplished is not so obvious. Merely assigning exclusivity of access does not ensure that individual resource owners will make optimal decisions about resource use. Indeed, how can the property system, with or without exclusivity of access, determine what optimal use is?

The answer most frequently espoused in modern Western legal thought in general, and in Anglo-American property law in particular, leans heavily on neoclassical economic theory of markets and the individual owner's ability to divest resources through market exchange (Buck 1998; Rose 2002). As discussed in chapter 3, in the theoretical efficiently competitive market where buyers and sellers are free to bargain over the exchange of goods and services, prices send signals to all market participants about the comparative economic gain to be had by engaging in different activities such as producing goods, owning land, and providing services. In the case of land and resources, individual owners acting purely out of self-interest will use relative prices as a guide to identifying the use of property that is most likely to maximize their economic position. If all owners do this, so postulates market theory, then the sum total of all their selfish efforts is to produce the maximum possible welfare position for the society as a whole. Markets, in other words, align personal interests with the general welfare. In theory, therefore, competitive markets will lead to the most efficient allocation of resources.

Of course, the presence of active competitive land markets need not result in the development of natural capital and the depletion of ecosystem services. If the demand for resource conservation is sufficient, the market will produce conservation of resources through nonuse status. *Nonuse* is thus a misnomer, in the sense that property would be put into conservation only if the market values that outcome over all other alternatives. Land trusts, for example, enter the private market to buy land, and then leave the land in conservation status rather than reselling it, because their members value conservation highly and are willing to bid against others who might put land to different uses. Local land trusts own conservation rights in over 5 million acres in the United States (McLaughlin 2005), reflecting the strong demand many people have for land conservation over other uses. Conservation, therefore, is every bit as much a use of land as is a shopping center.

Given that conservation is already a use of land valued in the market, one should expect ecosystem services to make a difference in market outcomes as well. Because ecosystem services clearly provide economic value to humans, the price of land where the natural capital that produces them is located and of the land where the services are delivered ought to reflect those values. From the perspective of market theory, it would make no sense to convert a forest into a shopping center if the owner of the land could derive more gain by leaving the natural capital intact. Neither the self-interest of the landowner nor the general welfare achieves its market-produced maximum potential if that happens. So why would it happen?

The simple answer often given for why such "market failures" appear is that markets don't function smoothly when property rights, assuming the state is willing and able to enforce them, are either unclearly defined or unwisely defined. When one considers how the market pricing mechanism works in theory toward achieving the optimum allocation of resource uses, it is readily apparent why well-defined and vigorously enforced private property rights are a necessary element for the market to operate to its fullest potential (Medema and Zerbe 2000). After all, who would pay for a right if there were no reasonable expectation that anyone else will recognize the right or that the state will enforce it? Consider again the coffee plantation example. The value of the coffee plantation in the market would be different in the case where the law is clear that landowners have a strongly enforced right to undisturbed pollination, versus in the case without such a right. In the latter case, the coffee plantation owner might represent to prospective purchasers that adequate pollination has not been a problem, but no purchaser could reasonably rely on there being a *right* to continued pollination from other properties, and thus no rational purchaser is going to pay for it. The price of the coffee plantation, therefore, will reflect the risk of diminished pollination to which purchasers believe the land is exposed, such as from the conversion of nearby forest tracts to grazing.

It does not exaggerate, therefore, to observe that without property rights, markets in resource allocation will not achieve satisfactory results, much less optimum results. It is equally true, however, that without markets, some private property rights are largely irrelevant. For example, if in the coffee plantation example the pollination right belongs to whoever owns the pollinator habitat, we saw in chapter 3 that the forest tract owner nonetheless will have great difficulty capitalizing on such a property right. One cannot easily herd wild pollinators to purchasers' properties! Transactions costs and free rider problems thus render pollination a positive externality the forest tract owner will find difficult to internalize through the market pricing mechanism. The law could define the pollination right as clear as day, and the state could stand ready to enforce it with all its sovereign power, but how could the forest tract owner charge others market prices to use it? In other words, the physical realities of a particular resource may make the third condition of private property—an efficient means for divesting the property interest—difficult to achieve.

Pointing to these preliminary concerns, some commentators quickly go so far as to suggest that the challenge is insurmountable, that the very nature of ecosystem services defies assigning property rights, and thus no markets in ecosystem services are possible (Fitzgerald 2005). Clearly, though, where natural capital and the ecosystem services flowing from it are well defined ecologically, geographically, and economically, it makes sense to think about how to configure private property rights in them in order to facilitate their optimum allocation through the market pricing mechanism. To be sure, doing so is not as easy as thinking about property rights in natural resource commodities such as timber. When considering a tree as a commodity, a default rule for assigning the property right is quite straightforward—it belongs to whoever owns the land on which the tree stands. There is nothing about the tree as a commodity that begs to invite any other property owner into the picture. As noted above, however, because the point of *origin* of ecosystem services—that is, natural capital—and the point of *delivery* of the ecosystem service benefits are often geographically and temporally distant, the configuration of property rights in them necessarily must take into account the relations between the relevant property parcel owners.

Private Property Rights and the Problem of Transboundary Resource Flows

The best starting place to assign property rights in trees as a timber commodity is with the owner of the parcel on which the trees stand, because there is no other person who has a more direct interest in the timber value of the trees. The parcel owner is most likely to be the person who best optimizes care for and harvesting of the trees as a timber resource. However, when we think of trees as natural capital from which ecosystem services flow, and also take into account that the value of the ecosystem services flowing from that resource base often materializes at distant times and locations, the baseline norm for allocation of the property rights is not as readily apparent (Salzman 2005). The complication is that in order to fully account for all the relevant property interests, rights must be assigned not only in the resource *stock* (natural capital in our case) but also in the resource *flow* (ecosystem services in our case) (Bouckaert 2000; Ostrom et al. 1999). And as the flow of ecosystem services often, if not usually, transcends the boundaries of defined private property parcels, this requires a mechanism for distributing the rights among a broad spectrum of property owners, all of whom can reasonably make some claim to them.

At one extreme, we could assign an absolute right in the natural capital and its associated flow of ecosystem services to the owner of the parcel on which the natural capital is located. Unlike the timber rights solution, however, this allocation of rights does not lead as obviously to smooth market outcomes and appropriate allocation of resources. We have seen already how transaction costs and free rider problems can confound the ability of the natural capital owner to secure the value of the ecosystem services through market transactions. And it may be equally as difficult for the state to enforce injunctive or compensatory remedies against property owners who "steal" services such as pollination from owners of pollinator habitat. But even if those problems can be effectively overcome, there is something unusual about the idea of property owners charging others for clean air, nutrient cycling, and pollination. As a purely normative matter, in other words, shouldn't owning property carry with it some security in the continued provision of basic ecosystem services, even if their source is from natural capital located on some *other* parcel of property?

If we decide that is the case, it would be easy to move to the opposite extreme and assign an absolute right in ecosystem services to the owner of the parcel on which they are delivered. This right could be enforced through either injunctive relief or compensatory damages, depending on whether a rule-based or liability-based approach is preferred. Yet this approach raises two independent problems. First, it would mean that in order to put the property on which the natural capital source is located to an alternative use—one that may even be more valuable to society as a whole—its owner would have to buy out the rights to the ecosystem services held by a potential multitude of other property owners. The transaction costs of such an endeavor in many settings would be daunting, and the end result of the landowner deciding not to undertake them could be inefficient use of land. Even more complicated, however, would be the question of how to allocate ecosystem service rights among the many different property owners who might all be in a position to alter, store, or consume services that would otherwise flow onto and then off of their respective properties. If everyone has an absolute right to transboundary ecosystem services such as flowing water and flying bees, how can anyone use them for private gain, much less sell them in the marketplace?

Unlike rights in commodities, therefore, assigning absolute rights in natural capital and ecosystem services is no quick answer to the problem of how to tap into the market pricing mechanism. Owners of parcels where natural capital is located are inseparably bound to owners of parcels where ecosystem services are delivered, and the latter are all inseparably bound to each other. The absence of *any* restriction on use, therefore, is likely to lead to inefficient resource outcomes, as what is optimal for the individual landowner may pose negative externalities to other landowners and be suboptimal from a broader social welfare perspective (Hsu 2003). Such distortions in resource allocation attributable to the presence of externalities have generally led to calls for a redefinition of property rights designed to "internalize" all costs and benefits in each landowner so as to reestablish the elegant matchup between private and social interests (Baumol and Oates 1988). Indeed, in theory the complete assignation of all conceivable property rights should eliminate all potential for externalities, positive and negative (Simpson 1998). The real question, then, is how do we define rights in natural capital and ecosystem services in a matrix of interconnected property interests?

As the doctrine of nuisance illustrates, the law has a time-honored answer for distributing property rights in situations where allocating absolute rights is problematic—simply demand that everyone act *reasonably*. But this balancing approach doesn't always present a good fit with the market either, for the market is based on the proposition that everyone acts *selfishly*. If owners of natural capital and users of ecosystem services must balance their respective self-interests with a duty to act reasonably toward each other, the efficiency of the market is put into question. Transaction costs are likely to be high, as selfish landowners will argue over who is acting more reasonably and turn to the state to decide between their competing claims. And it is seldom self-evident which competing land use is the more reasonable; indeed, as a legal concept, the balancing test for "reasonableness" in nuisance law has defied objective standards, leading that body of law to be described famously as an "impenetrable jungle" (Prosser et al. 1984, 264). Indeed, it may very well be that the gain to society from defining property rights to completely eliminate externalities is offset by the costs to the state of enforcing the rights and adjudicating disputes (Salzman 2005; Simpson 1998).

In general, however, there is no reason why a system of private property rights cannot rely on combinations of absolute rights and balancing tests. As complex as natural capital and ecosystem services are ecologically, geographically, and economically, it should be no surprise that they would also be complex as a matter of property rights. But if that is the case, why should we expect that only a system of *private* property rights can offer suitable solutions for the optimal conservation of resources? Shouldn't other kinds of property regimes be explored?

Other Kinds of Property Regimes

The classic theory of private property suggests that it is effective in producing good decisions about the use and conservation of resources—decisions that harmonize private individual interests and public welfare—because it facilitates open competitive exchange of resources in the market (Fitzpatrick 2006). We have seen, however, that natural capital and its ecosystem services present significant challenges for tapping into the market's elegant allocation mechanism. It might take an elaborate and vigorously enforced set of rights to make the market work in ecosystem services contexts. But this leads us back to the irony that this could only come about because the members of the relevant *community* would agree to be coerced by the *state*.

This necessarily begs the question why, if the community and the state must have a hand in the private property system to make it work as a means of efficient allocation of natural capital and ecosystem services, the community or the state could not do just as good a job as the individual owners, if not better, of making wise decisions about resource use and conservation? In other words, can't the problems of positive externalities, transaction costs, and free riders be effectively mitigated through property regimes that rely on group-owned property or state-owned property? Indeed, are there situations in which we *must* turn to property regimes other than individually owned private property? If so, what might those regimes look like?

Between open access and individual private property, Ostrom et al. (1999) place group property, which they define as differing from private property according to the ease with which any individual in the group can buy or sell a share of the resource. Property law provides many arrangements for group ownership, known in law as common property or concurrent property, such as co-tenancy, partnerships, corporations, and family-owned property. Cole (2002) explains that in common property arrangements the ownership group holds the right to exclude others—making it functionally private property for the group relative to outsiders—whereas within the group each individual has both rights and duties with respect to use of and investment in the resource. One potential downside of common property, it has been suggested, is that it might result in simply a smaller-scale version of open access if there are no well-defined rights and duties *within* the group of owners (Bromley 1991, 149). But if the conditions exist for society as a whole to establish private individual property regimes, there is no reason to believe that common property arrangements will necessarily devolve into miniatures of open-access property, because no more intragroup cooperation would be needed to avoid that result among the ownership group in a common property regime than among all the multitude of individual owners in a private property regime.

Indeed, if one expands the size of a cooperating common property group to the size of society as a whole, one has what Ostrom et al. (1999) call government property, more commonly referred to in law as public or state property. After all, a private property system assumes all members of society cooperate sufficiently to enlist the state to enforce the agreed-upon rules and liabilities. In a public property system, the cooperative effort merely replaces the individual owners with the state, and the state defines the rights of each citizen with respect to the resources and the rules and liabilities enforced against them. That shouldn't involve any more cooperation than is required to produce the private property system.

Hence the question of which property regime will work best really isn't settled by suggesting that private property systems require less mutual cooperation, for they involve at least as much cooperation as common property and public property systems. Rather, the question is which of these systems is more likely to lead to cooperation that makes the most sense with respect to resource use and investment. Using the market pricing mechanism discussed above as a benchmark, a strong case can be made that common property and public property systems are less likely to mesh well with market allocation forces than will private property. The selfish individual interests that make market pricing work so efficiently for private property systems are likely to find competition with some, if not many, of the group or government interests that act as decision drivers under the common property and public property regimes. On the other hand, the fact that there might be such things as group and government interests independent of individual interests suggests that the market might not account for everything of importance to people, such as normative goals of fairness and long-term distributional equity of resources (Raymond 2003). Indeed, Ostrom et al. (1999) suggest that there is strong empirical evidence that no single type of property regime works efficiently, fairly, and sustainably in all resource settings, and Fitzpatrick (2006) points to numerous examples of private property not living up to its promise compared to other property regimes. It is appropriate, therefore, to consider how well each of the three alternatives to open access has performed to date with respect to allocating use of and investment in natural capital and the flow of ecosystem services.

The Status of Natural Capital and Ecosystem Services under Existing Property Regimes

Cole's model of property rights (2002, 10) recognizes that most land in the real world is owned in hybrid configurations of rights. Some property is owned exclusively by an individual, or an organized group, or the government, but property may also be held by combinations of such entities, or by corporations, land trusts, tribal or local governments, and many other kinds of ownership arrangements, in ways that bridge these three strictly defined regimes. Law students learn, therefore, that property rights are like "bundles of sticks," and not every interest in a parcel of property necessarily comes with every "stick" (Duncan 2002). The right to exclude others from access is often recognized as the core right in property, but rights to income, to extract resources, to manage, and to possess free of term also contribute to the totality of interests associated with a particular parcel. One single entity—an individual, organized group, or governmental unit—need not hold absolute control of every conceivable interest in a parcel of land to be considered its "owner" for most practical and legal purposes, and it may be that individuals, groups, and governments each have a share in a particular parcel. For example, a land trust might purchase the right to develop a parcel in order to ensure certain resources are conserved, but an individual might retain the right to occupy and graze livestock on the land and a corporation may have rights to extract some of the subsurface mineral resources. Property is often held in such hybrids of individual, group, and government interests that evolve over time (Yandle and Morriss 2001), further complicating any examination of how best to manage property rights from the perspective of ecosystem services.

Moreover, even when a single owner holds all legally recognized rights in a parcel, it is never the case that the owner can use the property free of concern for other interests. As seen previously, property rights necessarily must take into account the relationships between properties and, in common and public ownership regimes, between the people involved in the group or public ownership pool. Nuisance law requires each individual landowner to adjust what a court deems "unreasonable" behavior toward other landowners, and public law, such as environmental regulation, may require the same of a landowner to protect the public interest. Likewise, individuals acting in group and public policy contexts assert their individual interests, which, depending on the rules of the group or government, may need to be taken into account by others sharing ownership interests. Cole thus concludes that "a property regime can only be *relatively* public or private" (2002, 13).

Nevertheless, because there is a relative spectrum, it is reasonable to expect that, for a particular natural capital resource or flow of ecosystem services, the gravity center of rights most closely associated with its management will often rest with a single identifiable owner entity, be it an individual, an organized group, or a governmental unit. There may be other rights in the property owned by other entities, but how the principal owner treats the natural capital or ecosystem service may depend on what kind of property regime it most closely resembles, that is to say, whether it is *relatively speaking* more private, common, or public. Hence, for purposes of examining the treatment of ecosystem services in the context of property rights, it is most convenient to consider the conventional typology of private property, common property, and public property as presenting the important distinctions for comparison. As it turns out, however, the distinctions may be purely academic, as none of the three regimes has performed so as to integrate natural capital and ecosystem service values into resource allocation decision making in any meaningful way.

Natural Capital and Ecosystem Services on Privately Owned Lands

The common law is widely regarded among legal scholars as remarkably adaptive and capable of evolving over time to receive new knowledge about the world and society (Epstein 1995; Fuller 1968); yet, while it is capable of rapid movement, it is seldom in a hurry to do so. Change generally comes slowly to the law of property, and thus as new as "natural capital" and "ecosystem services" are to ecology and economics, it comes as no surprise to lawyers that they are hardly on the tip of property law's tongue. This makes the project of describing the present state of property rights in natural capital and ecosystem services one of inductive interpretation rather than a direct reading of the pages of judicial opinions and statutory text. Nevertheless, there is ample content with which to work.

DEFINING NEW RIGHTS OR REDEFINING OLD RIGHTS?

The Tragedy of Ecosystem Services is conventionally described as due in large part to poorly defined private property rights—that is, it is unclear whether a landowner has the right to destroy natural capital and thus to be paid to refrain from doing so, or must conserve natural capital and thus comply with legally enforced remedies when seeking to destroy it. In the lexicon of property law, this view treats natural capital and ecosystem services as *res nullius*—assets that have not been appropriated through property rights because they either were previously unknown to the concerned community or were known but not yet allocated (Bouckaert 2000). In this view, more clearly defined rights will facilitate creating markets in ecosystem services (Salzman 2005).

Presumably, any community that has a well-developed private property rights system will have developed mechanisms for assigning rights in *res nullius*. Historically, for example, Western legal culture has favored a rule of first appropriation, under which the first person to occupy or possess the previously unowned resource is allocated the property right. Almost all American law students study the case of *Pierson v. Post*, in which the court used this rule to decide who owns wild animals (Bouckaert 2000).[3] Alternatively, rights in *res nullius* could be allocated by auction to the highest bidder. Economists and legal scholars have debated the relative advantages and disadvantages of these and other means of allocating rights in *res nullius*, but it would be moving too fast to declare natural capital and ecosystem services *res nullius* and explore those arguments now, for it is not at all clear that natural capital and ecosystem services are *res nullius* at all.

To be sure, some ecosystem services, such as pollination, present property rights questions that appear quirky, in that the question of ownership hasn't often been taken up directly in the law of property. Yet, as noted earlier, what is explicit at the surface of the common law is often revealed to be merely the tip of the iceberg after a deeper intuitive analysis. In fact, a plausible interpretation of the law of property rights is that the Tragedy of Ecosystem Services is not a consequence of poorly defined private property rights, or about rights incapable of being assigned, but rather stems from private property rights that are quite clearly delineated and need to be *redefined*.

THE ANTI-ECOSYSTEM BIAS OF AMERICAN PROPERTY LAW

In his little noticed but profoundly insightful article "The Anti-wilderness Bias of American Property Law," law professor John Sprankling (1996) convincingly demonstrates why American property law is anything but unclear about a landowner's discretion over the fate of natural capital and ecosystem services. Property law has traditionally been portrayed as silent or neutral on the question of what rights or duties a landowner has over undeveloped land on which natural capital is located. This "neutrality paradigm" supported the premise that property law neither encourages nor discourages property owners from destroying or degrading natural capital, meaning that their decision whether to do so must be seen as a voluntary act driven by rational economic behavior. Indeed, were this the case, it would be encouraging to the project of defining rights in natural capital and ecosystem services, for it would mean that the law would be filling the gaps of *res nullius*—improving the clarity of rights—rather than reorienting settled but no longer effective doctrine.

But a careful reading of the evolution of American property law from its English common law roots to its contemporary framework suggests it is not gaps that must be filled, but walls that first must be taken down. Sprankling's thorough historical analysis reveals that early American property law, as formulated through judicial opinions building the common law of property rights, embraced agrarian development as its central purpose and saw the nation's abundance of wilderness as essentially a license to tilt property law toward what Sprankling calls an "anti-wilderness bias." It was "an instrumentalist judiciary [that] modified English property law to encourage agrarian development, and thus destruction of, privately owned American wilderness" (Sprankling 1996, 520), and this was perceived as having no downside given the supply of undeveloped land the nation enjoyed. No less than the United States Supreme Court joined in this retooling of common law, as Justice Story observed in 1829 that "the country was a wilderness, and the universal policy was, to procure its cultivation and improvement."[4] The result was a body of law that actually *encouraged* destruction of natural capital and devalued its status in the market.

In one of his most striking examples, Sprankling traces the evolution of American property law on the doctrine of adverse possession, under which the long-term possessor of land can oust the true title owner of possession. The doctrine was a means of resolving title disputes in England, which lacked an organized title recording system, in the context of what was a densely agrarian landscape long before the development of American law. English common law, which early American courts adopted wholesale, required the adverse claimant, among other things, to have engaged in open and obvious activities likely to afford notice to a diligent owner, such as establishing residence on the land, cultivating it, or fencing in portions. Over time, however, American courts began systematically to promote development by modifying these requirements based on the nature of the land involved. Thus adverse possession of wilderness lands could successfully be established by infrequent, inconspicuous acts, such as occasional berry picking or taking of timber, that would likely have gone unnoticed by anyone, even an observant and diligent owner. This made it easier to establish adverse possession of wilderness lands through minimal development activity, and thus sent a clear message to landowners to develop their land first lest they lose it to interlopers.

Even the law of nuisance, the common law doctrine most attuned to the relationship between property owners, joined in the evolution of the anti-wilderness bias. English common law enforced a strict harm-based test for nuisance, under which any act that harmed the productive usefulness of other land could be deemed a nuisance. In America, however, the pro-development common law evolved so that the "reasonableness" of the harm mattered, and locality and circumstances became the criteria with which to measure what was reasonable. The result was that "all other things being equal, conduct was less likely to be enjoined as a nuisance if it occurred in a wilderness area than in another, more developed, locality" (Sprankling 1996, 554). One court, for example, went so far as to refuse to enjoin a dam that would have flooded a tract "so wet, marshy, and sour as to be worthless for agricultural . . . purposes."[5] Of course, as nuisance law systematically made it less likely a court would find harmful land uses a nuisance in wilderness areas than in developed areas, potential nuisance-causing land uses gravitated to undeveloped areas to reduce their exposure to liability.

As one might expect, the American West was where Sprankling found the anti-wilderness bias had penetrated deepest into property law. Because of England's dense crop and pasture land uses, English common law held to rigid lines on the doctrine of trespass, making stock owners liable for any damage their animals might cause to other landowners. By contrast, American law, particularly in the West, tore down the "invisible fence" of English trespass law and replaced it with a "free-range" standard under which stock could roam over private lands without creating trespass liability. By statute, many American states purported to reverse the English rule so as to facilitate agrarian development. Locating livestock near forested land or the prairie thus became viewed as a beneficial use of the adjoining natural resources. Although courts in New England states construed these statutes quite narrowly, elsewhere they prevailed under theories that the free-range standard had become the common law equivalent of "customary use" of undeveloped lands, in effect making privately owned wilderness open-access land for purposes of grazing. As the Ohio Supreme Court put it, "to leave uncultivated lands unenclosed was an implied license to cattle and other stock at large to traverse and graze them."[6]

Sprankling's assessment of American property law thus reveals why establishing markets in natural capital and ecosystem services will involve more than simply clarifying private property rights, as if the rules and liabilities are not already clear. In short, owning wilderness, which is where one would reasonably expect to find intact natural capital, is a *burden* to landowners under American property law. On balance, a landowner is better off developing natural capital to other uses, lest ownership be lost to an adverse possessor, lest nuisance uses locate in the vicinity for safe harbor from liability, lest stock owners graze their cattle there, and so on. Turning natural capital into an asset in the eyes of property law can only be hindered under this entrenched common law cloud.

Of course, today it would be unusual for a judge to characterize a wetland as a worthless tract of sour marsh. Modern perceptions of wilderness have raised it from wasteland status to an important public resource. For example, in upholding the regulation of development in wetlands under the Clean Water Act, the U.S. Supreme Court acknowledged that "wetlands may serve to filter and purify water draining into adjacent bodies of water . . . and to slow the flow of surface runoff into lakes, rivers, and streams and thus prevent flooding and erosion."[7] But this change of heart has largely been embodied through public legislation with its focus on the use of public lands or the protection of discrete resources. Notwithstanding the changes in public perception and the rise of public legislation aimed at protecting the environment, Sprankling (1996) found that the contemporary common law of property has remained stuck in its nineteenth-century anti-wilderness bias. He concludes,

> Modern courts have lost sight of the historical roots of our property law system. Although espousing prowilderness sentiments in good faith, the judiciary blindly applies most of the anti-wilderness doctrines of the past. Thus, individual disputes tend to be resolved in favor of wilderness exploitation. More importantly, the entire body of anti-wilderness opinions continues to exist, setting public norms for private conduct outside of the litigation arena. The accumulated precedents of the two centuries constitute a virtual common law of wilderness destruction that threatens the existence of privately owned wilderness sanctuaries. (569)

While Sprankling couches this phenomenon on the effect the common law's bias has on conservation of wilderness, the importance of undisturbed wildlands to the sustainability of dynamic ecosystems surely demands that the bias be reframed as one of "anti-ecosystem" dimensions.

REGULATORY TAKINGS AND THE REDISCOVERY OF THE COMMON LAW

Sprankling's account of the evolution of the common law of property rights, confirmed in other historical studies (Eagle 2004; Freyfogle 2006; Klass 2006; McElfish 1994; Purdy 2006), finds unmitigated support in the unlikely field of regulatory takings law. The tenacity of the common law's drift toward the anti-ecosystem bias meant that any meaningful protection of natural resources on private lands would have to come through public legislation. Although McElfish (1994) identifies many sporadic instances of such legislation in the states happening simultaneously with the common law's evolution in the opposite direction, no one could reasonably argue that a comprehensive body of statutory public law existed, even by the mid-1900s, to reverse the anti-ecosystem bias of the common law. The wave of federal environmental legislation beginning in 1970 did include laws with substantial impact on private land use, most notably the Endangered Species Act and the regulation of wetlands that has grown out of Section 404 of the Clean Water Act. But as that body of land use regulation expanded, the claim grew ever louder that its effect on land use cut so hard against the grain of settled common law property rights as to constitute a taking of property without just compensation in contravention of the Fifth Amendment to the Constitution.

Ironically, although this "regulatory takings" tension has not resulted in many successful litigation claims seeking compensation, it led eventually to a legal development that placed the pro-development common law in the role of gatekeeper for the validity of pro-environment legislation. In the U.S. Supreme Court's 1992 decision in *Lucas v. South Carolina Coastal Commission*,[8] Justice Scalia announced the majority's ruling that where a new land use regulation denies all economically beneficial or productive use of land—in that case a blanket prohibition of development in coastal dune areas—it must be treated as a per se taking of property for which just compensation is due under the Fifth Amendment. Justice Scalia's caveat was that just compensation would not be due if the regulation does "no more than [simply] duplicate the result that could have been achieved in the courts—by adjacent landowners (or other uniquely affected persons) under the State's law of private nuisance, or by the State under its complementary power to abate nuisances that affect the public generally. . . . "[9] In his concurring opinion, Justice Kennedy expressed concern with the idea that state regulation could go no further than duplicating the common law of nuisance without exposing itself to the now infamous "categorical taking" problem, for as he put it, "[c]oastal property may present such unique concerns for a fragile land system that the State can go further in regulating its development and use than the common law of nuisance might otherwise permit."[10] In other words, Justice Kennedy took it as a given, as Justice Scalia and the Court's majority also clearly did, that the common law of property does not protect the "fragile land system." Indeed, although leaving the final say to state courts, Justice Scalia surmised that "it seems unlikely that common-law principles would have prevented erection of any habitable or productive improvements on petitioner's land. . . . "[11]

In an effort to turn Justice Scalia's caveat into the exception that swallows the rule, some legal scholars have rediscovered the importance of the common law of property rights in the constellation of environmental law. For example, Blumm and Ritchie (2005) offer a comprehensive survey of common law doctrines that could, in some cases in their existing forms and in others only through some evolutionary judicial development, impose restrictions on the ability of a landowner to destroy natural capital and thus insulate public regulation that duplicates that effect from attack as a regulatory taking of property. Most of the doctrines they examine, however, relate to common law formulations of ostensibly superior *public* rights in land, which are dealt with later in this chapter. As for expressions of *private* property rights, the only doctrine Blumm and Ritchie describe as having yet departed from Sprankling's anti-wilderness bias thesis derives from a famous case in American environmental law, *Just v. Marinett County*,[12] in which the Wisconsin Supreme Court held that "[a]n owner of land has no absolute and unlimited right to change the essential natural character of his land so as to use it for a purpose for which it was unsuited in its natural state and which injures the rights of others." Although they and others (McElfish 1994) maintain that this "natural use doctrine" has firm roots in English common law and has been adopted by a few other American state courts, at best its contours remain hazy and its development nascent. Indeed, *Marinett County* and all cases endorsing it invoke the "natural use doctrine" only to defend public regulation of land from regulatory takings claims, not to adjust or define rights as between private property owners, so the doctrine's utility in overcoming the anti-ecosystem bias of the common law of property rights remains unproven. In short, notwithstanding considerable efforts by legal scholars to uncover property doctrine exceptions to Sprankling's thesis that would widen the caveat to the per se taking rule of *Lucas,* none has established itself in the existing body of common law.

TITLE TO LAND, BUT NO RIGHTS IN ECOSYSTEM SERVICES

Further evidence supporting Sprankling's evaluation of American property law is found in the absence of precedent for the proposition that landowners have rights in the continued flow of ecosystem services from other persons' lands. After all, such rights, if they were recognized and enforced, would be the antithesis of any notion that property law favors the development of natural capital. If Sprankling were wrong about the anti-ecosystem bias of the common law, therefore, one could reasonably expect to find precedent supporting a landowner's right inherent in title—that is, without formal contractual agreement or regulatory intervention—to some level of continued provision of ecosystem services flowing from natural capital found on other persons' lands. At the very least, under the assumption that such rights are presently unclear, one should expect to find the law silent on the matter. In fact, however, the property law of ecosystem services is the mirror image of the property law of natural capital—the common law is clear that there are no such rights inherent in title to land.

In this sense English and American common law are much closer in unison than is the case for the common law of natural capital. The kind of right that would require one landowner to refrain from interfering with the flow of ecosystem services to other lands is referred to in law as a negative easement (Dukeminier et al. 2006). The English common law recognized four negative easements inherent in title: the rights to stop other landowners from (1) blocking a neighbor's windows, (2) interfering with the flow of air in a defined channel, (3) removing artificial support for buildings, and (4) interfering with the flow of water in an artificial channel. Also, under the doctrine of "ancient lights," if a landowner has received light from across adjacent parcels for a sufficient period of time, a negative easement could arise. But English courts, cautious in general of attaching too many encumbrances to land, stopped there in establishing any more expansive negative easements as a matter of title. American courts accepted all but the ancient lights doctrine, which has been disavowed repeatedly,[13] and stopped there as well except in a few isolated and very limited instances (Dukeminier et al. 2006). American common law has also widely recognized that landowners must provide the lateral and subjacent support that an adjacent parcel would receive under natural conditions, imposing a general duty on landowners not to cause subsidence on other properties through excavation or withdrawal of groundwater.

Beyond this limited set of negative rights—that is, rights to prevent a person from doing something that injures another's land—American property law ventured no further. To be sure, American courts have been more generous than their English counterparts in recognizing the creation of negative easements *by agreement.* Land trusts routinely employ that mechanism to purchase (and not use) rights to develop land, leaving title and limited use rights in the seller (McLaughlin 2005). But as a matter of property rights in ecosystem services that inhere in title to land, there are none firmly established in common law, and this is no accidental gap in the law.

Property rights in ecosystem services thus are reduced to whatever might be successfully established on an ad hoc basis through nuisance claims. Legal scholars many decades ago suggested that a set of "natural rights" should guide nuisance law to protect a landowner's use of land in its natural condition, with one boldly claiming that "ownership of land insures far more than mere occupation and use of soil and vegetation on the surface of the earth. It protects the reasonable use of all the elements nature places on the surface" (*Stanford Law Review* 1948, 53). To date, however, few published judicial opinions have picked up on that thesis. One court in Texas found that cloud seeding unreasonably interfered with natural rainfall on the plaintiff's property, holding that a "landowner is entitled . . . to such rainfall as may come from clouds over his property that Nature, in her caprice, may provide."[14] In a more modern context, the Wisconsin Supreme Court found that interfering with the flow of light to solar panels could give rise to a nuisance claim given that "access to sunlight as an energy source is of significance both to the landowner who invests in solar collectors and to a society which has an interest in developing alternative uses of energy."[15] An exhaustive survey of case law reveals these are rare exceptions to the general rule—nuisance law has simply not developed, at least not yet, so as to provide any sense that a private landowner can secure continued provision of important ecosystem services.

THE COMMON LAW AS A WALL, NOT A GAP

As it stands today, therefore, American property law is not simply neutral on the question of private property rights in natural capital and ecosystem services but downright hostile to them, making it no wonder that neither finds much stock in the marketplace. To be sure, Justice Scalia acknowledged in *Lucas* that "changed circumstances or new knowledge may make what was previously permissible [under common law] no longer so,"[16] thus opening the door to the kind of evolution of law legal scholars such as Sprankling, Blumm, and Ritchie seek. At this time, however, the emerging knowledge about the ecological and economic value of ecosystem services has yet to gain discernible traction in that evolutionary process. Rather, knowledge of the economic value of natural capital and ecosystem services seems thus far to have budged the anti-ecosystem bias of American property law little, if at all.

The private landowner in such a system has no reason to think that conserving natural capital will be to his or her advantage; indeed, doing so may be a disadvantage. Likewise, the beneficiaries of ecosystem services flowing from natural capital on other persons' lands have no expectation based on our common law experience that they may protect those benefits through enforcement of property rights. Neither condition is the result of private property rights being "poorly defined." Rather, in the absence of intervening public legislation, we have been handed a clear set of rules from our common law system of property rights—landowners have almost total discretion over natural capital on land they own, with strong incentives to destroy it, and they have no rights in the continued provision of ecosystem services from land owned by others. There is no gap in private property rights to be filled, in other words, but rather a well-constructed wall to be taken down.

Group Ownership of Natural Capital and Ecosystem Services through Common Property

The classical model of private property in market analysis uses a rational *individual* owner out to maximize personal utility to depict how different rules and liabilities affect ownership behavior (Cole 2002). Of course, not all individuals derive personal utility the same way. Many property owners may not care to derive the maximum *economic* gain from property—some may value their property for conservation, or family heritage, or simple privacy more than for financially more lucrative uses—but each individual in a private property regime gets to make that decision for himself or herself and deal with the outside world on terms that maximize his or her utility. By contrast, when two or more individuals share ownership in the same interest in property, they must deal not only with the outside world, but also with each other.

For purposes of this chapter, the focus is on the limited question of how property law addresses the group ownership of property interests in an established private property system. There is an extensive body of literature, covered in chapter 6, dealing with how groups might manage common-pool resources by informal agreement, community norms, or otherwise when property rights are not well established or enforced by law. In a property law system that meets the foundational conditions of private property, however, common property in the formal legal sense is simply a way for a group of individuals to share an interest in property, there being no question what it means to own the interest and that the group is the owner. When all relevant property rights are well established and enforced, therefore, the only question group ownership introduces to the picture is what rights the members of the group have *relative to each other*. In other words, nothing about the fact that a group of individuals shares ownership of a property interest changes how the law recognizes or enforces that interest relative to owners of other parcels of property or, for that matter, relative to owners of other interests in the same parcel of property.

In the absence of either an agreement among the group members or a regulatory intervention, therefore, common property is a "commons on the inside, property on the outside" (Rose 1998). Hence, when all members of the ownership group agree in all respects on how to use the property interest, all that has been said above about American property law for private property with respect to individual owners applies equally in the common property setting—the group might as well be an individual as far as property law is concerned. To be sure, group ownership may prove advantageous for navigating through the private property regime by reducing the transaction costs of bargaining and of pooling resources. Land trusts take advantage of such economies of scale (Anderson 2004) and thereby can work to overcome the anti-ecosystem bias of American property law through sheer numbers and a unifying vision. But those advantages spring from the collective resources of the group members and the strength of their agreement, not from any special status of common property under property law.

When the members of the group do *not* all agree on how to use the common property interest, however, the law should provide means of deciding whose view prevails. In many instances this has little to do with property law or property rights directly. A corporation, for example, is owned by its shareholders, and thus property to which the corporation holds title is a form of common property (Cole 2002; Demsetz 2002). The same is true for a legally formed partnership. But when disagreements arise among the group members in such cases about how to manage the common property interest, fields such as corporate law and partnership law govern the decision-making process for the group and the relative power of each group member in that process. When disagreements over the use of such a group's common property arise, therefore, frequently this kind of legally enforced decision-making apparatus can be engaged to decide the dispute, and property law simply accepts the answer.

Hence, the real point of interest about common property arises when the group of owners cannot agree on how to use the property interest and there is no specialized body of law governing the group with which to resolve the dispute, meaning property law must supply the answer directly. In other words, if the ownership group is ungoverned—meaning it has no "law of the group" pursuant to other enforced bodies of law—how does property law resolve disputes over use of common property among the group members?

Demsetz's (2002) model of the "compactness" of the ownership group—the number and closeness of the persons involved—provides a useful insight on how this answer has played out in American property law. For the most part, as Demsetz points out, examples of large ungoverned groups owning common property are rare in the United States, particularly when compared to many other parts of the world. The classic cases of large-group collective ownership—the open field system of seventeenth-century Europe, the commonfield agricultural lands of the South American Andean highlands, the common forest and grazing lands of the Japanese Iriachi, and similar collective property regimes—generally involve modestly productive agrarian land uses. Highly articulated, state-enforced private property regimes would be costly to establish and enforce in such settings (Cole 2002); instead, as is discussed in chapter 6, the group often develops informal customs and norms to regulate behavior within the group and relies on group vigilance to enforce its collective management decisions against the outside world.

By contrast, property is put to highly productive uses in the United States, and resource allocation issues are often complex economically and socially (not to mention ecologically). Simply put, property in the United States has become too valuable, and its management challenges too complex, to expect that it would wind up being held in substantial amounts by large, amorphous, ungoverned groups of individuals. Rather, with the exception of the riparian water rights system discussed later in this chapter, large groups in the United States generally own common property through highly governed entities such as corporations, partnerships, land trusts, or governments, which are decidedly not "commons on the inside." Hence, given their well-defined rules of internal governance, whether such entities do or do not conserve natural capital in ways that maximize social welfare would have nothing to do with how clearly property law has defined the rights of the group members sharing ownership through common property.

Not surprisingly, therefore, American property law, when concerned with common property, focuses on developing rules that are most useful when the property is owned by relatively compact, close-knit, ungoverned groups, such as siblings, spouses, or neighbors. To the extent such small groups might disagree over uses of their common property interest, they are unlikely to present profoundly complex or intractable "commons on the inside" problems for property law. The small group size, the close-knit nature of the group members, and their ability to exclude outsiders make these arrangements closer to what economists refer to as "club property" (Buchanan 1965; McNutt 2000), to distinguish it from larger, more amorphous group-sized ownership patterns. In other words, modern American common property does not present the kind of large-group contexts that have drawn attention from many scholars in developing theories of collective resource management, most notably Elinor Ostrom and Daniel Bromley (summarized in Cole 2002, 110–29), which are taken up in chapter 6. The law of common property thus is fairly mundane in the United States and, if anything, follows the anti-ecosystem bias found generally to have evolved in the common law of private property owned by individuals.

For example, the common law property rule most relevant to common ownership situations in which there is no formal agreement between the group members, known as the doctrine of waste, was enforced in England mainly to preserve the status quo. Sprankling (1996, 534) explains that it "resolved disputes between competing interest holders by preferring existing uses to new uses." Particularly given England's wood-dependent economy and wood-scarce landscape, any substantial cutting of trees on forested land was considered waste, allowing the objecting co-owner to prevent his or her co-owners from doing so. In the early American context, the situation was quite the reverse— the landscape was tree-abundant and farm-scarce. The English version of waste would have impeded agricultural development, and thus the American courts soon deemed that "lands in general with us are enhanced by being cleared" and that it would "be an outrage on common sense" to apply the English doctrine.[17] This sentiment eventually forged the American "good husbandry" standard of waste, which permitted a co-owner to clear wilderness land for cultivation or grazing without fear of being found to have committed waste (Purdy 2006). Sprankling (1996, 569) surveys more recent case law to demonstrate that, while the number of cases decided pursuant to the common law doctrine has diminished considerably (likely because most co-owners today act through formal governing agreements), the courts remain committed to this approach, leaving "the modern law of waste . . . staunchly hostile to wilderness."

Common property regimes in the modern American property law context thus provide very little traction for the recognition of natural capital and ecosystem service values. Common property is owned primarily by groups that operate under formal group governance mechanisms, such as corporate law, partnership law, and contractual agreements. These group governance mechanisms are designed to produce a single answer from among the many members' voices, making such groups, for all practical purposes, individuals in the eyes of property law. As discussed in the previous section, the private property system leaves that universe of common property essentially at odds with the recognition of natural capital and ecosystem service values. What little common property does not fall in this category must resort to the common law for resolution of land use disputes between the co-owners, where the doctrine of waste replicates the same hostility toward conservation as do the common law doctrines for individually owned private property.

Public Ownership of Natural Capital and Ecosystem Services

The federal government owns more land in the United States than any other landowner—about 650 million acres spread through the fifty states, which is over one-third of the nation's landmass—and state, local, and tribal governments are significant landowners as well. Quantity, though, does not necessarily mean quality. True enough, the previous sections suggest that private property and common property systems in the United States have stunted development of property rights in natural capital and ecosystem services. But what reason is there to believe that public property systems, even in democratic societies, will perform any better?

Cole (2002, 20–44) and Gottfried et al. (1996) summarize the extensive body of literature positing many reasons why public ownership cannot necessarily be assumed to produce better resource allocation outcomes than a market-based private property system (and vice versa). Indeed, many commentators argue that public ownership is inherently likely to lead to *worse* outcomes (Anderson and Leal 1992; Posner 1998; Stroup and Baden 1983; Stroup and Goodman 1992). The crux of the argument boils down to the difference between the economic incentives private actors send and receive in the market and the institutional incentives elected officials and appointed bureaucrats send and receive in the political realm. Strong believers in the virtues of the market argue that political institutions will act as a poor second best to the market, whereas skeptics of the market, particularly those wary of distributional inequities, place more faith in democratic institutions. In short, it is a matter of whom one prefers to make decisions about resource management, private individuals and corporations, or bureaucrats and politicians.

The principal reasons presented for why public ownership is likely to fall short are the lack of profit motive, the ability of government to pass off the costs of poor management to the public, and the short-term planning horizon of election and budget cycles (Cole 2000, 298–305; 2002, 20–44). On the other hand, because it does not operate subject to market constraints, government can make policy decisions, such as preserving natural capital, that produce economic value not recognized in the market. In particular, the government need not compare ecosystem service values using the same private discount rate that market participants use in assessing alternative uses of a resource. Rather, the government can adopt a "social" discount rate that reflects a longer view toward sustainable resources, greater aversion to risk, and a greater appreciation of the complex nature of ecosystems (Arndt 1993; Markandya and Pearce 1991; Pearce et al. 1989, 132–52). Moreover, government is likely to face lower transactions costs than private entities in setting aside and managing large tracts of land necessary for maintaining ecosystem services dependent on landscape-level ecosystem integrity. There are, in other words, theoretically valid arguments on both sides.

The proof, as they say, is in the pudding. This chapter is focused on how effectively different property regimes have *in actual practice* integrated the value of natural capital and ecosystem services into resource allocation decisions. Despite the theoretical basis for believing that private property will effectively direct the market toward this end, supporters of private property in fact have little positive evidence to advance on its behalf in this regard. So what evidence is there either way with respect to public property? With virtually plenary power over the use and disposition of federally owned land granted to it under the Constitution (Appel 2001),[18] one could reasonably expect that the federal government, if it wished to, could produce significant gains in securing the integrity of natural capital and maintaining flows of ecosystem services on its lands and to others' lands. Alas, the story for public property is not encouraging either.

To be fair, the federal government has put much of its land into conservation status. Over 40 million acres are in protected wilderness status, over 90 million acres are wildlife refuges, and several other conservation categories of federal land management add millions of additional acres to what amounts to conservation as the sole or dominant use of the public land management unit. For the most part, however, the federal government has chosen to manage the lion's share of its lands under a policy framework known as "multiple use/sustained yield," or MUSY. Over 192 million acres of national forests, 83 million acres of national parks, and 260 million acres of unclassified public lands administered by the Bureau of Land Management are subject to this resource management model. As is discussed in more detail in chapter 5, the U.S. Forest Service's MUSY mandate for the national forests typifies its amorphous nature, defining multiple use as "the management of all the various renewable surface resources of the national forests so that they are utilized in the combination that will best meet the needs of the American people; making the most judicious use of the land for some or all these resources or related services over areas large enough to provide sufficient latitude for periodic adjustments in use to conform to changing needs and conditions."[19]

As Nagle and Ruhl (2006) explain in their survey of this and the other federal public land MUSY mandates, this "standard" vests considerable discretion in the land management agencies to decide what "best meets the needs of the American people." Courts, which in general must defer to agency discretion, have found virtually no toehold in MUSY with which to subject agency decisions to any meaningful scrutiny. Courts have held, for example, that in weighing different options for use of public lands subject to the MUSY mandate, the federal land management agencies are subject to no particular "accounting method" and may exclude consideration of the benefits of ecosystem services because of the "lack of certainty" in assigning quantitative value to them.[20]

Of course, the open-ended discretion of the MUSY mandate goes both ways, affording the land management agencies considerable latitude to integrate natural capital conservation and maintenance of ecosystem service flows into their MUSY decision-making calculus. That the agencies need not consider ecosystem service benefits does not mean they cannot. Historically, however, their track record has been repugnant to using ecosystem service values to guide public land use decisions. With all their MUSY-infused discretion to balance competing uses of public lands, the federal land management agencies have been subjected to pressure by the interests associated with the different uses to tip the scales in their respective favors. The result, on balance, has been to promote resource development, commodity extraction, and recreation—that is, the uses with values reflected in the market—against management of ecological functions that might secure natural capital and maintain ecosystem service flows (Cheever 1998; Feller 2001; Nagle and Ruhl 2006). Indeed, no significant study of the Forest Service, Bureau of Land Management, or other federal land management agency implementing a MUSY regime has evaluated the historical experience as an ecological success story. The Forest Service, for example, used MUSY in the 1960s to portray clear-cutting as a rational, scientifically sound forest management policy, leading Congress to manage the practice more closely through the National Forest Management Act of 1976 (Nagle and Ruhl 2006). The Forest Service has also been sharply criticized for selling timber harvest rights to private companies for well below market prices (Stroup and Baden 1983). Similarly, the Bureau of Land Management has for decades sold grazing permits to private ranchers at below-market cost, leading its lands to be evaluated as almost completely depleted of ecological value (Donahue 1999). As the U.S Supreme Court has documented, Congress, independent commissions, and even the Bureau of Land Management itself regard the history of public grazing as having left "vast segments of the public rangelands . . . in an unsatisfactory condition."[21]

Only recently have public land management agencies discovered ecosystem services and recognized that their management policies have left the value of ecosystem service flows out of the decision-making calculus. For example, delivering a speech on "Innovations in Land and Resource Governance" at the August 2005 White House Conference on Cooperative Conservation, Secretary of Agriculture Mike Johanns outlined a long agenda of reforms for agricultural policy and the national forests, one of the last of which addresses ecosystem services:

> Today, I am announcing that USDA will seek to broaden the use of markets for ecosystem services through voluntary market mechanisms. I see a future where credits for clean water, greenhouse gases, or wetlands can be traded as easily as corn or soybeans. We will collaborate with partners to establish a role for agriculture and forestry in providing voluntary environmental credits. I know that it's one thing to announce a new policy and quite another to achieve meaningful results. Therefore, we are creating a new Market-Based Environmental Stewardship Coordination Council. The Council will help to ensure that we produce a sound market-based approach to ecosystem services.[22]

The details and extent of this enlightenment remain to be seen. For now, one can only conclude that the history and current record of public land ownership in the United States suggest that public ownership as a property regime cannot claim any superiority to private ownership insofar as the integration of natural capital and ecosystem service values into resource allocation decisions is concerned.

Property Rights and Water

Cole (2002) reminds us that resources don't always fit tightly into private, common, or public ownership status, but rather the "bundle of sticks" representing the rights associated with a parcel of property is often found subject to an admixture of multiple owners and kinds of ownership. No property interest illustrates this reality better than water. Indeed, water rights law is so complex, it is often seen by lawyers as a separate field from property law. But rights to own and use water are in concept assignable as property, and in practice water finds itself strongly represented in all three kinds of ownership regimes. Water, therefore, offers an opportunity to evaluate how mixed property regimes work toward integrating natural capital and ecosystem service values into resource allocation decisions. In particular, because the distribution of water rights requires assigning rights in both resource stocks (lakes, watersheds, aquifers) and resource flows (rivers, groundwater flow), it provides a close analogy to the challenges of assigning property rights in natural capital and ecosystem services.

A Primer on Water Rights

Two core interrelated issues will drive the law and policy of ecosystem services in freshwater ecosystems—water quality and water quantity. As environmental protection laws and policies increasingly demand authority over both of those domains, they have run head-on into what in many parts of the nation is dearer than diamonds—water rights. Congress understood the potential for this collision at the dawn of the environmental legislation movement in the early 1970s, advising that one policy of the Endangered Species Act of 1973 be "that Federal agencies shall cooperate with State and local agencies to resolve water resource issues in concert with conservation of endangered species."[23] Alas, that has been easier said than done. Some background on water rights helps explain why.

Generally speaking, freshwater is far scarcer in supply west of the nation's 100th meridian, which has led to the development of two different, but equally rich, bodies of water law (Dellapenna 2002; Tarlock 2000a). Water in its natural watercourses is in all the states nominally owned by the states as a public resource, but *use* of the water is what matters, and the states have developed highly articulated legal regimes for distributing these "usufructuary" rights (Hayes 2003; Miano and Crane 2003). As one court put it, the state "acts as the trustee: it owns the resource, but holds it not for itself, but for the benefit of the public. . . . [I]t cannot sell, lease, or give it away, it can only make it available for use by others."[24]

In the East, where water is more plentiful, surface water has traditionally been allocated under the common law "riparian rights" system, under which water is treated essentially as common property of all owners of land underlying or bordering the water body, known as riparian lands. Under strict riparian rules, only riparian land may use the water, and the water may be used only on and for the riparian land, though for obvious reasons (e.g., public water supplies) many states have altered that rule to allow specified off-site uses. Also, although the amount of water that may be used was in early versions of the system limited such that the landowner had to leave the "natural flow" of the watercourse unimpaired, conventional riparian rights systems are based on a "reasonable use" standard that allows some reduction in natural flow so long as other riparian owners are not harmed. What constitutes a reasonable use depends on a variety of factors such as the purpose of the use, suitability of the use to the watercourse, its economic and social values, and the extent of harm to others. To summarize, several general characteristics define the classic riparian system:

1. Riparian rights are of equal priority.
2. The right is not quantified, but rather extends to the amount of water which can be reasonably and beneficially used on the riparian parcel.
3. Riparian rights are correlative, so that during times of water shortage, the riparian proprietors share the shortage.
4. Water may be used only upon that portion of the riparian parcel within the watershed of the water source.
5. The riparian right is part of the riparian land and cannot be transferred for use on other lands.

The classic riparian rights system, based as it is on the balancing test applied for reasonable use, can lead to significant uncertainty as to the quantity of water to which each riparian is entitled and how to resolve disputes among them (Miano and Crane 2003). The on-site use requirement can also constrain economic development. In general, moreover, the riparian rights system leaves water flows generally unregulated and unplanned. About half of the eastern states, therefore, have moved to a permit allocation system that, while preferring riparian uses, focuses primarily on the purpose of the use and facilitates a more orderly monitoring and enforcement of approved uses. Florida provides a well-developed example of this system (Fumero 2003), which is often referred to generally as "regulated riparianism" (Dellapenna 2002).

As for groundwater, the eastern states follow several different doctrines. The "rule of capture" or "absolute ownership" allows a landowner to withdraw unlimited supplies of groundwater from beneath the surface regardless of the consequences for other landowners drawing from the same source. Some states temper this under the doctrine of "correlative rights," which allows use only in proportion to the relative size of the surface estate unless using more would not injure other users from the same source. And about a third of the eastern states have adopted a "reasonable use" rule that recognizes proper beneficial uses, such as domestic supply, irrigation, and mining, and allows withdrawal as long as no injury is caused to other beneficial uses relying on the same source. As has happened with surface water, however, most of the eastern states have adopted a more formal permitting system for groundwater uses.

Although water in the East was a plentiful resource well into the twentieth

century, a growing urban population has demanded ever more supply, agriculture has increased the use of irrigation as insurance against drought, and environmental concerns have made water quantity an important factor in the management of water quality and wildlife habitat. Conflicts over water supply,
particularly between urban and agricultural users, have become more frequent
(Hayes 2003; Ruhl 2003).

In the West, where water supply has always been scarce and unreliable, allocation of surface water supplies generally follows an "appropriative rights" system that implements a rule of "first in time, first in right." Four core principles
guide this system:

1. Water in its natural course is the property of the public and is not subject
   to private ownership.
2. A vested right to use the water may be acquired by appropriation and application to a beneficial use.
3. The person first in time is first in right.
4. Beneficial use is the basis, measure, and limit of the right.

What matters under this approach, in other words, is appropriation of
water to a beneficial use, which defines a user's allocation from the particular
stream. The time of appropriation is of the utmost importance, with the "senior" appropriators who put water in a stream to use earlier than "junior" users
having a priority when water supplies are short. In dry periods, therefore, the
senior user takes as much as his or her allocation allows, the next most senior
user comes next, and so on until the supply is exhausted, meaning some junior
users may have no water at all.

Although it makes water rights more predictable than does the riparian system (which is why the western states abandoned riparianism early in their settlement), there are flaws in the appropriative rights system as well (Miano and
Crane 2003; Tarlock 2001). For example, senior uses may not be the most economically useful or efficient, but will nonetheless trump socially superior uses
in dry periods. The risk associated with junior rights can also deter investment
in watersheds that have unreliable supplies. New users may be able to purchase
rights from senior users, but many states severely restrict interwatershed transfers of water. Also, because senior users must maintain their beneficial use to
keep their rights intact, the appropriative system does not reward water conservation. In many watersheds, moreover, the quantity of appropriated rights and
the quantity of water supply were sometimes far from certain, a problem many
states address through an "adjudication" process requiring users to prove their
seniority and amount of beneficial use in a multiparty proceeding that results
in each user's rights being established through an "adjudicated right."

The appropriative rights system has evolved with many different nuances among the western states. For example, some states prioritize beneficial uses, usually with domestic and agricultural uses receiving the highest priority. And some states even recognize limited riparian rights. One of the most controversial issues, though, has been how to ensure that, for recreational and wildlife purposes, some minimum amount of water remains undiverted and not consumed, known as "instream flow" (Boyd 2003). The catch has been that the conventional appropriation system requires a beneficial use *and* a diversion. Some state courts have recognized instream flow as a beneficial use not requiring a diversion.[25] Most states have handled this through statutes allowing public agencies to "reserve" instream flow from appropriation by others under prescribed conditions and procedures (Mathews 2006; Neuman et al. 2006). Some states are also beginning to allow nongovernmental entities to lease instream flows (Tarlock 1991). Yet change has come slowly, and many commentators have emphasized that far more concerted and deliberate initiatives are needed for the law of instream flow to catch up with the ecological realities of western water resources (Benson 2006; Bonham 2006).

Groundwater rights are even more varied among the western states, ranging from a pure "rule of capture," to a rule of "correlative rights" assigning rights in an aquifer relative to surface area of the land, to a rule of "reasonable use," to a system basically the same as the surface water appropriative rights approach. Some states have responded to massive agricultural and urban overpumping of groundwater supplies by enacting regulatory codes, but in many arid states groundwater depletion is a growing concern (Hayes 2003).

Water Rights as a Mixed Property Regime

The foregoing summary of water law is a standard story told in countless law books, taken as a given by courts and law practitioners around the nation, and considered sacrosanct by those who believe they hold precious water rights. But in many respects the exact nature of water rights is ambiguous. Are they private property, common property, public property, or all at once?

Virtually all state water rights systems start with the premise that the state sovereign owns the water resources in the state, yet the previous description reveals that most of water law in the East and the West alike is about how private interests (as well as public entities) secure rights to *use* water. The West's appropriation system is decidedly private in context, particularly in its seniority mechanism for allocating rights during dry periods (Dellapenna 2002; Tarlock 2000a). The Utah Supreme Court in 1917 went as far as to proclaim that "the very purpose and meaning of an appropriation is to take that which was before a public property and reduce it to private ownership."[26] The East's riparian rights system, while not an example of formal common property ownership, is usually held out as a property regime relying on correlative sharing rights, and in that sense is an example of common property (Bouckaert 2000; Dellapenna 2002, 2004; Tarlock 2000a). But whether it is more private or group in nature, a riparian right is decidedly not public property. In both cases these systems create legally recognized rights in the use of water that are rights in the true property sense of the word, in that they enjoy protection from uncompensated governmental takings under federal and state constitutions (Hayes 2003). Hence, Benson (1997) observes that "state or public ownership of water has far more meaning on paper than in practice" and "water rights holders generally view the water they use as being their own, and they stress the private property nature of water rights" (375).

Tarlock (2000b) suggests that the gradual but unmistakable privatization of water rights "has gradually eroded the connection between humans and an actual physical space by making property a universal abstraction rather than a situation-dependent entitlement" (72). By treating water as a commodity up for private and common ownership grabs, water law has, in his view, detached rivers from their surrounding ecosystems. To be sure, the public interest in water has supported regulation of private use rights as discussed above, but Tarlock's point is that once water law severs water from ecosystems, treating it as a consumptive commodity, and so long as regulation focuses on water the commodity, there is little reason to expect markets in water to reflect natural capital and ecosystem service values not associated with the commodity value.

Yet public ownership of water, albeit pushed into the background of the water use rights system as a matter of practice, is nonetheless a potential constraint on the private system, the source of which is not simply the public interest but public ownership, and the effect of which is not merely regulation but the assertion of an ownership right. For the most part, however, this has been an unrealized potential, used judiciously at the outer edges of the system to address only the worst failures in the prior appropriation and riparian systems. Perhaps the most obvious example comes in the form of the so-called public trust doctrine, the lineage of which can be traced to Roman law's conception of water resources as *res communes*—a thing incapable of individual ownership and thus held in common (Duncan 2002).

The name is impressive, suggesting great possibilities, but the lodestar case of the public trust doctrine in the United States, at least for purposes of thinking about it as a tool of resource conservation, was no harbinger of a day when it might guide management of natural capital and ecosystem services. In the U.S. Supreme Court's 1892 opinion in *Illinois Central Railroad Co. v. Illinois*,[27] the Court held merely that Illinois could not sell fee interests in the land under Chicago Harbor to private developers, because

> [t]he state holds the title to the lands under the navigable waters. . . . It is a title held in trust for the people of the state that they may enjoy the navigation of the waters, carry on commerce over them, and have liberty of fishing therein, freed from the obstruction or interference of private parties.

That is as far as the Court took the public trust doctrine. Nevertheless, in a landmark legal article Professor Joseph Sax (1970), then on the law faculty at the University of Michigan, outlined an ambitious agenda for evolving the doctrine into the nation's bedrock source of ecosystem management law. Sax argued that "[o]f all the concepts known to American law, only the public trust doctrine seems to have the breadth and substantive content which might make it useful as a tool of general application for citizens seeking to develop a comprehensive legal approach to resource management problems" (1970, 474). But this never came to be. Why not?

One important reason is that the U.S. Supreme Court declined the invitation to take the doctrine there. As Ruhl (2005a) explains,[28] the Court has declined to expand the scope of the doctrine, either geographically beyond submerged lands or categorically beyond the trust uses described in *Illinois Central Railroad*. As far as the Court is concerned, the states may not alienate fee title in tidelands, shores, and other public trust lands in violation of the public trust doctrine, and that's as far as the courts have gone to constrain state behavior.

To be sure, many state courts have opined more broadly on the scope of their state's version of the public trust doctrine, in particular to extend it to water more generally, not simply the land under water (Blumm and Ritchie 2005; Ryan 2001). For example, one famous case from California, regarding the diversion of water from Mono Lake, ruled that "[t]he state has an affirmative duty to take the public trust into account in the planning and allocation of water resources, and to protect public trust uses whenever feasible."[29] This and other state cases like it, however, are mindful of the "publicness" of public trust resources, emphasizing uses such as navigation, fishing, and recreation, and not necessarily preservation or even active conservation of ecosystems, much less ecosystem services (Ryan 2001). It is true that an occasional state case suggests an ecologically oriented purpose to the doctrine. Perhaps the most noted case in this regard is from Wisconsin, in which the court found that the doctrine required that wetland areas be limited to uses consistent with "natural conditions."[30] Several more recent cases are variations on that theme (Blumm and Ritchie 2005; Lum 2003; Ruhl 2005a).[31]

Some commentators thus assert that the public trust doctrine is "definitely growing" as an ecosystem management tool (Lum 2003, 73). By and large, however, the state courts have declined to mobilize Professor Sax's vision of the public trust doctrine as a means of effective and broad judicial intervention in resource management policy (Ryan 2001). There is, simply put, no broad-based resource management duty to be found in the judiciary's version of the public trust doctrine, certainly not one that has reached private lands on which ecologically important natural capital resources are found.

In short, while it may be hard to detect any aversion in the case law to expanding the public trust doctrine into the domain of natural capital and ecosystem services, it is even harder to detect any sense of urgency or enthusiasm (Lahey and Cheyette 2002; Scott 1998). One rather obvious possibility for this lethargic approach is that, not long after Professor Sax suggested how the doctrine's latent power could be tapped, the legislative revolution of the 1970s unfolded to bring one after the other of comprehensive resource management laws into being. New federal legislation protecting wetlands, the coastal zone, and endangered species, as well as managing federal public lands, obviated the need for the Supreme Court to revisit the public trust doctrine, and the eventual blossoming of similar state legislation did the same at the state level (Lazarus 1986; Ryan 2001). Maybe the public trust doctrine could have become what Professor Sax envisioned in 1970 and what many commentators still hold out hope for, but with the surge of federal and state environmental legislation that transpired, who needed it?

Yet what goes around, comes around, and it may be that the environmental legislation of the 1970s, perhaps responsible for taking the wind out of Sax's view of the public trust doctrine, in fact sowed the seeds for its eventual reemergence or something like it. A major disruption of the settled body of water law has come in the form of environmental protection legislation (Arnold 2005). Simply put, the demands of federal environmental laws such as the Endangered Species Act, Clean Water Act, and National Environmental Policy Act, as well as their many state counterparts, include demands on water quantity. The description in the prologue of the conflict over the Apalachicola–Chattahoochee–Flint (ACF) River basin in Georgia and Florida demonstrates this phenomenon in the East. An acute example from the West is the water allocation conflict taking place in the Klamath River basin straddling the border between Oregon and California. When a severe drought struck the overappropriated basin in 2001, the federal agencies managing and monitoring operation of a water supply reservoir were forced to choose between maintaining instream flows for endangered fish species versus providing continued irrigation flows to farmers (National Research Council 2004b). The agencies chose the fish in 2001, returning over 200,000 acres of farmland to their natural arid conditions, then chose the farmers in 2003, which many believe contributed to a massive fish kill near the mouth of the river (Doremus and Tarlock 2003). In both years it was clear that the water rights systems of the two states had simply been outstripped in their capacity to accommodate all the water demands and had no effective mechanism to manage the competing interests.

The ACF and Klamath are just two among many examples of how settled expectations about water rights have been rocked in many states by the implementation of the environmental legislation regime (Alderton 2003–04; Hayes 2003; Miano and Crane 2003; Rossmann 2003; Tarlock 2001). The result has been to put water rights under a microscope, revealing in many ways that what was once thought of as a cohesive, self-contained property system is in fact an amalgam of several different property regimes, the workings of which may come a bit unglued when confronted by a force as powerful as environmental law. Particularly in the face of increasing scarcity of water, states have become more willing to assert, or at least consider asserting, their public ownership of water in response to the environmental effects of overuse that has so often been the result of private water rights systems (Andreen 2006; Deason et al. 2001; Emel and Brooks 1988; Hayes 2003). Private interests, on the other hand, have been no less willing to challenge governmental incursions on their use of water (Benson 2002; Leshy 2005; Parobek 2003). Litigation over water rights is heating up in many states as a result, and courts are having to mediate competing claims on water between government and private water users in an evolving legal landscape. Recently, for example, a California court held that locally imposed restrictions on groundwater pumping by agricultural users did not unduly interfere with their rights, and so did not constitute a taking of property, a ruling that surely rocked the agricultural community's conception of the scope of their access to groundwater.[32]

Unlike government regulation of land use practices, however, this imminent collision of interests is between two systems of *property* rights—public versus private—and not simply a question of how far government may regulate in the public interest. In other words, with respect to water, the fact that the government can assert a property interest rather than only a regulatory interest can alter the calculus of what consequences flow when the government attempts to assert that interest by adjusting the scope of private property interests in water (Roos-Collins 2005). As noted earlier, in the U.S. Supreme Court's 1992 decision in *Lucas v. South Carolina Coastal Commission,*[33] Justice Scalia hinged the question of government liability for takings of private property on whether regulatory land use restrictions duplicate constraints already imposed under the "background principles of nuisance and property law,"[34] in which case no liability would attach. Most important, he noted that "changed circumstances or new knowledge may make what was previously permissible no longer so,"[35] which many property law scholars take to mean that the background principles for purposes of government takings liability evolve dynamically with the changing contexts of appropriate land uses and property rights (Lazarus 2004; Sugameli 1999). The public trust doctrine, and public ownership of water in general, are just that sort of evolving background principle of property law (Duncan 2002; Kanner 2005). The rebalancing of water rights between public and private ownership interests, therefore, may be a case in which property law adjusts to new knowledge (for example, about ecosystem services) by arriving at new configurations of the relative balance of rights within the mixed property regime system (Rose 1990b).

Although some legal scholars do not agree that Justice Scalia meant to leave this door open or that going through it would be wise (Callies 2000; Callies and Breemer 2002), this is precisely the project Blumm and Ritchie (2005) have undertaken through their comprehensive survey of common law "background principles." Whereas the discussion above regarding nuisance law suggests that they have yet to identify a prospect for rebalancing private versus private-property interests, their work on the public trust doctrine and other sources of public ownership interests in water outlines the legal argument for moving the line in the public versus private water rights balance toward the public side. Indeed, the details of some states' water rights law reveals this as an emerging theme, Florida being just one example where riparian rights have given way in recent years in some degree to public rights (Proctor 2004). It remains to be seen whether this trend continues in Florida and comes to pass more broadly among the states and, if it does, whether it improves management of water resources. But the fact that it is a realistic prospect illustrates what may be the underlying advantage of a mixed property regime—that it may respond more adaptively and effectively to new knowledge than can a strictly private, common, or public property regime.

As it stands today, however, conflicts such as the ACF and Klamath suggest that the mixed property regime of water rights has yet to prove its adaptive power with respect to natural capital and ecosystem services (Andreen 2006). The private law systems of water rights—prior appropriation in the West and groundwater withdrawal rights in many of the states—have failed to prevent, if not contributed to, overdevelopment of water resources. The common property approach of riparian rights is strained to its limits as well. Regulatory interventions have fixed some of the flaws of each system, but still no coherent approach has emerged with respect to natural capital and ecosystem services. We have reached the point where, without technological advances, it is no longer possible to continue the pace of water development *and* retain an intact stock of aquatic natural capital necessary for continued delivery of valuable ecosystem services (McCool 2005). It may come down, as Blumm and Ritchie posit, to a redistribution of the rights themselves as between the different sectors represented in the mixed property regime, but for now that prospect is more a source of conflict than it is consensus. As Hayes (2003, 24) aptly concludes, the rising frictions between water rights and resource management "are exacerbated by the uncertainty of how public trust and public interest state water law principles, as well as federal law overlays, affect the scope of private water rights. Everyone is learning on the fly."

5 Regulation

Even if the common law's anti-ecosystem bias were stripped from the private property rights system, it may well be that ecological, geographic, and economic contexts would present significant barriers to developing an efficient set of private property rights governing natural capital and ecosystem services. Where privatization does not get the job done, the cue from Hardin's "Tragedy of the Commons" (1968) suggests that the work can be done through regulation, or as Hardin put it, through "mutual coercion, mutually agreed upon by the majority of the people affected" (1247).

Indeed, given the depth and breadth of environmental legislation in the United States, one could reasonably assume that regulation would have become a formidable engine for making decisions about natural capital and ecosystem services. As this chapter shows, however, one would be wrong to draw that conclusion. After exploring the nature of regulation and rationales for using it as a foundation for resource management decision making, this chapter surveys the evolution, modern status, and current trends of environmental regulation, tracing in each case the failure of this otherwise formidable source of law to address comprehensively and effectively the challenges of natural capital and ecosystem services.

What Is Regulation?

Regulation involves the exercise of *legislative* authority to control conduct, usually of many people or entities, through prescribed standards designed to promote the public interest (Pierce et al. 1999). In the United States, federal regulatory authority depends on powers enumerated in the Constitution, with the primary workhorse for environmental regulation being Congress's power to regulate commerce between the states, known as the interstate commerce power (Klein 2003), as well as the power to spend (Binder 2001) and to control federal public lands (Appel 2001). States derive general regulatory authority through the so-called police power, under which states may regulate to protect the public health, safety, and general welfare. Although debate continues over the precise scope and overlap of these respective federal and state powers, both sources of regulatory authority undeniably are expansive and reach a broad array of private activity, including the management of natural resources on private property.

Usually the legislature establishes regulatory standards in statutes that delegate implementation power to administrative agencies, as Congress has done through many statutes for environmental agencies such as the Environmental Protection Agency and the Fish and Wildlife Service. A rich and complex body of administrative law applies in the federal and state systems to govern how agencies exercise these delegated authorities. Administrative law provides for agency discretion to issue general regulatory standards in administrative regulations and to adjudicate specific cases arising under those standards, so long as the exercise of discretion is consistent with the statutory provisions. Opportunities for public participation allow interested parties to provide input to these agency decisions, and to seek relief in the courts when the agency acts inconsistent with constitutional or statutory requirements. Necessarily, therefore, administrative law also establishes a system for judicial review of the agency's conduct when such claims are brought.

The development of administrative law in the United States has put agencies in a position of tremendous power. Statutes seldom prescribe agency duties or authorities in specific terms, thus leaving wide latitude to agencies to "fill in the blanks." Public participation in standard setting and other regulation promulgation proceedings is generally limited to "notice and comment" opportunities, and in adjudicatory proceedings usually only parties with requisite "standing" may participate. Judicial review is in general guided by the overarching principle of deference to agency exercise of discretion, such that only "arbitrary and capricious" agency action—decisions clearly not grounded in the statutory directives or, in the case of adjudications, in the factual record. To be sure, successful challenges to agency decisions are not infrequent, but agencies by and large wield significant authority over virtually every aspect of life in America.

Paradoxically, the two principal justifications for regulation as practiced in the United States are to remove constraints to efficient operation of markets and to remove perceived inequities resulting from efficient operation of markets (Pierce et al. 1999). On the one hand, regulation can supply one or more of the factors necessary for an efficient market, such as more clearly defined property rights. On the other hand, society may conceive goals for social and economic equity that will not be produced by unrestrained private market outcomes, in which case regulation can supply the necessary intervention. In both

cases, regulation has proven to have pervasive influence over resource management decisions made by private property owners. For example, by prescribing
standards for pollution of air and water resources, regulation can internalize the
costs of pollution to the polluters and thus promote more efficient market production decisions. But regulation is used just as effectively to intervene in mar-
ket-driven decisions. For example, the power of states to zone land uses, dictat-
ing where industry and residential developments may or may not locate, has
been recognized since the 1920s precisely because the market produced results
society frequently found undesirable. Hence, as important as the common law
power of courts to define the scope of property rights has been, the power of
legislatures and agencies to regulate uses of property has been of unquestionably profound impact.

Indeed, Cole (2002) goes so far as to describe this exercise of regulatory
authority as an assertion of public property rights, in that regulation, while not
based on a claim of public ownership, "in effect . . . creates a mixed property
regime, comprised of both public and private property rights in the resources"
(29) (see also Yandle and Morriss 2001). Yet, although regulation unquestionably can and often does limit what private owners of property may do with
their resources, there are important distinctions between public property rights
and public regulatory authority that remain pertinent to questions of resource
management on privately owned lands. As the discussion of instream water
flows in chapter 4 illustrated, when the government acts to protect public property rights in a mixed private–public ownership property regime such as water,
its authority is based on the public's de jure ownership interest. By contrast,
when regulation limiting private property owners is not based on a public ownership interest in the resource, it must rest entirely on the public's regulatory
authority.

This distinction is not trivial. It is the difference between government acting to manage its share of a mixed ownership regime versus government regulating when it has no formal ownership share in the resource. That the public
does not always have a direct ownership share in a resource is a necessary consequence of having a private property system. To be sure, the public has an
interest in how all resources are managed, but if it cannot assert that interest in
particular cases based on a share of ownership, it must rest the exercise of power
exclusively on the general police power to protect health, safety, and the general
welfare. The government's authority to exercise the police power has never been
described as depending on assertion of a property right, hence neither should
its exercise be thought of as creating a property right, converting private property into a mixed property regime.

If we did not make any distinction between these two sources of power—if
the scope and nature of government authority were the same regardless of the

extent of its direct ownership interest in a resource—the government would be less inclined to "put its money where its mouth is." In other words, why bother incurring the expense and trouble of purchasing and managing a share in the ownership of a resource when the same ends could also be accomplished through sheer exercise of regulatory authority? To put it another way, if the government believes its interests in how a privately owned resource is managed are so paramount to the private owner's interests, let the government purchase the ownership share it needs to fulfill its interests. The concern, as should be apparent from the distinction so stated, is that if government does not ever have to pay in such cases, it will overregulate, or regulate inefficiently, as it is not having to bear the true costs of the impact of its regulations. This is frequently offered as the rationale for making the government pay just compensation to private property owners when regulation substantially interferes with the owner's property interests (Dukeminier et al. 2006).

There is far from consensus over the question of whether the government should *ever* have to pay for regulation that impedes private owners' resource management decisions and, if so, under what circumstances. The reality is that, given how the U.S. Supreme Court has construed the scope of the Fifth Amendment, most regulation of private property based on the government's police power is in no substantial danger of being characterized as a taking of private property for which just compensation is due (Dukeminier et al. 2006). Yet, the fact that this topic presents a great divide between competing conceptions of regulation suggests that, for the present purposes, the distinction between public *ownership* and public *regulation* is important to recognize. The exercise of the government's police power to regulate privately owned property interests, in other words, *is* distinct from the exercise of a government property right to protect its direct ownership interests in a mixed property regime.

Why Regulation?

Because regulation is not based on an ownership interest, the government's incentives are not the same as a true owner's, which means the market forces that in theory lead a private property owner to act efficiently toward the resource will not be in play. As chapter 4 explored, however, there is more than ample evidence that the private property market does not always produce efficient resource uses. So, the possibility that government regulation of private property might result in less than optimal efficiency of resource management does not mean it will necessarily perform worse than will the private (or public) property regime. If the choice is between market failure and regulatory failure, we should pick the least bad. Although the prevailing view is that private property markets will outperform government regulation in this sense, "existing 'empirical' studies do not demonstrate either that command-and-control regulations are inherently inefficient or that they are invariably less efficient than market-based alternatives" (Cole and Grossman 1999, 892). Coase (1960, 15–18) suggested that there is no necessary superiority in this regard, but rather that it will depend on the circumstances. Thus Cole (2002), focusing on pollution regulations, suggests that the choice between private property rights and regulation should favor "which, in the circumstances, would achieve exogenously set societal goals at the lowest total cost, where total cost is the sum of compliance, administrative, and residual pollution or consumption costs" (17).

In particular, just as the government can adopt a social discount rate in the management of publicly owned resources to better align decisions with non-market societal goals, such as intergenerational sustainability of natural capital stocks, so too can it base regulatory policy on a discount rate that reflects considerations the private market might underplay, such as the unpredictable nature of complex ecosystem dynamics (Pearce et al. 1989, 132–52). Chapters 3 and 4 explored the potential of private market decisions to account for ecosystem service values in a manner that fails to sustain maximum social welfare over the long run. Regulatory intervention can, in effect, force private actors to incorporate the social discount rate as a means of adjusting market outcomes in this respect.

As a theoretical matter, however, even if based on the private market discount rate, regulations should perform better than the market under Cole's test with respect to the production and conservation of public goods. As chapter 3 shows, where rivalry of consumption of a resource benefit is high and the costs of effective exclusion to the resource are low—meaning private property owners can easily limit access to the resource and allocate benefits among potential purchasers—private property regimes may be the more efficient approach. But where the benefit lacks rivalry of consumption and the costs of exclusion are high, the private property system will find it difficult to contain positive externalities and thus is likely to result in the benefit being undersupplied. These are the defining characteristics of public goods, and "the traditional solution to the problem of underproduction of public goods is government intervention" (Bell and Parchomovsky 2003, 10).

A pure public good is inexhaustible and thus subject only to undersupply, not to overexploitation. But as chapter 3 also showed, where the benefit exhibits some rivalry of consumption, user loads may reach the point of congestion, at which point users acting "rationally" will, in the absence of some exogenous constraint, move the resource toward the collective ruin Hardin predicted. As Bell and Parchomovsky (2003, 12) postulate, conservation is likely in the long run to lose out as a use for such "impure" public goods, because it represents a use that is necessarily incompatible with *all* exploitive uses of the resource. Hence, regulation may turn out to be the most efficient solution for undersupply *and* overexploitation in the case of impure public goods, which is to say for many natural capital and ecosystem service contexts. Indeed, because regulation can force beneficiaries of a resource to pay for any positive externalities they otherwise would reap for free (such as through taxes) and can limit access to the resource (such as by physical or other constraints), Holly Doremus has observed that "regulation can create markets for public goods" (2003a, 220). And "the possibility exists government, by direct provision, may outperform the market" in terms of ensuring through regulation that the efficient amount of the good or service is supplied and used (Randall 1983, 137).

Whether regulation, in these and similar circumstances, lives up to this potential is an empirical question that is for the most part beyond the scope of this examination. The threshold question of concern here is whether regulation has in fact been *effective* at accounting for natural capital and ecosystem service values. If it has been more so than the private property system (which chapter 4 shows has not been at all effective in this regard), that would place the burden of proof on the private property system to demonstrate it could turn its record around and, ultimately, be at least as effective as, and more efficient than, the government's exercise of regulatory authority. On the other hand, if regulation has been no more effective than private property at getting a handle on natural capital and ecosystem service values, we would need to experiment with both, as well as with other institutions (see chapter 6), to test their relative performance under Cole's criterion.

Alas, the history and current state of environmental regulation suggests that we are in this "starting from scratch" position. Despite its claims to superior efficiency in the public goods context, the preponderance of the evidence weighs heavily toward the conclusion that regulation has been ineffective at incorporating natural capital and ecosystem service values into natural resource decision making.

The Rise and Stall of Environmental Common Law

Most comprehensive treatments of the evolution of environmental regulation begin with the common law as setting the stage for its development (Elliott et al. 1985). In particular, over time the nuisance doctrine developed into a powerful means of protecting the environment (Hylton 2002), so much so that a leading environmental law treatise observes that

> [t]here is no common law doctrine that approaches nuisance in comprehensiveness or detail as a regulator of land use and technological abuse. Nuisance actions reach pollution of all physical media—air, water, land, groundwater—by a wide variety of means. Nuisance actions have chal-

lenged virtually every major industrial and municipal activity that today is the subject of comprehensive environmental regulation. (Rodgers 1994, 112–13)

As Ruhl (2005a) has explained,[1] this tradition reached its apex almost a century ago, when the U.S. Supreme Court decision in *Georgia v. Tennessee Copper Co.*[2] suggested that the common law doctrine of public nuisance could play an important and innovative role in pollution control. Public nuisance involves claims brought by the sovereign or select private interest on behalf of the public welfare, whereas private nuisance claims relate only to the injuries suffered by the plaintiff. After agricultural landowners in Tennessee were unsuccessful in private nuisance actions brought in state court at stopping harmful air emissions from copper smelting plants located in the eastern reaches of the state, Georgia used a public nuisance theory of liability to sue the companies in its sovereign capacity on behalf of its citizens.

Georgia's claim fell on sympathetic ears in the Supreme Court. The Court was "satisfied by a preponderance of the evidence that the sulfurous fumes cause and threaten damage on so considerable a scale to the forests and vegetable life, if not to health, within the plaintiff State" as to justify an injunction.[3] Indeed, in a later remedial decree,[4] the Court, much like a modern administrative agency, required the company to keep daily records of its operations, to submit to court-appointed inspectors, to meet performance standards for emission rates, and to comply with maximum total daily emission loads. Although the Court later relaxed some of the limits during wartime, ultimately the case had a technology-forcing effect as the fear of liability led the industry to develop a new smelting process that allowed reclamation of the sulfur (Percival et al. 2003).

Indeed, because the injunctive remedy in public nuisance cases can involve judicial regulation of the nuisance-causing activity, the public nuisance doctrine is the genesis of Cole's theory that all environmental regulation is an assertion of public *property* rights. He points to early British judicial explanations that "in private nuisance the injury is to individual property, and in cases of public nuisance the injury is to the property of mankind,"[5] and to American cases basing public nuisance remedies on injury to "the right to clean air"[6] (2002, 33–34). As suggested earlier, however, it is neither necessary nor accurate to go this far. In some public nuisance cases the sovereign might in fact be prosecuting injuries to public property, but in many cases the government is simply protecting the private interests of many of its citizens in one suit. Thus courts have made it clear that public nuisance claims do not depend on asserting public ownership interests, nor do they in any way establish such interests.[7]

Nevertheless, although public nuisance is not based on *ownership,* it is entirely appropriate to recognize the *regulatory* effect of public nuisance. Pursuant to public nuisance injunctive relief, countless cases through history have resulted in regulation or full prohibition of resource-harming activities. Given this judicial power, public nuisance doctrine could serve as an effective means of regulating private decisions about the exploitation of natural capital. Recently, for example, in *Palazzolo v. Rhode Island* a Rhode Island court pointed to ecosystem service values as a basis for finding that state agencies did not commit an uncompensated taking of property when they denied a permit application Anthony Palazzolo submitted to authorize him, under state wetlands laws, to fill and develop part of a salt marsh located adjacent to a tidal pond. As discussed in chapter 4, when government through legislation or administrative regulation denies a landowner the opportunity to engage in land uses that would have been restricted under common law in the first place, no regulatory taking can be found to have occurred. The court reasoned this to be the case pursuant to the public nuisance doctrine:

> [The] development has been shown to have significant and predictable negative effects on Winnapaug Pond and the adjacent salt water marsh. The State has presented evidence as to various effects that the development will have including increasing nitrogen levels in the pond, both by reason of the nitrogen produced by the attendant residential septic systems, and the reduced marsh area *which actually filters and cleans runoff.* The Court finds that the effects of increased nitrogen levels constitute a predictable (anticipatory) nuisance which would almost certainly result in an ecological disaster to the pond. . . . Because clear and convincing evidence demonstrates that Palazzolo's development would constitute a public nuisance, he had no right to develop the pond as he proposed. Accordingly, the State's denial to permit such development cannot constitute a taking.[8]

Yet there is a limit to how far public nuisance can be expected to carry the cause of accounting for natural capital and ecosystem services. Even in its core target zone of pollution control, confidence in the effectiveness of nuisance doctrine has waned over time. The death knell to any hopeful thinking about private nuisance in this respect came in 1970 in the famous case of *Boomer v. Atlantic Cement Co.,* in which New York's highest court declined to enjoin a cement plant's air emissions, ruling instead that a damages remedy, previously not available under New York law, was the more efficient approach. While the case is known mostly for that shift in remedial doctrine, the court's rationale for backing off of injunctive relief for private nuisance claims sent a loud message to legislatures that their help was needed. The court warned,

It seems apparent that the amelioration of air pollution will depend on technical research in great depth; on a carefully balanced consideration of the economic impact of close regulation; and of the actual effect on public health. It is likely to require massive public expenditure and to demand more than any local community can accomplish and to depend on regional and interstate controls.

A court should not try to do this on its own as a by-product of private litigation. . . . This is an area beyond the circumference of one private lawsuit. It is a direct responsibility for government and should not thus be undertaken as an incident to solving a dispute between property owners.[9]

The date of the *Boomer* opinion, not coincidentally, marks the advent of the wave of federal legislation regulating air, water, and land pollution. So it is no surprise that law students are taught today, quoting from the leading environmental law casebook (Percival et al. 2003, 871), that "there is wide agreement that private nuisance actions alone are grossly inadequate for resolving the more typical pollution problems faced by modern industrialized societies."

For similar reasons, it is probably expecting too much to think that public nuisance will become the champion of ecological protection. Like air pollution, many of the positive externality and public good problems plaguing effective management of natural capital and ecosystem services are either so ubiquitous, or function at so grand a scale, that the reasoning the *Boomer* court used to caution what to expect from private nuisance seems equally appropriate for public nuisance. For example, in *Connecticut v. American Electric Power,* several states recently sued a collection of electric power companies to enjoin the defendants' emissions of carbon dioxide and other "greenhouse gases" alleged to cause global warming. The states argued that "the natural processes that remove carbon dioxide from the atmosphere are now unable to keep pace with the level of carbon dioxide emissions," and that the power companies therefore are "liable for contributing to a public nuisance, global warming." The court dismissed the lawsuit, however, on the ground that "resolution of the issues presented here requires identification and balancing of economic, environmental, foreign policy, and national security interests" that are "consigned to the political branches, not the Judiciary."[10]

Hence, while cases like *Palazzolo* suggest that nuisance doctrine can integrate natural capital and ecosystem service values into the balancing of interests in particular settings, a prospect considered more fully in chapter 18, cases like *American Electric Power* suggest it is unlikely that this judicially administered common law doctrine can alone do the work, or even carry the lion's share, of establishing a comprehensive framework governing decisions about natural capital and ecosystem services. Moving beyond the common law form of regulation based on judicial power, to the more familiar form of regulation based on legislative power, thus seems as necessary and appropriate for managing natural capital and ecosystem services as it was for managing pollution.

The Emergence and Evolution of "Command-and-Control" Environmental Regulation

Accordingly, histories of environmental regulation in the United States frequently identify several stages of growth in regulatory scope and power beginning with its eclipse of the common law. Percival et al., for example, suggest the following progression:

1. The Common Law and Conservation Era: Pre-1945
2. Federal Assistance for State Problems: 1945–1962
3. The Rise of the Modern Environmental Movement: 1962–1970
4. Erecting the Federal Regulatory Infrastructure: 1970–1980
5. Extending and Refining Regulatory Strategies: 1980–1990
6. Regulatory Recoil and Reinvention: 1991–present (2003, 85)

In his epic history of environmental law, Lazarus points out that this road from common law roots to an established regulatory infrastructure by the 1980s was not an easy one:

> Environmental law beat the odds during the 1970s, 1980s, and 1990s. Its supporters overcame massive institutional and political obstacles to develop a comprehensive series of federal environmental protection statutory programs. Environmentalists defeated repeated efforts to reduce environmental laws' reach and stringency, and environmental protection laws steadily became, notwithstanding or perhaps because of these challenges, more comprehensive, far-reaching, demanding, and pervasive than ever. (2004, 167)

Yet, as the stages of development suggest, with its success environmental law also sowed the seeds of a counterrevolution. Conventional environmental regulation indeed reached far, was substantially more stringent than the common law, and had pervasive impact on property rights. By the 1980s this impressive regulatory infrastructure was considered in many circles as a top-heavy, centralized, uncreative, ossified "command-and-control" system of decision making that neither produced more efficient markets nor corrected for perceived inequities of the market. After picking the low hanging fruit of pollution control through end-of-the-pipe discharge limits, environmental law began to look increasingly inept at tackling the more complex challenges of nonpoint source agricultural water pollution, invasive species, climate change, and habitat loss

(Hirsch 2001, 2004; Ruhl 2000; Stewart 2001, 2003). It is fair to say that by the mid-1990s "virtually everyone . . . agree[d] that our historical command-and-control approach [was] inefficient and inadequate by itself to carry us where we still need to go" (Thompson 1996, viii). Indeed, to the extent "where we still need to go" includes accounting more completely for natural capital and ecosystem service values, the track record of command-and-control environmental regulation bears out this assessment.

Conventional Environmental Regulation and Ecosystem Services

Chapters 3 and 4 explored why a private landowner's decision about whether to convert land with intact natural resources to other uses is unlikely to take into account ecosystem service values the natural capital stock is capable of supplying. This chapter so far has opened the door to the question whether, if private property markets do not adequately take ecosystem service values into account, regulatory programs can and should fill the gap. If regulation as practiced in the United States has indeed done so, this section explores three regulatory programs that one would reasonably expect to be prime examples: wetland resources protection, coastal resources protection, and forest resources protection.

These three regulatory programs represent three significantly different implementation frameworks (Ruhl 1995). Wetlands are regulated nationally under the Clean Water Act, even on private lands, but many states have also adopted parallel regulatory authorities that in some cases exceed federal regulation in stringency, geographic scope, or both. Coastal resources protection, by contrast, is not the subject of direct, comprehensive federal regulation. Rather, under the Coastal Zone Management Act the federal government has established national goals for coastal protection and provided incentives for states to adopt measures, including regulation of private lands, designed to meet the national goals. The federal government is even more removed from the picture in forest resources protection. As chapter 4 discusses, the federal government owns and manages the vast national forest system as a public property regime, but it has promulgated neither direct regulation of private forests nor a cooperative program designed to induce states to do so. Some states, however, have established regulations governing private forestry practices, and all have regulations governing management of state-owned forests.

To provide a snapshot of the existing implementation of these programs, all state legislation and administrative regulations relating to each program were evaluated and ranked according to the following scale:

1. Makes minimal or no effort to protect the resource

2. Acknowledges the general importance of the resource to humans and wildlife, but makes no reference to ecosystem functions or ecosystem service values
3. Acknowledges the ecosystem functions and/or ecosystem services provided by the resource as a reason for or purpose of the regulatory program, but does not expressly incorporate either into implementation standards or authorities
4. Includes implementation standards and authorities that expressly incorporate ecosystem functions into resource management decision making, but not ecosystem service values
5. Includes implementation standards and authorities that expressly incorporate ecosystem service values into resource management decision making

Evaluation under these criteria necessarily involves subjective interpretation in many cases, as few statutory or administrative regulation texts mention "natural capital" or "ecosystem services" as such. Moreover, statutory text, as noted earlier, is often broad and may include grand proclamations in a "legislative findings and purposes" section that are noticeably absent in provisions establishing regulatory authority. In the end, moreover, what matters is how statutory and administrative regulations are implemented and enforced, which may appear to be a far cry from what is written in the text. Nevertheless, an evaluation of what is written in the pages of legislative statutes and administrative regulations provides at least a sense of the general "state of mind" in which legislatures and agencies have approached the question of how to integrate natural capital and ecosystem service values into environmental law.

Wetland Protection Laws

Wetland protection in the United States is carried out through parallel, and to a large extent overlapping, federal and state authorities. The federal program applies throughout the nation to wetland areas within the constitutional and statutory scope of the Clean Water Act. Not all states provide state wetland protection programs, and of those that do, not all extend regulatory protection to all intrastate wetland areas. Overall, however, to the extent activities in or affecting wetland resources are regulated by federal authority, state authority, or both, ecosystem service values are not part of explicit decision-making mechanisms.

The Federal Wetland Protection Program

At the federal level, section 404(a) of the Clean Water Act (CWA) authorizes the secretary of the army, through the Army Corps of Engineers (the Corps), to "issue permits for the discharge of dredged or fill material in the navigable waters of the United States at specified disposal sites."[11] The U.S. Supreme Court has construed this authority to extend to wetlands adjacent to or otherwise having a substantial physical, chemical, and biological effect on navigable waters, but not to isolated wetlands.[12] Pursuant to section 404(b)(1) of the CWA, the Environmental Protection Agency (EPA) must promulgate substantive permitting standards focused on environmental factors, known as the "404(b)(1) Guidelines," which the Corps must follow in administering the permit program.[13] Thus, under the CWA, and subject to specified exceptions, wetlands may be filled only if a permit is granted in accordance with the 404(b)(1) Guidelines (Pifher 2005). These permits, known ubiquitously as "404 permits," "wetland permits," or "Corps permits," have become the cornerstone for federal protection of wetland resources (Williams and Connolly 2005).

Many routine land development activities require and receive a 404 permit, and, along the way, permit applicants and the agencies often confront the issue of "mitigation" as one of the conditions the developer must satisfy in order to obtain the permit (Gardner 2005; Wilson and Thompson 2006). The 404(b)(1) Guidelines provide extensive descriptions of wetlands values that the Corps should consider in assessing potential mitigation requirements. As Ruhl and Gregg (2001) explain, although the guidelines do not specifically mention the full scope of ecosystem service benefits supplied by wetlands, the guidelines provide clear regulatory authority to consider ecosystem service values, such as those derived from the water purification function that wetlands provide.[14]

Initially, the Corps and the EPA "clashed over the proper role of mitigation in the . . . permitting process" (Veltman 1995). However, in 1990 the Corps and the EPA signed a memorandum of agreement—the *Mitigation Guidance*—clarifying how wetlands mitigation will be administered under the 404(b)(1) Guidelines (Department of Army and Environmental Protection Agency 1990). The *Mitigation Guidance* divides mitigation into three phases—avoidance, minimization, and compensatory mitigation—and requires that those phases be conducted sequentially. With respect to compensatory mitigation generally, the *Mitigation Guidance* requires that it be used only for unavoidable adverse impacts that remain after all appropriate and practicable minimization has been required, and it expresses preferences for on-site mitigation and for wetlands restoration (as opposed to wetlands creation). It also requires (at 9,212) that "functional values" be examined in connection with those determinations.[15] Thus the *Mitigation Guidance* simply requires that functional value be examined and compensation provided—preferably on-site—for unavoidable adverse impacts.

This declaration of purpose is strong, but the methods to achieve it are not well defined. The Corps declared a goal of no overall net loss of values and functions, but the methods used to determine whether this goal is being met are only broadly described. Although the *Mitigation Guidance* pays homage to the idea of "functions and values" in numerous instances, it never defines these essential terms and there is no express recognition of wetland ecosystem service values in the mitigation calculus. Overall, these open-ended provisions have led some observers to describe the *Mitigation Guidance* as providing the Corps "virtually unfettered discretion in determining whether a just compensation for destroyed wetlands has been achieved" (Veltman 1995, 673–74).

This history of regulatory inattention to wetland ecosystem services means little headway has been made in calculating the *economic* values of wetland ecosystem services (Boyd and Wainger 2002b). For example, in *Measuring the Benefits of Federal Wetland Programs* (1997), Paul Scodari summarized the literature addressing the theoretical use of wetland assessment methodologies to generate economic values, including ecosystem service values. He concluded that the theory is lacking at both ends of the mitigation process—that is, it fails to offer viable methods for assessing wetland functions and services for purposes of developing the currency, and it fails to provide a valuation method for purposes of mitigation. He also found that, in practice, "wetland functional assessments produce measures of functional indices that are only suggestive of the capacity of wetlands to provide certain important outputs" and thus "limit our ability to develop estimates of wetland protection benefits" (54). In the absence of more informative and reliable assessment methods, Scodari concludes that valuation theories and estimates have necessarily been "based on flawed procedures that calculate measures that are, to varying degrees, inconsistent with the economic concept of value" (58).

In 2006 the Corps and the EPA proposed a new regulation governing wetland mitigation (Department of Defense and Environmental Protection Agency 2006) that is built around a "watershed approach" and gives express attention to ecosystem services. The proposal defines "services" to mean "the benefits that human populations receive from functions that occur in aquatic resources and other ecosystems," and defines "values" as meaning "the utility or satisfaction that humans derive from aquatic resource services" (15534–36). Going beyond mere definitions, the proposed rule states that "in general, compensatory mitigation should be located within the same watershed as the impact site, and should be located where it is most likely to successfully replace lost functions, services, and values" (15536). The proposal notes that compensatory mitigation might be sited away from the development project area, but that in such cases "consideration should also be given to functions, services, and values (e.g., water quality, flood control, shoreline protection) that will likely need to be addressed at or near the areas impacted by the permitted project" (15536). Yet nowhere does the proposal suggest how such "consideration" is to be made—how to define and measure lost and replacement services, and how to weight that analysis in the final regulatory decision regarding the amount and location of mitigation. Hence, while the proposal would, if adopted, improve the federal program's recognition of ecosystem service values, it would not by its terms provide the implementation framework necessary to give content to the regulatory program. It would be up to the agency, through subsequent implementation practice, to forge that policy.

STATE WETLAND PROTECTION PROGRAMS

Many states have adopted wetland regulation statutes and administrative implementing regulations. Yet, while the majority of these programs include implementation standards and authorities that expressly incorporate ecosystem functions into management decision making, they do not go as far as to use ecosystem service values as explicit decision-making criteria. The Virginia statute, for example, makes it unlawful to engage in activities "that cause significant alteration and degradation of existing wetland acreage or functions."[16] A substantial minority of states, moreover, fall below even this standard. For example, Nebraska's administrative regulations recognize that "wetlands serve a multitude of important functions,"[17] with a long list provided, but no specific regulatory standards or requirements are linked to those functions. Even fewer states go the full way toward including implementation standards and authorities that expressly incorporate ecosystem service values into management decision making. An example of this rare breed is from Connecticut. The state's statute recognizes that unregulated activities can "disturb the natural ability of tidal wetlands to reduce flood damage . . . [and] to absorb silt,"[18] and the state's administrative implementing regulations require that permit issuance be based on, among other considerations, "the environmental impact of the proposed action, including the effects . . . [on] natural capacity to . . . control sediment, to facilitate drainage, and to promote public health and safety."[19] Table 5.1 categorizes each state based on its wetland statutes and regulations in place in 2005.

TABLE 5.1. Evaluation of state wetland resource protection statutes and regulations. (Refer to categories listed at end of table.)

State/ Stage	Statutes					Regulations				
	1	2	3	4	5	1	2	3	4	5
Alabama		X							X	
Alaska	X							X		
Arizona				X			X			
Arkansas				X					X	
California				X					X	
Colorado	X								X	
Connecticut					X					X

(continues)

TABLE 5.1. *Continued*

State/ Stage	Statutes					Regulations				
	1	2	3	4	5	1	2	3	4	5
Delaware				X					X	
D.C.	X					X				
Florida				X					X	
Georgia		X							X	
Hawaii	X					X				
Idaho		X					X			
Illinois				X					X	
Indiana				X					X	
Iowa			X					X		
Kansas		X							X	
Kentucky		X					X			
Louisiana					X		X			
Maine			X						X	
Maryland				X					X	
Massachusetts		X								X
Michigan				X					X	
Minnesota				X					X	
Mississippi			X						X	
Missouri		X					X			
Montana		X						X		
Nebraska		X						X		
Nevada				X			X			
New Hampshire				X					X	
New Jersey				X					X	
New Mexico	X						X			
New York					X					X
North Carolina				X					X	
North Dakota			X				X			
Ohio				X					X	
Oklahoma			X					X		
Oregon				X					X	
Pennsylvania	X								X	
Rhode Island				X						X
South Carolina			X						X	
South Dakota	X							X		
Tennessee		X						X		
Texas			X						X	
Utah	X								X	
Vermont				X						X
Virginia				X					X	
Washington				X					X	
West Virginia	X								X	

State/	Statutes					Regulations				
Stage	1	2	3	4	5	1	2	3	4	5
Wisconsin				X					X	
Wyoming				X					X	

Categories:
1. Minimal or no effort to protect the resource
2. Acknowledges the general importance of the resource to humans and wildlife, but makes no reference to ecosystem functions or ecosystem service values
3. Acknowledges the ecosystem functions and/or ecosystem services provided by the resource as a reason for or purpose of the regulatory program, but does not expressly incorporate either into implementation standards or authorities
4. Includes implementation standards and authorities that expressly incorporate ecosystem functions into resource management decision making, but not into ecosystem service values
5. Includes implementation standards and authorities that expressly incorporate ecosystem service values into resource management decision making

Coastal Resource Protection Laws

The Coastal Zone Management Act (CZMA)[20] is the nation's primary foundation for beach and coastal area conservation. The CZMA authorizes the Department of Commerce to administer a federal grant program to encourage coastal states to develop and implement coastal zone management programs for the purpose of protecting, developing, and enhancing coastal zone resources, which include wetlands, floodplains, estuaries, beaches, dunes, barrier islands, coral reefs, and fish and wildlife and their habitat. The coastal states include any bordering an ocean or the Gulf of Mexico, Long Island Sound, or the Great Lakes. The coastal zone within these states includes coastal waters and adjacent shorelands, islands, transitional and intertidal areas, salt marshes, wetlands, and beaches. The objectives of the grant program and related CZMA provisions are to improve the management of the coastal zone resources within those states, which necessarily includes managing private and public land development actions in the coastal zone.

Two features set the CZMA apart from many other federal natural resource management laws, such as the wetlands program discussed earlier. First, it relies heavily on states to implement national policy through state-designed coastal management programs (CMPs) that establish land management decision-making frameworks. Second, it obligates federal agencies to implement their respective actions in a manner consistent with state CMPs. The result is a form of resource management regime that is quite decentralized but reaches a broad array of actors and actions. Thus one of the congressional findings supporting the CZMA was that

> [t]he key to more effective protection and use of the land and water resources of the coastal zone is to encourage the states to exercise their full authority over the lands and waters in the coastal zone by assisting the states, in cooperation with Federal and local governments and other vitally affected interests, in developing land and water use programs for the coastal zone, including unified policies, criteria, standards, methods, and processes for dealing with land and water use decisions of more than local significance.[21]

The CZMA thus outlines a national policy on coastal zone management but allows states to devise their own plans for fulfilling those goals and to use state law to implement them. To receive federal approval, which triggers the requirement of federal agency consistency, a state CMP must describe "permissible land uses and water uses within the coastal zone" and "the means by which the state proposes to exert control over the land uses and water uses . . . including a list of relevant State constitutional provisions, laws, regulations, and judicial decisions."[22] This legal framework must provide for "adequate consideration of the national interest involved in planning for, and managing the coastal zone, including the siting of facilities . . . which are of greater than local significance,"[23] and it must include "procedures whereby specific areas may be designated for the purpose of preserving or restoring them for their conservation, recreational, ecological, historical, or esthetic values."[24]

As long as the national goals are satisfactorily addressed, the states have considerable latitude in the design of their land use management frameworks. The CZMA lays out three general schemes from which the states can choose: (1) local implementation of the state CMP and state-promulgated standards, subject to state review; (2) direct state regulation; and (3) state review of state and local decisions for consistency with the state CMP. The intensity of land use regulation can vary under any of these approaches, and some states go well beyond the minimum necessary scope of regulation to implement the CZMA's national goals, albeit others do not. Very few do so in a way that extends the decision-making criteria of the program expressly to natural capital and ecosystem services.

State CMPs are generally regarded as being effective in making contributions to the national objectives (Good 1998). Yet, although about one-third of the state coastal regulation programs include implementation standards and authorities that expressly incorporate ecosystem *functions* into management decision making, they do so without express mention of ecosystem service values. For example, the Texas administrative regulations mandate that waterfront construction shall not "interfere with the natural coastal processes,"[25] but no specific mention of ecosystem service values is included. A substantial majority of states, moreover, go no further than acknowledging the ecosystem functions and/or ecosystem services provided by coastal resources as a reason for or purpose of the regulatory program, and many do not even go that far. The New Jersey coastal protection statute, for example, declares that the estuarine zone "protects the land from the force of the sea, moderates our weather, . . . and assists in absorbing sewage discharge by the rivers of the land,"[26] but neither the statute nor the administrative regulations make ecosystem functions or ecosystem service values an explicit criterion for issuance of approvals. A few states do go the full way toward including implementation standards and authorities that expressly incorporate ecosystem service values into management decision making. Florida, for example, establishes a "coastal construction control line" and requires anyone wishing to engage in construction seaward of the line to demonstrate, among other things, that the construction will not reduce "the existing ability of the [beach and dune] system to resist erosion during a storm," and that any man-made frontal dune system designed to mitigate for such effects "shall be constructed to meet or exceed the protective value afforded by the natural frontal dune system."[27] Similarly, the Massachusetts policies for review of federal agency actions for consistency with the state program includes a criterion of protecting "the beneficial functions of storm damage protection and flood control provided by natural coastal landforms."[28] Table 5.2 categorizes each state based on its coastal resource protection statutes and regulations in place in 2005.

TABLE 5.2. Evaluation of state coastal resource protection statutes and regulations. (Refer to categories listed at end of table.)

State/Stage	Statutes					Regulations				
	1	2	3	4	5	1	2	3	4	5
Alabama		X					X			
Alaska		X					X			
California			X						X	
Connecticut		X				X				
Delaware		X							X	
Florida			X							X
Georgia			X				X			
Hawaii		X					X			
Illinois		X					X			
Indiana	X					X				
Louisiana				X			X			
Maine		X					X			
Maryland				X			X			

(continues)

TABLE 5.2. *Continued*

State/ Stage	Statutes					Regulations				
	1	2	3	4	5	1	2	3	4	5
Massachusetts				X						X
Michigan		X							X	
Minnesota		X					X			
Mississippi		X					X			
New Hampshire			X						X	
New Jersey			X					X		
New York			X							X
North Carolina			X						X	
Ohio		X						X		
Oregon			X						X	
Pennsylvania	X						X			
Rhode Island		X							X	
South Carolina			X						X	
Texas		X							X	
Virginia		X							X	
Washington		X						X		
Wisconsin		X							X	

Categories:
1. Minimal or no effort to protect the resource
2. Acknowledges the general importance of the resource to humans and wildlife, but makes no reference to ecosystem functions or ecosystem service values
3. Acknowledges the ecosystem functions and/or ecosystem services provided by the resource as a reason for or purpose of the regulatory program, but does not expressly incorporate either into implementation standards or authorities
4. Includes implementation standards and authorities that expressly incorporate ecosystem functions into resource management decision making, but not into ecosystem service values
5. Includes implementation standards and authorities that expressly incorporate ecosystem service values into resource management decision making

Forest Protection Laws

As discussed in chapter 4, the federal government has a complex set of statutes and regulations for governing the national forest system under a public ownership property regime. No federal statute directly and comprehensively regulates private forested lands or forestry practices, however, thus leaving it to the states to enter that field. Very few states have done so through exercise of regulatory powers, though most have established regulations governing use of state-owned public forests.

FEDERAL FOREST REGULATIONS

As the nation's largest single owner of forest lands, the federal government's decisions about forest management have lasting effects on the ecosystem values of our nation's forests. Of the federal land management agencies, the U.S. Forest Service controls the largest holding of federal forests—192 million acres of land in 42 states, the Virgin Islands, and Puerto Rico—through its jurisdiction over the National Forest System. The system is composed of 155 national forests, 20 national grasslands, and various other lands under the jurisdiction of the secretary of agriculture (the secretary). The vast majority of national forest acres are located west of the Great Plains, though some other states have significant holdings.

The discussion in chapter 4 included the national forests as one example of a public ownership property regime in the United States that has failed to account effectively for natural capital and ecosystem service values. Of course, by virtue of the fact that they are publicly owned, the national forests would not be expected to assume whatever advantages private property markets can offer to that purpose. In the absence of strong social norms, therefore, the public property manager must resort to regulation of private users to control access to and exploitation of the public resource. The poor performance of the national forests and other major federal public land ownership regimes discussed in chapter 4 must therefore be attributable, at least in part, to the regulatory decisions the federal government has made about access to and use of national forest resources.

As Nagle and Ruhl (2006, 435–41) explain in their history of national forest policy,[29] the first glimmer of a national forest management policy was seen when the American Forestry Congress of 1882, meeting in Cincinnati, created the American Forestry Association to cooperate with federal and state governments toward formulating a definite policy for managing public forest lands. In the following two decades, several states created state-level forestry agencies to regulate fires, encourage timber culture, and, in some cases, promote forest conservation, preservation, and extension. While in most cases the motivating force behind these laws was security of timber supply, laws such as one in New York enacted in 1885 were among the first truly comprehensive forest management policies in America. The New York law established a system for designating, maintaining, and protecting state forests, complete with a state forest commission, wardens, forest inspectors, foresters, and other staff.[30]

Federal policy witnessed a similar trend after the 1873 meeting of the American Society for the Advancement of Science appointed a committee to present to Congress a plan for the extension and preservation of forests, providing the impetus for a flurry of additional studies and proposals and even several federal laws promoting timber culture and the collection of forest statistics. The seeds of today's Forest Service were also planted in the Department of Agriculture through creation of the Division of Forestry, which began with one employee and an annual budget of $2,000. By the late 1800s, though, federal forest policy remained a complete muddle. While promoting the extension of forestlands by subsidized plantings and other culture programs (albeit often in areas not suited to trees of any kind), the federal government was at the same time disposing of vast tracts of prime forestlands into private possession.

To resolve this inconsistency, a "rider" provision to the 1891 General Revision Act, known as the Forest Reserve Act of 1891, authorized the president to establish forest reservations on federal lands.[31] Fortunately, not all the federal lands had yet been given away or sold, and Presidents Harrison and Cleveland withdrew extensive areas in the western states from sale or entry and declared them national forest reserves. But much remained uncertain: western interests were quite bitter over the turn of events, there was debate over the actual authority of the president under the 1891 law, and there were no monies appropriated to manage what, by 1896, amounted to millions of acres of national forests. Congress resolved the situation with the passage of the Forest Service Organic Act of 1897 (Organic Act),[32] which ratified the presidential reservations and authorized administration of the national forests, then called forest reserves, through a federal agency.

The Organic Act marks the beginning of the development of comprehensive federal forest policy and administration. Although it made no mention of biodiversity, ecosystems, or even wildlife, it provided that the national forests should be established "to improve and protect the forest within the boundaries, or for the purposes of securing favorable water flows, and to furnish a continuous supply of timber for the use and necessities of citizens of the United States."[33] For the latter purpose, the Organic Act authorized the Department of the Interior, then (after 1905) the Forest Service in the Department of Agriculture, to

> cause to be designated and appraised so much of the dead, matured or large growth trees found upon such national forests as may be compatible with the utilization of the forests thereon, and may sell the same. . . . Such timber, before being sold, shall be marked and designated, and shall be cut and removed under the supervision of some person appointed for that purpose by the Secretary. . . .

Gifford Pinchot, who became head of the Forest Service in 1898, envisioned the national forests as primarily a timber supply resource, and for nearly seventy years the Forest Service interpreted the Organic Act as allowing widespread extraction of timber. Indeed, after World War II, housing construction demands placed tremendous new pressure on the nation's timber supply, and on the Forest Service. Clear-cutting became a common practice on private forestlands, thus depleting private timber supplies and causing the timber industry to pressure the Forest Service to increase the yield from national forests. The agency met this demand, but by doing so fueled a conflict between timber harvesting and another demand that boomed after the war—recreation. As clear-cutting became common in the national forests, so too did the previously uncommon instance of public criticism of Forest Service decisions.

Congress nevertheless gave the Forest Service basically a free hand in all such matters of national forest policy, intervening only once to enact the Multiple Use and Sustained Yield Act of 1960 (MUSYA). MUSYA expanded the purposes of national forest management from water flows and timber supply to include "outdoor recreation, range, timber, watershed, and wildlife and fish purposes."[34] Recognizing that "some land will be used for less than all the resources," MUSYA requires that the five multiple uses, which Congress deliberately named in alphabetical order, be treated as coequal and managed "with consideration being given to the relative values of the various resources." The statute describes the core mandate of multiple use as

> the management of all the various renewable surface resources of the national forests so that they are utilized in the combination that will best meet the needs of the American people; making the most judicious use of the land for some or all these resources or related services over areas large enough to provide sufficient latitude for periodic adjustments in use to conform to changing needs and conditions. . . .

Conservation and recreation interests opposed the legislation; the agency and timber industry actively supported it. Critics of the Forest Service charged that the agency had elevated timber extraction above the other uses and exercised widespread clear-cutting without due regard to the Organic Act, and would continue to do both under MUSYA. Indeed, for all practical purposes MUSYA codified precisely the policy discretion the agency sought (and argued it had even without MUSYA). After MUSYA, the law of national forests explicitly recognized the breadth of the agency's discretion. While courts demanded that the agency give "due consideration" to each of the multiple use components,[35] in the final analysis most courts agreed that "the decision as to the proper mix of uses within any particular area is left to the sound discretion and expertise of the Forest Service."[36] MUSYA's multiple use mandate was essentially rendered directionless, leaving it to the agency to decide where to go and providing no meaningful legislative or judicial check on the path chosen.

As Congress and the courts continued to afford the Forest Service wide latitude in setting policy after MUSYA, the agency began experimenting with planning as a way to resolve multiple use conflicts, requiring each national forest to develop a land use plan. Yet the agency used the end products as a vehicle to portray its policy of clear-cutting as not merely a capitulation to the powerful timber industry but the result of a rational, scientifically sound policy decision-making process. Neither congressional appropriations nor agency will would have supported any other outcome, thus leaving the growing recreational and conservation interests looking in through the window.

A significant blow to the agency came in 1975, however, as the clear-cutting age came to a screeching halt when the court in *West Virginia Division of Izaak Walton League v. Butz*[37] used plain English meanings to interpret the Organic Act's "designated," "marked," "dead," "mature," and "large" terms to prohibit widespread clear-cutting in most circumstances. The court simply noted that the Organic Act referred only to "dead, matured, or large growth" trees as eligible for harvesting, and that the statute required the Forest Service to designate and mark the trees before removal. MUSYA did not alter that basic starting point, so, the court concluded, clear-cutting is illegal. In modern terms, this simply did not compute for the agency or the industry. As environmental groups seized the moment and filed suits around the nation to extend the court's reasoning, the Forest Service and timber industry immediately sought congressional action to clear up what the agency could and could not do with respect to timber extraction policy.

Of course, what the agency sought and fully expected it would receive was legislative nullification of the court's opinion, but by this time criticism of the Forest Service had crept into Congress, focused initially through the so-called Bolle Report, commissioned by Senator Lee Metcalf of Montana, and the Church Commission hearings held in 1971 before the Senate Subcommittee on Public Lands of the Committee on Interior and Insular Affairs. At the request of Senator Metcalf, Arnold Bolle led a team of academics from the University of Montana in 1970 to study Forest Service practices. Their report was sharply critical of the Forest Service's land management practices, concluding that the agency overemphasized timber production and thus undermined the multiple use mandate. At the Church Commission hearings the next year, numerous distinguished witnesses testified in those hearings as to the environmental harm Forest Service policies had caused. Amid the emerging broad attention to environmental affairs that took hold in Congress during the early 1970s, this testimony proved critical in convincing Congress that the agency required more explicit direction. The result was the National Forest Management Act of 1976 (NFMA).[38]

Adding to rather than replacing the Organic Act and MUSYA, the NFMA prescribed a set of substantive standards and planning requirements for the Forest Service. In general, it restricts timber harvests to only those national forestlands where "soil, slope, or other watershed conditions will not be irreversibly damaged" and that could "be adequately restocked within five years after harvest." In particular, clear-cutting and other even-aged management techniques are specifically addressed and restricted by standards that, while loose, were more than had appeared in previous law. Also, making the NFMA particularly relevant to the question of ecosystem services, the statute requires the Forest Service to "provide for diversity of plant and animal communities."[39] These and other standards are to be coordinated for each national forest through individual "land and resources management plans" that require public input and are subject to judicial review. Hence, although it was not without its detractors, the NFMA unquestionably charted a new direction for national forest policy, one in which, for the first time in Forest Service history, biodiversity values had to be taken into account.

Nevertheless, although the NFMA provided vastly more detail to guide Forest Service policy than had the Organic Act and MUSYA, it left many more questions than it answered. A rich history of administrative interpretation and litigation helps fill in the details of such issues as where timber can be harvested under the "irreversible damage" standard, when clear-cutting is allowed, whether forest plans have been properly compiled, and, in particular, what follows from the "diversity of plant and animal communities" standard.

The Forest Service first implemented that statutory provision in 1982 through its regulation known as the Planning Rule.[40] Pursuant to that regulation, the diversity mandate, as well as all other concerns the Forest Service must consider under its multiple use mandate, is factored into each national forest's land and resource management plan, or LRMP. Preparation of an LRMP is the first step in resource allocation within a national forest. In the case of timber harvesting, the LRMP outlines generally where, when, and under what conditions harvesting can occur. The Forest Service then authorizes harvesting in particular locations by selecting a timber sale area and preparing an environmental assessment subject to public review and comment. The agency must consider the environmental consequences of each sale and must determine that a decision to sell in a particular area complies with the LRMP. Only then can the agency award a timber harvest contract.

During the 1990s in particular, environmental groups pressed hard on the Forest Service to emphasize the biodiversity side of the agency's forest management mandate and deemphasize the use of national forests for timber extraction. The groups initiated litigation challenging numerous LRMPs. Although the suits largely failed in achieving the intended overhaul of Forest Service policy by judicial decree, the effort scored modest successes in some courts. The relentless pressure the groups placed on the agency, coupled with recommendations from an independent body of experts (Brooks et al. 2002), had by the end f the decade produced a proposal for change by the agency under the heading of "ecological sustainability." In 2000, the agency issued a broad policy prescription regarding this new focus (Department of Agriculture 2000, 67516):

> The concept of sustainability has become an internationally recognized objective for land and resource stewardship. In 1987, the Brundtland Commission Report (The World Commission on Environment and Development) articulated in "Our Common Future" the need for intergenerational equity in natural resource management. The Commission defined sustainability as meeting the needs of the present without compromising the ability of future generations to meet their own needs. During the last twenty years, the world has increasingly come to recognize that the functioning of ecological systems is a necessary prerequisite for strong productive economies, enduring human communities, and the values people seek from wildlands.
>
> Similarly, the Forest Service and scientific community have developed the concepts of ecosystem management and adaptive management. Scientific advances and improved ecological understanding support an approach under which forests and grasslands are managed as ecosystems rather than focusing solely on single species or commodity output. Indeed, ecosystem management places greater emphasis on assessing and managing broad landscapes and sustaining ecological processes. Ecosystem management focuses on the cumulative effects of activities over time and over larger parts of the landscape. Planning and management under ecosystem management also acknowledges the dynamic nature of ecological systems, the significance of natural processes, and the uncertainty and inherent variability of natural systems. Ecosystem management calls for more effective monitoring of management actions and their effects to facilitate adaptive management, which encourages changes in management emphasis and direction as new scientific information is developed. In accord with ecosystem management, regional ecosystem assessments have become the foundation for more comprehensive planning, sometimes involving multiple forests and other public land management units. . . .
>
> Taken together, ecosystem management, scientific reviews, and collaboration enable the Forest Service to identify key scientific and public issues and to target its limited resources on trying to resolve those issues at the most appropriate time and geographic scale. Based on these changes in the state of scientific and technical knowledge, the Forest Service's extensive experience, and a series of systematic reviews, the Forest Service has concluded that the planning rule must be revised in order to better reflect current knowledge and practices and to better meet the conservation challenges of the future. Indeed, while the 1982 planning rule was appropriate for developing the first round of plans from scratch, it is no longer well suited for implementing the NFMA or

responding to the ecological, social, and economic issues currently facing the national forests and grasslands.

These lofty goals, however, have yet to produce a regulatory framework
accounting for natural capital and ecosystem service values. In revised Planning
Rule regulations the agency adopted in 2000, the agency defined "sustainability" with reference to forest ecosystem "services" but failed to define ecosystem
service valuation metrics, procedures, or standards:

> To understand the contribution national forests and grasslands make to
> the economic and social sustainability of local communities, regions,
> and the nation, the planning process must include the analysis of eco-
> nomic and social information at variable scales, including national,
> regional, and local scales. Social analyses address human life-styles, cul-
> tures, attitudes, beliefs, values, demographics, and land-use patterns,
> and the capacity of human communities to adapt to changing condi-
> tions. Economic analyses address economic trends, the effect of national
> forest and grassland management on the well-being of communities and
> regions, and the net benefit of uses, values, products, or services pro-
> vided by national forests and grasslands. Social and economic analyses
> should recognize that the uses, values, products, and services from
> national forests and grasslands change with time and the capacity of
> communities to accommodate shifts in land uses change. Social and
> economic analyses may rely on quantitative, qualitative, and participa-
> tory methods for gathering and analyzing data. (67576)

The Planning Rule the agency adopted in 2005 to replace the 2000 rule
(which the agency described as too complex) did not go even this far (Department of Agriculture 2005). Rather than defining sustainability, at least in part,
by reference to ecosystem service values, the 2005 rule commits to little, simply recognizing that

> consistent with the Multiple-Use Sustained-Yield Act of 1960 (16
> U.S.C. 528–531), the overall goal of managing the National Forest Sys-
> tem is to sustain the multiple uses of its renewable resources in perpetu-
> ity while maintaining the long-term productivity of the land. Resources
> are to be managed so they are utilized in the combination that will best
> meet the needs of the American people. Maintaining or restoring the
> health of the land enables the National Forest System to provide a sus-
> tainable flow of uses, benefits, products, services, and visitor opportuni-
> ties. (1059)

As it stands today, therefore, regulation of uses of the national forest system
resources lacks any coherent framework accounting for natural capital and

ecosystem service values (Williamson 2005). To drive that point home, as mentioned in chapter 4, it was only in 2005 that the agency suggested it would turn its policy focus toward forging such a framework.

STATE FOREST REGULATIONS

As does South Carolina, most states, if they take any position at all with respect to private forested lands, establish programs for voluntary "best management practices," help coordinate federal subsidies for private forest conservation, and regulate only a short list of activities, such as prescribed fires.[41] And many states, such as Minnesota, manage their public forestlands using the same "multiple use/sustained yield" mandate[42] that has vexed the U.S. Forest Service in its management of national forests. Within these two frameworks, one for private and the other for public forestlands, few states do more than recite the ecosystem service benefits of forest resources, and many do not go that far. Even Oregon, which has some of the most comprehensive regulations governing private forestry practices, provides no coherent approval criteria based on ecosystem service values.[43] Table 5.3 categorizes each state based on its forest regulations, whether covering public lands, private lands, or both, in place in 2005.

TABLE 5.3. Evaluation of state forest resource protection statutes and regulations. (Refer to categories listed at end of table.)

State/ Stage	Statutes					Regulations				
	1	*2*	*3*	*4*	*5*	*1*	*2*	*3*	*4*	*5*
Alabama			X				X			
Alaska				X		X				
Arizona		X				X				
Arkansas		X						X		
California				X				X		
Colorado				X			X			
Connecticut				X		X				
Delaware	X					X				
D.C.			X					X		
Florida			X			X				
Georgia			X			X				
Hawaii			X			X				
Idaho		X						X		
Illinois			X					X		
Indiana		X				X				
Iowa		X					X			
Kansas					X			X		

State/Stage	Statutes					Regulations				
	1	2	3	4	5	1	2	3	4	5
Kentucky		X						X		
Louisiana			X			X				
Maine			X			X				
Maryland			X				X			
Massachusetts	X						X			
Michigan		X				X				
Minnesota		X				X				
Mississippi		X						X		
Missouri		X					X			
Montana			X				X			
Nebraska		X								
Nevada			X			X				
New Hampshire		X					X			
New Jersey		X				X				
New Mexico		X				X				
New York	X					X				
North Carolina		X						X		
North Dakota	X						X			
Ohio	X						X			
Oklahoma		X				X				
Oregon				X					X	
Pennsylvania	X						X			
Rhode Island	X					X				
South Carolina			X			X				
South Dakota	X					X				
Tennessee		X						X		
Texas	X							X		
Utah	X					X				
Vermont			X				X			
Virginia		X						X		
Washington			X			X				
West Virginia	X							X		
Wisconsin			X						X	
Wyoming			X			X				

Categories:

1. Minimal or no effort to protect the resource
2. Acknowledges the general importance of the resource to humans and wildlife, but makes no reference to ecosystem functions or ecosystem service values
3. Acknowledges the ecosystem functions and/or ecosystem services provided by the resource as a reason for or purpose of the regulatory program, but does not expressly incorporate either into implementation standards or authorities
4. Includes implementation standards and authorities that expressly incorporate ecosystem functions into resource management decision making, but not into ecosystem service values
5. Includes implementation standards and authorities that expressly incorporate ecosystem service values into resource management decision making

As these examples of wetland, coastal, and forest resource management programs illustrate, one will search in vain for an example of a state regulatory program, regardless of which approach it takes, that accounts for natural capital and ecosystem service values effectively, much less efficiently. Indeed, when considered together (see Table 5.4), the overall story from these three core programs is that environmental regulation has recognized natural capital and ecosystem service values, if at all, mainly as a justification for regulation, not as a standard for regulatory decision making.

Neither property rights nor regulation, therefore, has established a positive record in this regard. Chapter 6 turns to the final institution upon which we might rely for this purpose—social norms.

TABLE 5.4. Summary of state performances in statutes and regulations for protection of wetland, coastal, and forest resources. (Refer to categories listed at end of table.)

	Statutes			Regulations		
State	Wetland Resource Protection	Coastal Resource Protection	Forest Resource Protection	Wetland Resource Protection	Coastal Resource Protection	Forest Resource Protection
Alabama	2	2	3	4	2	2
Alaska	1	2	4	3	2	1
Arizona	4		2	2		1
Arkansas	4		2	4		3
California	4	4	4	4	4	3
Colorado	1		4	4		2
Connecticut	5	3	4	5	1	1
Delaware	4	2	1	4	4	1
D.C.	1		3	1		3
Florida	4	4	3	4	5	1
Georgia	2	4	3	4	2	1
Hawaii	1	3	3	1	2	1
Idaho	2		2	2		3
Illinois	4	2	3	4	2	3
Indiana	4	1	2	4	1	1
Iowa	3		2	3		2
Kansas	2		4	4		3
Kentucky	2		2	2		3
Louisiana	5	5	4	3	2	1
Maine	3	3	4	4	2	1
Maryland	4	5	4	4	2	2
Massachusetts	3	4	1	5	5	2

State						
Michigan	4	3	3	4	4	1
Minnesota	4	2	3	4	2	1
Mississippi	3	2	3	4	2	3
Missouri	2		3	2		2
Montana	2		4	3		2
Nebraska	2		3	3		
Nevada	4		4	2		1
New Hampshire	4	4	3	4	4	2
New Jersey	4	4	3	4	3	1
New Mexico	1		3	2		1
New York	5	3	2	5	5	1
North Carolina	4	3	3	4	4	3
North Dakota	3		2	2		2
Ohio	4	2	2	4	3	2
Oklahoma	3		3	3		1
Oregon	4		4	4		4
Pennsylvania	1	1	2	4	2	2
Rhode Island	4	2	2	5	4	1
South Carolina	3	4	4	4	4	1
South Dakota	1		2	3		1
Tennessee	2		3	3		3
Texas	3	3	1	4	4	3
Utah	1		1	4		1
Vermont	4		4	5		2
Virginia	4	2	3	4	4	3
Washington	4	2	4	4	2	1
West Virginia	1		2	4		3
Wisconsin	4	2	4	4	4	4
Wyoming	4		4	4		1

Categories:
1. Minimal or no effort to protect the resource
2. Acknowledges the general importance of the resource to humans and wildlife, but makes no reference to ecosystem functions or ecosystem service values
3. Acknowledges the ecosystem functions and/or ecosystem services provided by the resource as a reason for or purpose of the regulatory program, but does not expressly incorporate either into implementation standards or authorities
4. Includes implementation standards and authorities that expressly incorporate ecosystem functions into resource management decision making, but not into ecosystem service values
5. Includes implementation standards and authorities that expressly incorporate ecosystem service values into resource management decision making

6   Social Norms

Whatever the configuration of property rights and regulations governing natural capital and ecosystem services, they by no means provide the final word on how society manages its resources. Social norms—customs, tacit agreements, ways of getting along—can plug gaps in property rights and environmental regulations and work end runs around the walls they might erect to sensible resource management. There is considerable evidence from around the globe that this occurs in common-pool resource contexts, but not much of it is from the United States. In fact, limited domestic examples exist of social norms filling voids left by property law and public legislation or overcoming obstacles they have presented to sustainable resource allocation. Not surprisingly, as this chapter shows, no meaningful headway in that regard has been made with respect to natural capital and ecosystem services.

Why Social Norms?

The rules of private property and regulation are formal, institutionalized directives that have the effect of making people cooperate, even if they don't want to. Of course, people had to cooperate to make the rules of private property and regulation in the first place. Cooperation is often born of the social bonds and norms that define informal "rules of the game" within a community, what Pretty (2003) describes as "the mutually agreed upon or handed-down drivers of behavior that ensure group interests are complementary with those of individuals [and] give individuals the confidence to invest in the collective good" (1913). Known holistically as "social capital" (Krishna 2002; Putnam 2000), these collections of social norms build relations of trust, actions of reciprocity and exchange, common rules, sanctions, and connectedness to and between social networks and groups (Pretty 2003). There is evidence that people generally share an abstract norm favoring protection of human health and the environment (Vandenberg 2005), thus social capital built up from that norm could provide local and regional communities the source of cooperation necessary to achieve sustainable management of natural capital and ecosystem service values.

In a society acculturated to strong private property rights and pervasive regulation of land use, all of which, presumably, result from and are consistent with shared social norms, one nonetheless might reasonably ask what further role social capital could have with regard to resource management decisions. One might think the job of social capital is done once it builds and supports the formal rules. Yet the converse question is equally reasonable to pose: Given the extensive amount of cooperation needed for social norms to produce the formal private property and regulatory regimes, why bother going that far— why not simply rely on the force of social capital? The answer to both questions is that they are not either-or options. A "smart" society will choose whichever blend of these sources of cooperation provides the most efficient cost-effective way of managing the resource consistent with the collective interest.

Social capital is a viable candidate in this regard. Because "social capital lowers the transaction costs of working together, it facilitates cooperation" (Pretty 2003, 1913). Indeed, social capital could be so cost-effective in this sense that people find formal rules of private property and regulation unnecessary, or even go so far as to ignore them. Robert Ellickson (1991), for example, documented the rules early American whalers devised and tacitly agreed upon for determining which ship could claim a harpooned whale. The rules were so effective that no formal legislation was ever needed, and Ellickson could find few examples of litigation over whale possession. Even more compelling was his now famous study of rural landowners in Shasta County, California, who, notwithstanding and in contradiction to a governing set of state legislative measures, developed a set of social norms covering the liability of a rancher for the trespass damages caused by his or her roaming livestock. Members of the community had to cooperate on many different matters, such as water supply, controlled burns, social events, and volunteer fire control, thus building extensive social capital over time. This stockpile of cooperation, which Ellickson (at 624) boiled down to simple neighborliness, provided a low-cost means of resolving potentially divisive issues such as the damages caused by wayward cattle.

Ellickson pointed to these examples as evidence that while transactions costs are the paramount concern with regard to facilitating efficient exchange, law is not necessarily the best means of reducing them. Indeed, Coase (1960) also used cattle trespass liability as an example for his theory of efficient bargaining and transactions costs, discussed in chapter 3, through which he suggested that the legal system should play the central role in avoiding unnecessary transaction costs by assigning liability to the party in the best position to implement cost-effective precautions. California had attempted to do just that through its formal legal rules, but, as Ellickson demonstrated, social norms found a better, less costly, more collectively endorsed way of assigning liability.

Of course, building the kind of social capital Ellickson detected in Shasta County may require a long and expensive investment of social resources. Yet, once built, these "relations of trust lubricate cooperation, and so reduce transaction costs between people" (Pretty 2003, 1913). Individuals in trusting relationships need invest less in monitoring others, can engage in longer-term exchange with expectations of mutual gain, and develop the sense of reciprocal obligations. Moreover, it is likely that once people have invested so heavily in shared community norms, they are less prone to deviate from them than they might be from externally imposed rules. Indeed, property rules and land use regulations can certainly change behavior, but if they do not also promote changes in personal attitudes, it is likely that people will revert to their old ways if the rules or regulations end or are not enforced (Pretty 2003). Not surprisingly, therefore, resource management theory has focused increasingly on the potential of social norms to form cooperative, sustainable management regimes, particularly in the context of common-pool resources.

Social Norms and Common-Pool Resources

Recall from chapter 3 that common-pool resources are characterized by the two defining properties of excludability and rivalness: (1) it is costly through legal or physical barriers to exclude individuals from gaining access to the resource, and (2) the benefits any individual derives by consuming the resource subtracts from the benefits available to others (Ostrom 2000a; Ostrom et al. 1999). The degree of excludability and rivalness (or subtractability) define the common-pool resource management challenge, with the combination of low excludability and high rivalness presenting the hardest cases for sustainable resource management (Buck 1998).

Elinor Ostrom has been the most influential social scientist in forging the theory and study of common-pool resources in the context of social norms, with several purposes in mind. First, she has worked to show that Garrett Hardin's "Tragedy of the Commons" (1968), discussed in chapter 3, is an oversimplification of the "commons" (Deitz et al. 2003; Ostrom et al. 1999). It is by now widely accepted that Hardin's theory of the rational resource user works better in describing the reasons for resource depletion in true open-access resource contexts than it does in cases where a person, group, or government is capable of enforcing some level of excludability for the resource (Rose 2002).

Second, Ostrom has stressed the need to distinguish conceptually between common-pool resource systems and common property regimes (Ostrom

2000a). The former concept describes the characteristics of a particular resource that exhibits the characteristic of nonexcludability, whereas the latter describes the characteristics of a particular configuration of property rights. They share the term *common*, but common-pool resources can be managed through private property, common property, or public property ownership regimes, and common property rules can be employed to define ownership in resources that do not meet the characteristics of common-pool resources. As Ostrom puts it, "there is no automatic association between common-pool resources and common property regimes" (2000a, 338).

Third, Ostrom has devoted the bulk of her empirical work to demonstrating that the two solutions conventionally posited for the Tragedy of the Commons—private property and public regulation—are neither the only solutions nor in many cases the best solutions (Ostrom et al. 1999). Rather, social norms can give rise to self-organizing, community-based institutions that provide the excludability necessary to ensure limited access and the incentives necessary to ensure sustainable resource management. In other words, social capital can provide an effective common property regime where private property and public legislation regimes fail.

To advance her theoretical and empirical work, Ostrom (1990) developed a set of design principles for identifying when social institutions could produce an effective and stable common property management regime for a common-pool resource:

1. *Boundaries.* The boundaries of the resource and the access rights of users are well defined.
2. *Rules.* The rules of appropriation and provision are well suited to local ecological, economic, and institutional conditions.
3. *Participation.* Users subject to the rules participate in the decision-making process.
4. *Monitoring.* Users effectively monitor compliance with the management rules through monitors accountable to the users.
5. *Enforcement.* The users enforce a graduated system of sanctions for noncompliance with the rules.
6. *Conflict resolution.* Cost-effective mechanisms exist for conflict resolution between users.
7. *Recognition.* External governmental authorities recognize the common property management regime.
8. *Integration.* In the case of large-scale common-pool resources, the various common property regimes, governmental authorities, and other relevant institutions are organized in multiple layers of nested enterprises, with small, local common-pool resource units at the base.

Ostrom's theoretical framework has been increasingly influential among legal scholars studying natural resources management. In particular, Carol Rose (2002) has fused Ostrom's political theories with property law in her own work on community-based management regimes, and Dan Cole (2002) has drawn heavily from Ostrom's work in his exploration of property regimes and pollution. As valuable as this body of work has been, however, it also reveals the limits of applying Ostrom's theories to resource management in the United States. Rose and Cole use Ostrom's work chiefly to provide comparisons to alternative property regimes and regulatory instruments, such as tradable pollution rights (Rose 2002), rather than to suggest that social norms are likely to provide robust property regimes for resource management in the United States.

This is not surprising, because almost all of the empirical examples of social norms leading to effective common property regimes come from *outside* the modern experience of the United States. Much of the work supporting and building on Ostrom's theories focuses on irrigation systems and other group resource management efforts in modern developing nations or from norm-based common property regimes that defined historic pastoral communities (Ostrom et al. 1994). Cole's (2002) work on property regimes, for example, summarizes England's open-field system, the Andean mountain common-field agriculture tradition, and the Japanese Iriachi common lands.

To be sure, Ostrom does not advocate that social norms will or always should provide the dominant context of common property regimes, or that they will necessarily do so in the absence of private property, public property, or regulation. Her theory is nuanced and accounts for variability in common-pool resources, cultures, land productivity, economic conditions, and a variety of other physical and social attributes (Ostrom 2000b; Ostrom et al. 1999). Yet, as such, her theory also explains why, in a nation with a growing population, hotly competing land and resource uses, and high land and resource values, one might expect that supporting formal private, common, and public property regimes and a far-reaching regulatory system could be well worth the cost to individuals and society alike. Conversely, one might expect in such situations that the transaction costs of building and relying on social norm regimes for managing common-pool resources could be quite costly in comparison (Aviram 2004; Cole 2002).

This is not to say, of course, that social norm arrangements for common property regimes "fall outside modernity" (Agrawal 2002), as if they are too simple for the resource management challenges of modern developed nations. Rather, Ostrom's design principles contemplate property regimes that are likely often far too complex and sophisticated for modern Western land use markets, which, as Carol Rose has observed, demand a limited universe of "off-the-rack forms of property" to facilitate easy market transfer and thus "sharply discourage efforts to create more complicated forms of property" (2002, 247).

Hence examples are few and far between of resources in the United States that both meet the characteristics of common-pool resources and have not been substantially assigned through state-enforced property rights or comprehensively regulated through public legislation, or both. The most prominent example, which few studies of common-pool resources fail to mention, is from James Acheson's classic work (1988) on the "harbor gangs" of the Maine lobstering industry. Acheson studied the complex set of practices lobstermen developed in the mid-1900s for regulating access to and use of the common-pool lobster fisheries of Maine's inshore coastal harbors. The resulting property regime meets all of Ostrom's design parameters, including integration with a public legislation regulatory regime.

Lobstering in Maine has long required a state license, but this is just a minimum entry requirement. For decades, groups of lobstermen in each harbor formed harbor gangs to which anyone wishing to take lobsters from the harbor had to be admitted. Each gang had a well-defined territory and set of rules, and violations of any gang's territorial boundary or fishing rules were punished by a set of escalating sanctions, starting with verbal abuse and culminating, for repeated violations, in destruction of the violator's traps.

The result, as Acheson has chronicled (Acheson 1988; Acheson and Steneck 1997; Acheson et al. 1998), was an extremely effective, stable, and sustainable common property regime that was based on decentralized social norms, was not officially recognized by the state, and in fact relied on illegal property destruction as its means of enforcement. By limiting the number of lobstermen working in any territory, the harbor gang system allowed a sustainable fishery without having to set trap or catch limits.

The state's policy toward the system was one of implicit recognition but official denial. While state conservation laws established minimum and maximum size limits and restricted taking of lobsters with eggs, the state did not until recently limit the number of fishers, traps, or catches; yet, the state would prosecute persons caught enforcing gang boundaries by destroying traps (Acheson and Brewer 2000). Clearly, the harbor gang system worked in the shadow of the law because the state allowed it to work, but the state allowed it to work because the harbor gangs made it work.

Nevertheless, Acheson's more recent research (Acheson and Brewer 2000) traces the erosion of some of the harbor gang practices and the increasing conflict over territory boundaries that led eventually to intervention through public legislation. Technological advances over time effectively allowed any lobsterman to set more and more traps, which many began doing, and many part-time fishers began moving to lobstering as a full-time effort as other fisheries (e.g., scallops) became less profitable. Some lobstermen also began fishing farther offshore, where territories are harder to delineate, while nonetheless demanding incorporation into the traditional system. In response, several island communities petitioned the state to defend their fishing grounds from these incursions. The state established conservation zones around the islands within which the state would enforce limited access and the islanders would abide by strict conservation rules.

Finally, in 1995 the state enacted a law to address both the trap congestion and boundary disputes issues (Acheson and Taylor 2001). The Zone Management Law divides the entire coastal fishery into zones with formally mapped territorial lines and requires licensed fishers within each zone to vote on rules governing trap numbers and fishing times. Although zones and trap limits have been set, there has also been increasing conflict over territorial boundaries and increasing fishing in far offshore grounds (Acheson and Brewer 2000; Acheson and Taylor 2001). Consistent with Ellickson's (1991) theory of social norms, which predicts they will arise when they are comparatively more efficient than regulation as a means of maximizing overall social welfare, Acheson's work supports the conclusion that the harbor gang practices were largely about competition for access to lobsters, with conservation a secondary concern, and that the norms underlying them were primarily designed to address resource distribution conflicts efficiently (Acheson and Brewer 2000; Acheson and Taylor 2001).

The Status of Social Norms for Natural Capital and Ecosystem Services

Despite its long history of success, the harbor gang system seems "exotic and unusual in a modern country" (Acheson and Brewer 2000, 2). Indeed, it is difficult to conceive of how such a strong social norms system could be replicated in the United States across the full spectrum of ecosystem services.

To begin with, a driving problem in many ecosystem services management contexts is *not* in the nature of a common-pool resource. Rather, many ecosystem services flow from natural capital located on land that is privately owned. While the resource flow of the service benefit itself might exhibit the low excludability and high rivalness associated with common-pool resources, the resource stock responsible for the benefit is often locked up in private property and thus, as chapter 4 explains, is decidedly excludable. The resource management problem is not that the owner of natural capital cannot exclude others from access, but that, as chapter 3 explains, the benefits of the resource stock are positive externalities the owner cannot capture as value in the market. In other words, the beneficiaries of ecosystem services do not overconsume them; rather, the owners of natural capital undersupply them.

Moreover, even when the flow of ecosystem services exhibits common-pool resource qualities, or is on public land managed for high public access, one of the paramount design parameters of successful social norm regimes—well-defined boundaries—is likely to be missing in many instances. As chapter 2 explains, the geography of ecosystem services is such that their benefits are often felt across extensive spatial and temporal distances. A large riparian wetland, for example, might provide flood control benefits more than a hundred miles downstream and many days or weeks after a surge of snowmelt in mountains yet dozens of miles upstream of the wetland. The transaction costs of building social norms in such far-flung settings are likely to be insurmountable, suggesting that reliance on social norms rather than property rights or public legislation would not often be the wealth-maximizing strategy (Ellickson 1991).

Even so, in some cases ecosystem service management issues can be described as affecting a community of landowners in well-delineated spatial and temporal boundaries. For example, the service benefits that forest and riparian habitat provide to water quality can be reliably identified within well-defined watershed boundaries that are often associated with close-knit social communities, the members of which share a fair understanding of the ecosystem service dynamics (Adams et al. 2005; Ruhl et al. 2003). Even though the forest and riparian habitat may be in extensive private ownership, it might be perceived as unreasonable, or at least not very neighborly, for landowners to neglect or destroy their natural capital resources at the expense of other landowners in the community. For example, in their study of resource management in the Cache River watershed of southern Illinois, Adams et al. (2005) observe that

> underneath and alongside . . . formal governing bodies, numerous formal and informal institutions of "civil society," ranging from state-chartered corporations and organizations to customary associations and social orderings have more-or-less formalized rules governing their behaviors. In a locality such as the Cache, where many individuals live within widely ramifying sets of kin and other long-standing, multi-generational relationships, these informal governing rules often override formal laws. The overlapping jurisdictions of formal and informal institutions create a governing terrain in which "custom" can be as significant as formal procedures. (332)

If one were looking for evidence of social norms arising to manage natural capital and ecosystem services, watersheds provide an obvious starting point.

Alas, there is little evidence of the kind of norms regime Ostrom would define as stable and robust. For example, in his comprehensive examination of "watershed councils" across the United States, Peter Lavigne (2004) finds none that would meet Ostrom's design parameters. Watershed councils are organizations, found in almost all the states, that are alliances of government and nongovernment representatives devoted to advocacy of watershed management. Although they have existed in one shape or form for decades, in the 1990s the emerging policy focus on ecosystem studies and ecosystem management prompted a rejuvenation of their spirit and purpose. This, one might have reasonably hoped, could have provided the impetus for a robust social norm regarding management of natural capital in watersheds. Yet Lavigne describes nothing like that.

Lavigne's survey identifies significant differences between modern watershed councils in different regions of the nation. In the East, they are usually private nonprofit watershed protection advocacy associations composed of dues-paying members. As independent advocacy organizations, they provide policy communication, support other groups in their area, and pursue their advocacy agenda through lobbying and, in some cases, litigation. Watershed councils in the Midwest generally follow this model, with an increased focus on "multi-stakeholder" inclusion and educational mission. In the West, by contrast, Lavigne found that many watershed councils depend heavily on public funding. While ostensibly multi-stakeholder in composition, they are, in practice, usually stacked with representatives from resource user categories, such as farmers and ranchers and the federal, state, and local government agencies that serve them. As a result, Lavigne concludes, watershed councils in the West have been largely ineffective and exclusive.

Even in the East and Midwest, however, Lavigne finds that the effectiveness of watershed councils in accomplishing their goals of ecosystem management is still hit-and-miss. By their nature they can only hope to shape norms, not to set and enforce them. Their boundaries may be clearly defined, their composition inclusive, and their status recognized and endorsed by the state, but they cannot mandate rules and impose sanctions for violations. They are, in short, a far cry from the watershed equivalent to the harbor gangs of Maine.

The Crowding Out of Norms

What explains the absence of well-developed social norm regimes for managing natural capital and ecosystem services in ideal settings such as watersheds, where the management boundaries are clear and the relevant management community is often close-knit? The answer may be that American property law and public legislation have simply crowded out room for social norms to develop. As Elinor Ostrom recognizes, "externally imposed rules tend to 'crowd out' endogenous cooperative behavior" (2000b, 147).

Conservation policy in the United States, with its sweeping and intensive array of regulatory regimes at all levels of government set against a wide-reaching private property regime, seems to have been ripe for such an effect. In his epic history of modern environmental law in the United States, *The Making of Environmental Law* (2004), law professor Richard Lazarus charts the rise of environmental legislation, observing that "a snapshot of our nation's environmental laws in January 1970 with those today starkly reveals a dramatically changed landscape. . . . Today, there are comprehensive and stringent pollution control and natural resource management laws, and corresponding agencies responsible for their implementation and enforcement, in the federal government, all fifty state governments, and an increasing number of tribal authorities" (xiii). He concludes that today's environmental protection regulatory regime is "more comprehensive, far-reaching, demanding, and pervasive than ever" (167). As chapter 5 explains, our regulatory system does not in any significant way account for natural capital and ecosystem service values, but it may be so comprehensive in all other respects as to have suppressed the emergence of social norms in or beyond its shadow as a means of taking those values into account.

Indeed, in watersheds like the Cache, it seems unlikely that social norms could withstand the juggernaut of public legislation and all the funding, discretion, expertise, and power it vests in governmental authorities, much less override or supplant it. The federal government and, even more so, state and local governments are increasing their legislative attention to watershed management (Ruhl et al. 2003; Tarlock 2002), thus transferring increasing scope and power to regulatory institutions. Adams et al. (2005) found, for example, that in the Cache "the strong influx of resources to government agencies . . . act[ed] as a transfer of power to state and federal agencies from local governments and their polities. This undermin[ed] the possibility of creating a strong civic culture" (334). Similarly, Lavigne (2004) found that the traditional legal structures of the West governing resource management, which in many ways promote resource use over conservation, continue to exert a "stranglehold" on resource policy and "heavily influence the structure of the watershed councils and limit their effectiveness" (315–16).

One is hard pressed to find counterexamples to suggest that social norms are providing a robust, stable, and effective means of managing natural capital and ecosystem services in the United States. What matters in the United States has far more to do with who owns the land and what the government has done to regulate it, leaving little room for social norms to fill gaps or scale walls. In short, to the extent that private property regimes and public legislation in the United States do not adequately account for ecosystem service values, it seems they have more than adequately prevented social norm regimes from doing so in their place or despite them.

We close part II, therefore, with a rather bleak message—that none of the institutions usually relied on to facilitate sustainable resource allocation has worked effectively toward that end with respect to natural capital and ecosystem services. We turn next in part III to a series of case studies, some of which illustrate this failure, but others of which suggest the merits of exploring innovative institutional approaches. The lessons learned from these case studies help build the model for legal and policy design outlined in part IV.

Part III Empirical Case Studies in Ecosystem Services Law and Policy

Much of the discussion in part II was intended to provide a broad theoretical understanding of why property rights, regulation, and social norms thus far have failed to integrate natural capital and ecosystem service values. Because it is important to ground theory in application, part III presents a series of nine empirical case studies that explore the causes and consequences of the lack of attention property rights, regulation, and social norms have given to natural capital and ecosystem service values. The case studies focus first on the application of ecosystem services to individual parcels of land (chapter 7) and to the hydrologic cycle (chapter 8). They then explore the realm of agricultural land use and watershed management through the Conservation Reserve Program (chapter 9) and the National Conservation Buffer Initiative (chapter 10) as important existing ecosystem service subsidy programs, the shift from crop-based (amber) subsidies to ecosystem service–based (green) subsidies in the United States and the European Union (chapter 11), and how these policies affect the economy and ecosystem service provision of a typical agricultural watershed (chapter 12). Part III then investigates the successes, failures, and potential of market-based instruments for encouraging investment in natural capital and the consequent delivery of ecosystem services in the realm of wetland mitigation banking (chapter 13) and tradable pollution permits (chapters 14 and 15).

7 An Odyssey on 6,000 Acres: Pre-1670 to 2006

How do we balance the shifting nature of societal valuation of different ecosystem services reflecting cultural or societal norms of the moment with the science-based ecological knowledge that some ecosystem services might be more significant than others for the long-term viability of critical ecological functions? This reflects a question posed by Firey more than forty years ago in *Man, Mind, and Land* (1960) regarding the possibility of there being a lack of congruence among optima of what he called resource systems when viewed from ecological, ethnological, and economic perspectives. The changing nature of ecosystem services and their valuation from the same landscape area can be demonstrated using a 6,000-acre tract of land, known locally as Dry Hill with 2.2 miles of high and low dune frontage along Lake Michigan and its history from just prior to European settlement to the present. Dry Hill is located approximately 35 miles south of Traverse City, Michigan, along the Benzie–Manistee county line.

Before European settlement in this portion of the former Northwest Territories, this area was dominated by a mixture of northern mesic forests—sugar maple, birch, basswood, beech, hemlock, with some oak and white pine, along with open, grass savannas (see Cole et al. 2003). The land provided Native Americans with a stream of ecosystem services, including game for food and clothing; plants for food and medicine; materials for building shelter and transport and for heating; and clean water that not only provided for human needs but supported a diverse fishery along the shores of Lake Michigan and the many inland lakes fed by the streams and groundwater originating from Dry Hill (Harrison 2005). This diverse and complex landscape supported a population of subsistence-dwelling Native Americans. Up until the contacts with Father Jacques Marquette (Baulch, n.d.), other French missionaries, and European traders began in the latter half of the 17th century, other than the impacts of the Native Americans, there were not substantial changes in the land cover and in the intensity of land uses in the area. The historical record suggests that the flows of ecosystem services remained relatively constant in terms of types, quality, and quantity. This situation began to change even during the late 1600s as there was an increase in the population of Native Americans in the area as a consequence of European settlements further east forcing a westward movement of displaced Native Americans (Stearns 1997).

The land was surveyed in the 1830s (Arcadia 2006), and the first sawmill in the area was set up just to the north of the parcel of interest in 1851 by Harrison Averill (Howard 1929). The object of this timber operation was the harvest of white pine—a product, as Stearns (1997) points out, resulting from the "Little Ice Age" and its associated climatic pattern. According to Stearns, at this period of time, "white pine was considered the only species worth logging in quantity. Easily worked, light and strong, white pine floated well and so was readily transported from the woods to the mill" (13). Settler reports from 1854 indicate that between Manistee to the south and Traverse Bay to the north the area "still had only five white families with homes, two 'bachelor roosts,' and a number of Indian farms" (Howard 1929, 31).

The harvest of white pine expanded rapidly in the area, sustaining the small lumber-company towns of Watervale (1890–93) to the immediate north and Burnham (1880–1906)[1] just to the south of the parcel of interest. These towns cut timber from their own and surrounding land, sawed the timber into lumber, and shipped the lumber out on schooners that moored along piers jutting out into Lake Michigan. Lumber was generally shipped to Chicago. In addition to white pine, as the composition of the remaining forest changed, other species were harvested; for example, hemlock was often cut for lumber as well as for its bark, which was used in tanning. The small company towns lasted until the timber was exhausted (the white pine harvest peaked between 1890 and 1910) or until the next major financial crisis (1893 in the case of Watervale) (Howard 1929). In the case of Burnham, its buildings were either taken down or moved to the closest more permanent town, Arcadia.

On Dry Hill, cutover land or natural grasslands were converted by European settlers to diversified farms. The sandy soils were not well suited to supporting a prosperous grain-based agriculture, but the moderating effects of Lake Michigan on the area's climate made it ideally suited for diversified fruit production. Consequently, after the decline of the timber industry and its exploitation of the forests, the period of 1900 to the late 1960s saw two parallel tracks in the development and use of the ecosystem services from Dry Hill. On the one hand, there was the development of a fruit-based agriculture—sweet and tart cherries, peaches, apples, and other fruits—as well as a diverse, recreation-based economy, which included resorts, camps, and summer cottages nestled along pristine inland lakes; nonconsumptive bird-watching, hiking, biking, and cross-county skiing; and hunting and fishing. In short, the economy of the area became intertwined with the flow of ecosystem services that supported a unique mix of recreational services and activities or that provided a unique microclimate in combination with soils and topography that made fruit-based agriculture profitable. The economic viability of both activities rose and fell with the general economy. Over time, the mix of fruit changed as did the nature of tourism.

A major change came to the parcel in the late 1960s when individuals began working through the area buying up individual tracts of land and gradually consolidating them into a single 6,000-acre parcel with 2.2 miles of frontage on Lake Michigan and extending an average of 4 miles inland to the east of the lake. The local newspapers soon revealed that the purchasing agents had been working for Consumers Power Company of Jackson, Michigan, which was acquiring the property in order to construct a pump-storage facility for electricity generation (*Benzie County Patriot* 1969). Given the high dune above Lake Michigan, Consumers Power was proposing to construct an artificial lake on top of the dune and extending to the east. At the base of the dune, the power company would construct a facility that would take "excess" power in the electrical grid and use it to pump water from Lake Michigan up the dune and into the artificial lake. Then, when there was a "deficit" in the grid due to peak electrical demand, the water flow would be reversed, the water flowing in large pipes down the dune would turn the pumps (now turbines) and generate power for the grid to meet the demand. Hence, if we can argue that topographic differences yield ecosystem services, Consumers Power was proposing a wholesale shift in the type, quantity, and quality of ecosystem services provided by Dry Hill. The services that supported the recreational and agricultural activities in the area would no longer be provided to the same extent as they had been. The socioeconomic activities resting on these ecosystem services would be either eliminated or significantly changed based on the size and scope of the proposed pump-storage facility and the ensuing disruption of those ecosystem services. The residents of the region held their collective breaths, anticipating consequences for their individual livelihoods as well as the socioeconomic vitality of their local communities.

Gradually, Consumers Power started leasing the farmland back to some of the fruit growers. The large tracts of second-growth timber and dune areas continued to be used as an "unofficial" public recreation area for locals and people vacationing in the area during all seasons. The ecosystem services that had given the area value before Consumers Power acquired the property continued to be enjoyed by the public. However, there was a major difference concerning the uncertainty of future access and use: when would Consumers Power exercise its right to develop the property? When would the private decisions of Consumers Power impact the ecosystem service benefits that had been enjoyed and continued to be enjoyed as positive externalities by those living in and visiting the surrounding area? Partly as a consequence of the environmental problems Consumers Power experienced at its pump-storage facility 60 miles south at Ludington, Michigan, as well as other machinations within Consumers Power (now CMS Energy Corp.), the development of their facility on Dry Hill was postponed and has never materialized. Within the local community, however, there were rumors of possible large-scale mega-developments, such as an upscale, integrated resort with condominiums, a golf course, and a private airfield, that would significantly change the character of the region. The large size of the tract and its extensive frontage of both high and low dunes on Lake Michigan made it a prime candidate for the type of mega-developments that were beginning to take place along the lake's eastern shore. There were also concerns about the impact on CMS's balance sheet of a large 6,000-acre asset that was not "performing" and the pressure within the company to do something productive with it.

Given the presence of a number of threatened and endangered species, and the attractiveness of the land to maintaining the scenic beauty of undeveloped western Michigan, the environmental community was interested in protecting this tract of land and its ecosystem services. Consequently, starting as early as 1994, groups of concerned citizens petitioned the State of Michigan through its Natural Resources Trust Fund to attempt to acquire the land from CMS and protect its unique character and what were by then recognized as its significant ecosystem services. These efforts were unsuccessful, given the unwillingness of CMS to consider selling the property. By 2003, what was happening to Dry Hill in an area undergoing rapid development for recreation-based second homes, in one of the fastest-growing areas in Michigan, was a question that had taken on greater urgency (Trowbridge 2001). However, the cast of players had changed. Instead of unorganized groups of concerned citizens, the Grand Traverse Regional Land Conservancy (GTRLC) was now in place and provided the legal, scientific, technical, and political acumen necessary to direct a capital campaign and coordinate support among diverse stakeholders in order to (1) acquire and protect the parcel from significant commercial development, and (2) plan and manage how a large tract of land with significant ecosystem services would be used into the foreseeable future. Additionally, CMS now appeared willing to sell the parcel.

As part of GTRLC's many applications to state agencies, foundations, and private individuals for the more than $30 million needed to acquire this parcel and two associated tracts of land, it developed a list of the ecosystem services and a management plan regarding (1) how those ecosystem services would be used and how their integrity would be maintained into the future, and (2) how the land would be allocated to public use or returned to private ownership with its development rights retained by GTRLC (see GTRLC 2003). Based on its reconnaissance studies, GTRLC divided the parcel into 752 acres of dune and woods west of Michigan Highway M22, 891 acres of woods east of M22, 609 acres of forestry for sustainable management east of M22, 700 acres of grasslands or restoration grasslands east of M22, and 3,000 acres of mixed farmland—crop, fruit, and woods east of M22.

Given these different parcels and the literature on ecosystem services (see Costanza et al. 1997; de Groot et al. 2002), the following were provisionally identified as relevant ecosystem services: gas and climate regulation (CO_2, NO_x, CH_4); water quality, regulation, and supply; erosion control, soil formation, pollination, waste treatment and detoxification; biological control; food production; genetic resources; wildlife habitat; and nutrient cycling. While it is exceedingly difficult to place economic values on these ecosystem services, in the extended Table 2 accompanying Costanza et al.'s 1997 article, there is an extensive review of the valuation studies that have been done of various ecosystem services on a biome-by-biome basis. Additionally, data from the Conservation Reserve Program's rental rates and de Groot et al. (2002) give some additional values for some of the services under consideration. These services were valued very conservatively based on figures from the literature, calculated on a per acre basis, expanded by the relevant acreages to capture some of the effects of scale (see Gottfried et al. 1996), and capitalized by 5 percent interest rate (a relevant rate in 2003). The capitalized value is $25.7 million—close to the value being asked by CMS. This is the value before giving any consideration to development potential for homesites or consumptive or nonconsumptive recreational benefits. By 2010, the annual direct, tourist-based benefits from the 6,000 acres not undergoing development but remaining in their current state (dune, forest, grasslands, mixed agriculture, with noninvasive public access) were estimated to be $4.4 million per year with total direct and indirect economic activity of $12.2 million (GTRLC 2003).

By mid-2006, GTRLC was able to successfully conclude a $33 million campaign to acquire the subject property from CMS as well as two associated parcels. GTRLC is now actively engaged in developing a management plan for the land, including the restoration of the 700 acres of grasslands, the development and implementation of a sustainable forestry management plan for the woodlands, and the selling of the agricultural land to area farmers. When sold to local farmers, the agricultural land no longer has its associated development rights, so it cannot be developed in the future. Consequently, this land, with its unique microclimate and soils that make it ideal for fruit production, will remain available for such activity into the foreseeable future. The GTRLC is constructing a 15-mile hiking trail through the land so that visitors can see the diversity of this working rural landscape. Given the size of the tract of land and its proximity to Lake Michigan with its persistent winds, options are being reviewed for a site for alternative energy production.

Six thousand acres is a large tract in any community. This tract yields a diverse flow of ecosystem services that are critical for the sustained health and welfare of the community. The social institutions that determine or regulate which ecosystem services are produced, in what amount, and who has access to them are also critical to the community. In the case of our subject tract, this institutional environment has gone from an open-access resource, or a possible common property resource with the Native Americans, to private property being held by a large number of individual property owners, to a single corporate owner, and now to a nongovernmental agency charged to operate in the public's interest. Consequently, GTRLC is embarked on a process of involving the community in its decision-making process while it maintains its dedication to protecting the integrity of the critical ecosystems under its stewardship. From pre-1670 to the present, our understanding of ecosystem services and how dependent our survival is on their continued availability in adequate amounts and quality has expanded, especially in the last twenty-five years. As GTRLC embarks on developing its plans for this parcel, it not only needs to be mindful of protecting the integrity of the ecosystems under its stewardship, it needs to be aware of how the services from these ecosystems influence the health and welfare of the community—not just "on" Dry Hill but surrounding it, because many ecosystem services flow beyond the ecosystems that provide them.

While Dry Hill has unique characteristics and a unique history, so also does every tract of land. What we learn from this case study is that the ecosystems services provided by any area of land change over time as the physical environment evolves, as our knowledge about ecosystems expands, and, even more powerfully and dynamically, as society changes. As the ethnic and cultural identity of inhabitants changes over time, as population densities wax and (less commonly) wane, as the role that a place plays in the regional, national, and global economy evolves, as the rules governing property ownership are modified and as the identity and objectives of the specific landowner(s) are altered by the sale of a property, the ecosystem services provided by the tract, and thereby enjoyed by the local population, change with them.

8   Water: Blue, Green, and Virtual

We began this book with a discussion of water allocation in the Apalachicola–Chattahoochee–Flint River basin of Alabama, Florida, and Georgia, where direct human withdrawals of water are in competition with the water requirements of an estuarine ecosystem. Here we take a more systematic and conceptual look at human uses of the hydrologic cycle utilizing the ecohydrology approach developed by Swedish hydrologist Malin Falkenmark, in combination with the concept of "virtual water" developed by British geographer Tony Allan. We then bring to bear these broad hydrologic approaches to the law and policy of ecosystem services in the context of rights to water within various components of the hydrologic cycle.

Blue, Green, and Virtual Water

When you go to a typical textbook on water resources management and turn to the chapter on human uses of water, or to the U.S. Geological Survey Web site on "water uses," what you get is a treatment of water withdrawals—water that is pumped from rivers, lakes, or aquifers and transported through pipes to an end use. The first column of Table 8.1 provides these data for the United States in 2000 (Hutson et al. 2005), showing that irrigation and thermoelectric power are neck and neck in the race to be the foremost freshwater use. Thermoelectric power would win that race if saline seawater withdrawals were included; salt water serves just fine in cooling down a power plant but not for irrigating crops. Public municipal water supplies—the first water use most of us would think of—are a distant third, followed by commercial and industrial water users with independent systems.

The second most common approach to quantifying human uses of water is to consider water "consumption"—meaning that the water is evaporated or transpired to the atmosphere and therefore not locally available for further use. Using this criterion, irrigation emerges as by far the leading water use with over

Table 8.1. U.S water in uses related to Falkenmark's concepts of direct and indirect blue water and green water uses.

Water Use Sector	Estimated 2000 Freshwater Withdrawals (consumption[b]) (mgd[c])	Percent of Freshwater Withdrawals (consumption)	Percent of Human Freshwater Consumption	Percent of Human and Ecological Freshwater Consumption (incl. runoff to ocean)
Irrigation	153,000 (107,000)	39.5 (82.9)	22.0	2.5
Thermoelectric generation	152,000 (1,520)	39.3 (1.2)	0.3	0.0
Public supply	48,500 (9,700)	12.5 (7.5)	2.0	0.2
Self-supplied commercial & industrial	20,700 (4,140)	5.3 (3.2)	0.9	0.1
Aquaculture	4,150 (4,150)	1.1 (3.2)	0.9	0.1
Self-supplied domestic	4,030 (806)	1.0 (0.6)	0.2	0.0
Mining	2,250 (22)	0.6 (0.0)	0.0	0.0
Livestock	1,980 (1,980)	0.5 (1.5)	0.4	0.0
Total direct blue water	387,000 (129,000)	100.0 (100.0)	26.5	3.1
Rain-fed agriculture[a] (direct green)	0 (357,000)		73.5	8.5
Total direct human use	387,000 (486,000)		100.0	11.6
Transpiration by ecosystems (indirect green)	1,964,000			46.8
Aquatic ecosystems (indirect blue)	1,750,000			41.7
Total precipitation	4,200,000			100.0

Source: U.S. water withdrawals are from the Hutson et al. 2005. All others are first-order approximations by the authors based on Hutson et al. 2005 and USDA 2000.

[a]See calculation in Table 8.2. Does not include pastures and rangeland for grazing livestock.
[b]Estimate based on consumption rates of 70 percent for irrigation; 1 percent for thermoelectric power and mining; 20 percent for public supply, self-supplied commercial and industrial, and self-supplied domestic; and 100 percent for aquaculture, mining, and livestock (Dziegielewski 2005).
[c]Mgd = millions of gallons per day.

80 percent of all consumption, given the high rates of return flows for thermo-electric cooling and public water supplies (Table 8.1).

Falkenmark and Rockstrom (2004) argue, however, that this "blue water" approach misses the most important human uses of water that occur before rainfall reaches a water body or in that water body itself. Crops transpire large amounts of water as a requirement of their growth process (Table 8.2). This represents direct human use of "green" water, a use that in the United States, taking a first-order approximation, equals nearly all withdrawals of "blue" water. Use of green water constitutes nearly three times all consumption of blue water. Moreover, this is a substantial underestimate because it does not include rainfall that supports pastures and rangelands for livestock production, which is difficult to quantify but probably exceeds transpiration by crops. In general, meat requires at least five times as much water per pound than do plant foods. Therefore, the primary direct human use of water is for food production.

Falkenmark argues that it is arbitrary to include water for food production as water use only if that water is withdrawn from a water body or aquifer and applied to crops as irrigation and to exclude it if the water supporting crops and pastures comes in the form of rain. When green water is included, meat-rich and sometimes overabundant European and North American diets require 1,700–1,800 m^3 of water per capita per year (1,230–1,303 gallons per day) to produce; largely vegetarian and sometimes inadequate African and Asian diets require 600–900 m^3 per year (434–650 gallons per day). In comparison, North Americans withdraw 240 m^3 per capita per year (174 gallons per day) for domestic use, consuming about 48 m^3 per year (35 gallons per day).

TABLE 8.2. First-order approximation of rain-fed cropland transpiration.

Crop	m^3/ha/yr	Hectares Harvested in 2000 (thousands)	Total Use (m^3/yr) (millions)
Barley	3,560	2,105	7,494
Corn	3,170	29,546	93,661
Potatoes	3,290	548	1,803
Sorghum	2,600	3,102	8,065
Soybeans	3,400	29,553	100,480
Sugar beets	3,990	566	2,258
Wheat	3,740	21,460	80,260
Cotton	6,210	5,471	33,975
Other[a]	3,743	83,719	313,360
Total		176,070	641,356
Less irrigation (million m^3/yr)			-147,928
Rain-fed crop total			493,428
Rain-fed crop total (mgd)			357,242

Sources: Chapagain and Hoekstra 2004, USDA 2000.

[a] Taken as mean of all crops listed above.

It is also fruitful to take the ecohydrologic approach further and consider water for ecosystem services. Water defines aquatic ecosystems and coastal ecosystems such as estuaries that provide a very high level of ecosystem services per unit area (Costanza et al. 1997). Moreover, terrestrial ecosystems also depend on water, and, as a general trend, the more water they receive, the more ecosystem services they generate per hectare; wetlands and tropical rain forests lead the list, deserts finish last. Viewed from this perspective, the essential water resource is not stream flow or aquifer storage but total precipitation (Oki and Kanae 2006). Like photosynthesis, precipitation is therefore a fundamental ecosystem function that is critical to the creation of ecosystem services. By delivering freshwater, precipitation is a foundation of most ecosystem services delivered by terrestrial and aquatic ecosystems. In fact, on a regional basis, human population density is highly correlated with annual rainfall, up to a threshold where rainfall becomes excessive. Transpiration by ecosystems, including grazed areas, supports plant growth and makes possible the ecosystem services they provide. Falkenmark terms this "indirect green water use" and it is the leading use of water. This is followed by "indirect blue water use," water flow that supports the services of aquatic ecosystems. "Direct green water use" comes third. Blue water withdrawals constitute only 9 percent of U.S. freshwater use, and blue water consumption only 3 percent. Figure 8.1 captures these relationships in the journey of water falling as rain to return either to the atmosphere or to the ocean.

Virtual water is a component primarily of direct green water use. In *The Middle East Water Question: Hydropolitics and the Global Economy* (2001), Tony Allan of Kings College of London clearly illustrates that food imports have expanded the water resources available to populations in the Middle East and are, in fact, how that region has been able to support increasing populations despite extremely limited water supplies. Importing food allows water-short countries to forgo consumptive irrigation and reserve their water for domestic, commercial, and industrial uses, and, perhaps, support of aquatic ecosystems. Given that 1,000 metric tons of water or more, most of it from rain rather than from irrigation, is transpired in producing 1 ton of grain, grain imports essentially transport rainfall from water-rich to water-poor regions, and do so efficiently—in the 1,000-fold concentrated form of grain. In this sense, grain is an efficient way to transport rain as virtual water rather than as blue water. While Allan and other authors focus on the world's arid regions as virtual water importers, the United States is the world's largest virtual water exporter. Chapagain and Hoekstra (2004) have quantified virtual water trade on a global basis (Table 8.3), showing that North America is the primary virtual water exporting region at 162 billion gallons per day, more than tripling the exports

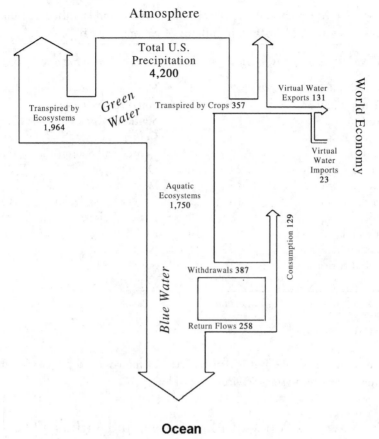

Figure 8.1. Blue, green, and virtual water. The annual flow of water in the United States in billions of gallons per day showing the relationships among precipitation, green water, blue water, virtual water, withdrawals, and consumption as human uses of water. The width of arrows is roughly proportional to the volumes of water. Rainfall on grazing lands is included in Transpired by Ecosystems.

of any other region. The United States exports 131 billions of gallons per day (bgd) with net exports of 108 bgd. This is direct green water made available to regions that purchase U.S. crops, livestock, and some other water-intensive manufactured goods. Note that the United States exports about the same quantity of green virtual water that it consumes from blue water withdrawals. Each is about equivalent to the flow of the Ohio River.

From the conceptual analysis and first-order approximations illustrated above, it is apparent that direct human use of water drawn from streams, rivers, lakes, and aquifers is only a small component of the hydrologic cycle that pro-

TABLE 8.3. Mean annual virtual water imports to and exports from major world regions, 1995–1999, in billions of gallons per day.

Virtual Water Imports	Bgd	Rank	Virtual Water Exports	Bgd
Central and South Asia	142	1	North America	162
Western Europe	76	2	South America	50
North Africa	36	3	Southeast Asia	49
Middle East	30	4	Central America	28
Southeast Asia	29	5	Central and South Asia	22
Central America	24	6	Oceania	22
South America	15	7	Western Europe	20
North America	13	8	Former Soviet Union	13
Eastern Europe	8	9	Eastern Europe	9
Former Soviet Union	6	10	Middle East	8
Southern Africa	6	11	North Africa	4
Central Africa	2	12	Southern Africa	3
Oceania	1	13	Central Africa	1

Source: Chapagain and Hoekstra 2004.

vides value to people. But is that reflected in the laws and policies governing water and the ecosystem services it provides?

The Law and Policy of Blue, Green, and Virtual Water

As observed in chapter 5, regulated riparian and prior appropriation are very different legal systems governing water resources. Nevertheless, they have something in common; they both focus on withdrawals of blue water and largely ignore all other components of the hydrologic cycle illustrated in Figure 8.1. Briefly, regulated riparianism, the system in eastern states where precipitation generally exceeds evapotranspiration, requires a permit to withdraw water from a public water source. Permits are granted by the state on the basis of "reasonable use." Prior appropriation, the system in western states where potential evapotranspiration sometimes exceeds precipitation and where water is frequently a limiting factor in economic development, grants permanent rights to withdraw water on the basis of "first in time, first in right." Thus nineteenth-century pioneers raced to capture water rights through hydraulic mining and later irrigation, putting waters to "beneficial use"—a criterion that support of aquatic ecosystems fails to meet. Much has been written on the need to reform these water allocation systems on economic efficiency and environmental grounds, but that is not where we want to take this discussion. Rather, what is the law and policy of green water, of virtual water, and of water for ecosystem services?

If rainfall is the essential water resource and ecosystem service, then who has a right to it? Rain falls as the weather sees fit, often providing too much and then none at all when it would be most beneficial to crops or public water supplies. A strong low pressure system in July 2005 distributing an inch or more of rain over a broad swath of the Midwest might have increased the value of crop yields by a billion dollars; but it didn't happen, and 2005 stands as a year of poor yields due to drought. A prevailing rainfall regime and corresponding ecological and agricultural potential is appurtenant to the land unit that normally receives it. Favorable rainfall regimes raise the value of land, especially farmland, but the landowner has no right to the continuation of the rainfall regime that obtained when the land was purchased. Climate change, which, for example, has been shown to increase intense rainfall events while also intensifying droughts, can change the value of the land, but the landowner has no recourse.

On the other side of the coin, using the ecohydrology approach, we can see that land management decisions also considerably influence water quantity. As a general rule, oceanborne moisture can penetrate several hundred kilometers into continents only if ecosystems such as forests and wetlands transpire much of the rainfall back into the atmosphere. For example, the forests of the southeastern United States receive fairly high rainfall and transpire a considerable portion of this rainfall back to the atmosphere where it is carried by low pressure systems as water vapor further inland. The rain that supports crop production in the Midwest is therefore considerably enhanced by transpiration occurring on private forested lands to the south and east that recycle moisture originally evaporated from the Gulf of Mexico or elsewhere. In fact, the corn and soybean fields of the Midwest transpire moisture that supports the wheat fields of the Great Plains. Thus these lands are providing a valuable, though uncompensated ecosystem service. Conversely, recurrent damaging droughts in the Sahel savanna zone of Africa can be partly attributed to deforestation along the West African coast and consequent reduction in transpiration (Falkenmark and Rockstrom 2004). In none of these instances do landowners have any obligation to maintain certain levels of transpiration from their land. Nor are they rewarded if they do so.

In arid and semiarid lands, transpiration is often viewed as a problem, limiting runoff and the availability of blue water. Vigorous removal of exotic tree species is practiced in South Africa to increase stream flow (World Resources Institute 2000). Salt cedar, mesquite, and Russian olive are viewed as pests in the U.S. Southwest because of their high rates of transpiration. Removal of phreatophytes (deep-rooted plants that get their water from the water table or the layer of soil just above it) is good for aquatic ecosystems and junior appropriators, but limits the services of carbon storage and wildlife habitat in order to increase blue water flow. Cloud seeding, if it were ever to become technologically and economically effective, is fraught with legal implications because it essentially takes the service of rainfall away from downwind landowners. Water harvesting (capturing rainfall from roofs through gutters into cisterns for indoor and outdoor use) is a legal means to capture green water for blue water uses. There is no legal system governing the effect of landowners' decisions in transforming precipitation into green or blue water.

Although virtual water constitutes an effective exportation of large quantities of green water, many jurisdictions pose, or have attempted to pose, restrictions on the transport or export of blue water. For example, in *Sporhase v. Nebraska,* an injunction was served by the state of Nebraska denying Sporhase the right to transport water from Nebraska into neighboring Colorado on the grounds that Colorado had no reciprocal agreement allowing transfer from Colorado to Nebraska. The U.S. Supreme Court disallowed Nebraska's action as unconstitutional interference with interstate commerce, yet granted special status to water's unique features as a basis of community health and welfare (Henderson 2002). In response to construction of the Chicago Sanitary and Ship Canal, which transfers Chicago's wastewater from the Great Lakes basin to the Mississippi basin, and to counter grand schemes to transport Great Lakes water to the Southwest, laws have limited transport of water outside of the Great Lakes–St. Lawrence basin (Henderson 2002). Canada has likewise consistently opposed transfers of water across the border despite having the greatest surplus of water resources in the world. The language accompanying these cases reveals an obsession with control of blue water similar to territorial impulses that often apply to land. Yet, green virtual water is transported and exported in large quantities and without restriction or, in general, opposition. In fact, it is a way to capture comparative advantages in world trade made possible by the abundance of the ecosystem service of precipitation in favored geographical locations.

This case study shows that portions of the hydrologic cycle that are critical for ecosystem service provision lie to a considerable extent outside the law and policy regimes that govern water allocation. For example, there are few incentives for landowners to manage transpiration on their lands for the benefit of downstream aquatic ecosystems, downwind farmers, or anyone else. Water that evaporates or transpires to potentially fall as rain downwind, and water that runs off to provide sustenance to aquatic ecosystems or damaging floods is incidental to land use decisions that are made for reasons that may be completely unrelated to the hydrologic cycle and the ecosystem services that it provides. As the case studies that follow will show, this disregard of the effects of land use on the hydrologic cycle represents the central challenge of managing watersheds for ecosystem service provision.

9 The Conservation Reserve Program 1985–2006: From Soil Erosion to Ecosystem Services

As a consequence of the world food crisis in the early 1970s and rapidly expanding farm exports, U.S. Secretary of Agriculture Earl Butz exhorted American farmers to plant fence row to fence row (Mayer 1982). In response to his exhortation and high commodity prices, a large amount of marginal agricultural land was brought into production. When world demand fell for agricultural commodities and their prices fell in turn, income fell for American farmers while the marginal land they brought into production earlier continued to contribute to soil erosion, sedimentation, deteriorating water quality, and other environmental problems (Clark et al. 1985).

As the 1980s began, the U.S. farm economy was in a depressed state, with many farmers facing the prospects of bankruptcy while the federal government made large income-support payments to farmers. At the same time, the Soil Conservation Service, now known as the Natural Resources Conservation Service (NRCS), was analyzing the data from the 1977 National Resources Inventory (NRI), which allowed it to identify the areas of the country that were the biggest contributors to the twin problems of soil erosion and sedimentation (American Farmland Trust 1984). Based on nearly 70,000 sample points providing reliable data at the state level (Harlow 1994), the NRI provided information on soil capability, land use, and conservation needs. The 1977 NRI resulted in the first nationwide estimates of the location and extent of water and wind erosion. Analysis of the NRI data showed that soil erosion on private cropland was highly concentrated on a small percentage of the cropland: cropland eroding at 10 tons or more per year accounted for about 11 percent of the cropland and 53 percent of the nation's sheet and rill erosion (American Farmland Trust 1984).

The 1977 NRI and succeeding NRIs authorized under the Soil and Water Resources Conservation Act of 1977[1] resulted in a process that initiated the collection of extensive "data on the quality and quantity of soil, water and related resources, including fish and wildlife habitats"[2] (Natural Resources Conservation Service 2001), essentially becoming a tracking mechanism for assessing the status of ecosystem services related to working agricultural landscapes. The NRI also collects data on the agricultural practices used by producers that have a direct bearing on the viability of ecosystem services that are produced in conjunction with traditional agricultural commodities—what is now referred to as multifunctionality (Organization for Economic Cooperation and Development 2001). The NRI data focused the attention of researchers on the impacts of soil erosion on off-site damages and the potential impacts on on-site damages related to productivity losses (English et al. 1984; National Research Council 1986) as well as on policy options to deal with these problems (for an example, see American Farmland Trust 1984).

These events came together in fashioning Title XII, the Conservation Title, of the 1985 Food Security Act, or the farm bill (Public Law 99-198). One of the provisions of Title XII was the Conservation Reserve Program (CRP), which was designed to remove "marginal," highly erodible land from crop production through a ten- or fifteen-year contract during which the federal government essentially rented the landowner's cropping rights to the land. Landowners who successfully entered their land into the CRP through a competitive bidding process would also be eligible for 50 percent of the cost of establishing permanent vegetative cover on the land. Given the precarious finances of many farm operators, the CRP rental payments provided the landowner and farm operator with a predictable cash inflow that could prove useful in servicing outstanding debt.

While the hearings leading up to the 1985 Food Security Act and the Conference Report itself do not use the language of ecosystem services, the intent of the 1985 legislation, as well as the subsequent extensions and reauthorizations of the CRP in the 1990, 1996, and 2002 farm bills, is clear: to the extent possible, marginal agricultural land and land with high environmental value would be retired from crop production in exchange for an annual rental payment. Other parts of the successive farm bills over this same period of time had as one of their goals to encourage the adoption by farm operators of agricultural practices that were as benign as possible to the environment and the ecosystem services it provides—for example, programs such as Conservation Compliance, Environmental Quality Incentives Program, and Conservation Security Program.

The long-term goal authorized for the CRP was to idle approximately 40–45 million acres (about 10 percent) of the nation's cropland. Other goals for the CRP as authorized in 1985 included the following: (1) reducing soil erosion, (2) protecting soil productivity, (3) reducing sedimentation, (4) improving water quality, (5) improving fish and wildlife habitat, (6) curbing production of surplus commodities, and (7) providing income support for farmers (Agricultural Stabilization and Conservation Service 1986). The 1985 Food Security Act was signed by President Reagan on 23 December 1985, and the first national sign-up for the CRP took place from 3 March to 14 March 1986. Land with erosion rates greater than 2T or 3T,[3] depending on the offered parcel's Land Capability Class, was eligible. In the first sign-up, 9,407 contracts were approved for 753,668 acres at an average annual rental rate of $42.06 per acre with an average reduction in soil loss of 26 tons per acre (Osborn et al. 1995).

Through the end of the fifth sign-up in July 1987, the eligibility requirements for the CRP were focused on soil erosion with some emphasis on cover management. Beginning with the sixth sign-up in February 1988, the eligibility criteria expanded significantly to include special provisions (1) if the offered land was going to be planted to trees, or (2) if the offered land was between 66 and 99 feet wide and was adjacent to a water body. In both instances, the enhanced ecosystem service benefits to be derived from either the forested or riparian lands were a significant addition to those already accruing to the nation as a consequence of improved permanent vegetative cover on marginal land, reduced soil erosion, and enhanced water quality.

In sign-up period eight (February 1989), the eligibility criteria were expanded once again to include land with evidence of scour erosion caused by out-of-bank water flows or cropped wetlands. These expanded criteria pertained for the ninth sign-up in July and August 1989. In 1989, Marc Ribaudo estimated that the present value of the water quality benefits from the CRP program ranged from $3.5 billion to $4 billion. Ribaudo essentially endeavored to place a monetary value on the enhanced flows of ecosystem services related to water quality that had become available to society as a consequence of land being placed in the CRP. He included such services as reduced flows of contaminants (chemicals, sediments, and nutrients; increased dissolved oxygen; cooler water temperatures, etc.) that directly and indirectly impacted a range of human activities, such as recreation, commercial fishing, navigation, water storage, and drinking water supplies.

Over its life, the CRP has evolved into a program with an environmental focus that is broader than the programmatic foci of soil erosion on highly erodible agricultural land and the reduction of the resulting sedimentation that were the primary thrusts of the program in the mid-1980s. Beginning in March

1991 with the tenth sign-up, land offered by landowners for inclusion in the CRP through its competitive bidding process was rated based on an Environmental Benefit Index (EBI). The EBI reflected the interest of the nation and the Congress to have the CRP and its payments for land retirement secure a larger suite of environmental benefits than just a reduction in soil erosion. Consequently, the EBI was constructed to reflect a larger set of environmental benefits, many of which we now refer to as ecosystem services. The factors used in constructing the EBI, the weights assigned to the factors, and how rental rates might be used to adjust EBI scores have been areas of policy debate (Cattaneo et al. 2006; Feather et al. 1999; Heimlich and Osborn 1994). The EBI used in the tenth, eleventh, and twelfth sign-ups was constructed of seven coequal conservation and environmental goals: surface water quality improvement, potential groundwater quality improvement, preservation of soil productivity, assistance to farmers most impacted by conservation compliance,[4] encouragement of tree planting, enrollment in Hydrologic Unit Areas identified by the USDA's Water Quality Initiative, and enrollment in established conservation priority areas (Osborn et al. 1995).

As the structure of the EBI changes, the flow and regional location of the ecosystem services derived from the land retired from agricultural production change as does the flow of federal dollars in rental payments to owners and operators of the farmland who are successful in placing their land in the program. With annual CRP rental payments amounting to $1.765 billion, changes in the formula directing where these payments are made can have significant economic and political ramifications. In the early 1990s, with CRP acres concentrated in the Great Plains and western Corn Belt, one of the policy debates leading up to the 1996 farm bill and the reauthorization of the CRP centered around the EBI and whether the focus of the program was going to be on providing environmental benefits that were also supportive of upland wildlife habitat in the plains, as had been the case earlier, or if the CRP would be "tilted" through the EBI to provide greater water quality benefits in the Midwest.

Generally the EBI is constructed of a number of factors, with each factor comprising a number of subfactors. Each factor is assigned a number of possible points. The number of possible points assigned to a given factor, and how these are partitioned among its respective subfactors, reflect the relative weight or importance given to the factor in the overall construction of the EBI. For example, these dominant factors were used in constructing the EBI for the thirty-third general sign-up of the CRP from March 27 to April 14, 2006: N1—Wildlife Factor (0 to 100 points, with three subfactors including wildlife cover); N2—Water Quality Benefits from Reduced Erosion, Runoff, and Leaching (0 to 100 points, with three subfactors, including groundwater quality and surface water quality); N3—Erosion Factor (0 to 100 points, scoring based on potential for land to erode); N4—Enduring Benefits (0 to 50 points, scoring based on the likelihood conservation practices will remain on the land after the CRP contract period); N5—Air Quality Benefits from Reduced Wind Erosion (0 to 45 points, with five subfactors including wind erosion impacts and carbon sequestration); and N6—Cost (point adjustment for amount of cost share for installing practices and amount of CRP payment relative to allowable maximum) (Farm Services Agency 2006a).

During the recently completed thirty-third general sign-up of the CRP, 1.4 million acres of land were offered by farmland owners to the USDA. Of these offered acres, the USDA selected 1 million acres that had a minimum EBI score of at least 242. This score reflected the factors and the related ecosystem services listed above: reduced soil erosion, enhanced surface water and groundwater quality, conservation practices with long-term enduring benefits, enhanced air quality, and wildlife enhancement. The average EBI for the acres accepted by the USDA for the thirty-third CRP sign-up was 284, and the average annual per acre rental was $53.44 (see Berry 2006). With the conclusion of that sign-up, there are more than 36.7 million acres in the CRP. Under the 2002 Farm Bill, the USDA is authorized to enroll 39.2 million acres in the CRP.

As of July 2006, there were 735,494 contracts totaling approximately 36.7 million acres in the CRP (Farm Services Agency 2006a). Currently under the CRP, there is a provision for the continuous sign-up of lands that meet specific environmental objectives. These lands include "filter strips, riparian buffers, grass waterways, field windbreaks, shelterbelts, living snow fences, salt-tolerant vegetation, shallow water areas for wildlife, and wellhead protection"(Wiebe and Gollehon 2006, 176–77). There are 2.5 million acres enrolled in the CRP through the continuous sign-up provisions. Given the material presented in the NCBI case study, filter strips, riparian buffers, and grass waterways are providing a rich and diverse range of ecosystem services relative to their size. Under the continuous sign-up provisions, the USDA enrolls partial fields that are going to be converted to uses with large environmental benefits (Economic Research Service 1997). Consequently, if the landowner is willing to accept the maximum productivity adjusted payment as calculated by the Farm Services Agency (FSA), the offered acres are automatically accepted. The acres in the continuous sign-up portion of the CRP have an average annual rental rate of $88.71 per acre compared to land in the general CRP that has an annual rental rate of just $43.65 per acre. In a state like Illinois, the comparison is $80.08 per acre per year for the general CRP and $131.66 per acre per year for the continuous CRP (Farm Services Agency 2006b). The difference between the two payments reflects the increased flow of ecosystem services and their value from the land placed in the continuous CRP as well as the opportunity cost to the producer of placing this land in the program.

Another option under the continuous sign-up provision of the CRP is the Conservation Reserve Enhancement Program (CREP). Through CREP, the federal government gives state governments the option to partner with the USDA to address specific environmental problems in their jurisdictions. Using state funds, the state augments the rental payments to landowners for parcels of land from specific regions that help remedy the identified environmental problem(s). Consequently, a landowner would receive a higher combined payment of both state and federal funds. For example, in Illinois the restoration of the Illinois River is a state priority. One of the problems in the river's watershed has been excessive soil erosion and sedimentation. Consequently, the state entered into a CREP agreement with the FSA and NRCS regarding the placement of riparian buffers along specific tributaries of the Illinois River. Landowners and producers could place their land in the CREP under the provisions of the continuous CRP. However, rather than obtaining the average rental payment of $131.66 mentioned above, they qualified for an average CREP-enhanced payment of $159.47 (Farm Services Agency 2006b). There are currently thirty-seven CREP agreements between the USDA and twenty-nine states.

Starting in October 2004, producers could offer to enroll large wetland complexes and playa lakes located beyond the 100-year floodplain. These lands provide a range of water-related ecosystem services—filtering runoff, groundwater recharge, and flood pulse mitigation, as well as wildlife habitat and recreational opportunities. This option was in addition to a number of existing wetland provisions in the CRP. One of the intents of these wetland provisions is to provide expanded protection to the nation's limited supply of wetlands while enhancing their flow of ecosystem services. There are a little over 2 million acres currently in CRP practices related to wetland protection (Farm Services Agency 2006b).

Since its inception in 1985, the CRP has been modified from a program dealing primarily with soil erosion to one of the nation's premier programs retiring agricultural lands from production while protecting and enhancing the flow of diverse ecosystem services from working agricultural landscapes. Estimates of the value of services provided by the CRP vary; however, NRCS estimated prior to 2003 the monetized CRP benefits related to activities like hunting and recreation were $1.4 billion per year (Johnson 2005). This is in addition to nonmonetized values related to improvements in surface and ground water quality (Ribaudo 1989) and restored wetlands. According to Johnson (2005), FSA reported that compared to 1982 erosion rates, CRP has reduced erosion by more than 440 million tons of soil per year on the acres enrolled in the program. Johnson further reported figures from NRCS on the acres of wildlife habitat established (3.2 million acres), the reduction in the amount of nitrogen applied to the landscape (681,000 tons), and the reduction in the amount of phosphorus applied (104,000 tons). NRCS also reported that the CRP has sequestered 16 million metric tons of carbon per year.

Each time the program has been up for reauthorization or there have been arcane questions regarding whether or not its budget is in the USDA's budget baseline (Gullo 1994; Zinn 1994), the continued flow of these ecosystem services has been placed at risk. Over its twenty-year history, in rural America, the CRP has emerged as the primary vehicle for providing a range of ecosystem services related to surface water and groundwater quality, wildlife habitat, recreation, carbon sequestration, and flood mitigation, among others. However, as federal budgets tighten, continued funding for the CRP and the benefits it provides might well be questioned; they have in the past (Farm Services Agency 2003). Given these benefits and the fact that in 2007 and 2008 over 60 percent (21 million acres) of the existing CRP contracts will expire (Wiebe and Gollehon 2006), the implications of the program for the future flows of ecosystem services from rural areas that benefit the entire nation are significant.

IO The National Conservation Buffer Initiative: Ecosystem Services from Riparian Buffers

Riparian zones lie adjacent to surface freshwater features such as streams, rivers, lakes, and marshes and are ecologically connected to them. These areas are relatively modest in width—up to 100 feet, more or less, depending on topography, soils, and vegetation. As such, vegetated riparian areas can act as effective buffers for the adjoining water bodies, intercepting surface runoff, wastewater, subsurface flow, and deeper groundwater flows from upland sources. They thereby reduce the movement of associated nutrients, sediments, organic matter, pesticides, and other pollutants into surface waters and groundwater recharge areas (Welsch 1991). Riparian areas yield a range of ecosystem services related to the maintenance and enhancement of water quality—filtering of sediments, nutrients, pathogens, pesticides, and toxins in runoff (Chase et al. 1995; Fischer and Fischenich 2000; Hartung and Kress 1977; Peterjohn and Correll 1984; Waters 1995) and reducing soil erosion and providing sediment control (Castelle et al. 1994; Waters 1995) among other services.

However, ecosystem services related to water quality are only a portion of the services provided by riparian areas. Given their relative size in the landscape, riparian areas have the potential to provide a vast array of critical ecosystem services. These are summarized in Table 10.1 according to de Groot (1992) and his framework of ecosystem functions: *regulatory functions* (the capacity of ecosystems to regulate essential ecological process and life support systems through biogeochemical cycles), *habitat functions* (the ability of an area to provide space and the means to satisfy physical needs of humans, flora, and fauna), *productive functions* (the ecosystem goods and services that are produced naturally without alteration of natural processes by humans, but humans must spend time and energy to harvest the goods or use the services),

TABLE 10.1. Riparian ecosystem functions and services.

Number	Ecosystem Function	Ecosystem Services	References
1. Regulation Functions			
1.1	Gas regulation	Role of riparian ecosystem in biogeo-chemical cycles. Provides clean breathable air.	Wilson et al. 2005.
1.2	Climate regulation	Influence of land cover and biological-mediated process on climate. Influence terrestrial and stream temperature, human health, recreation, and crop productivity. Thermal refuge for aquatic species.	Wilson et al. 2005, Collier 1995a, Cunjak 1996, de Groot et al. 2002, Waters 1995, Wegner 1999, Woodall 1985.
1.3	Disturbance prevention	Influence of ecosystem structure on dampening environmental disturbance, such as flood attenuation, ice damage control, stream bank stabilization, maintaining channel morphology. Biological control mechanisms.	de Groot et al. 2002, Fischer and Fischenich 2000, Platts 1981, Postal and Carpenter 1997, Wegner 1999, Williams 1986.
1.4	Water regulation	Role of riparian cover in regulating runoff and stream flow. Infiltration and maintenance of stream flow.	Lowrance et al. 1984, Williams 1986.
1.5	Water supply	Filtering, retention, and storage of freshwater. Riparian buffers filter sediments, nutrients, pathogens, pesticides, and toxics in runoff. Infiltration of surface water that helps maintain baseflow. Water supply and groundwater recharge.	Chase et al. 1995, Fischer and Fischenich 2000, Hartung and Kress 1977, Peterjohn and Correll 1984, Waters 1995.
1.6	Soil retention	Role of vegetation root matrix and soil biota in soil retention. Reduce soil erosion and sediment control.	Castelle et al. 1994, Waters 1995.
1.7	Soil formation	Weathering of rock, accumulation of organic matter. Maintenance of topsoil and soil fertility.	de Groot 1992.
1.8	Nutrient regulation	Storage and recycling of nutrients such as N and P and organic matter. Contribution of organic matter to stream from adjacent vegetation.	Barling and Moore 1994, de Groot 1992.
1.9	Waste treatment	Role of riparian vegetation and biota in removal or breakdown of xenic nutrients and compounds. Storage and recycling of human waste.	Castelle et al. 1994, de Groot 1992.
1.10	Pollination	Role of biota in pollination.	de Groot 1992.

Number	Ecosystem Function	Ecosystem Services	References
2. Habitat Functions			
2.1	Refugium function	Suitable living space for wild animals and plants. Woody debris in the stream provides habitat and shelter for aquatic organisms. Terrestrial riparian ecosystem provides habitats for amphibians, mammals, and birds. Habitat for natural communities, and rare, threatened, and endangered species. Provide travel corridors for migration and dispersal.	Allan 1995, Chase et al. 1995, Hammond 2002, Kaufmann 1992, Keller et al. 1993, Naiman and Rogers 1997, Wegner and Fowler 2000, Verry et al.2000.
2.2	Nursery functions	Suitable reproduction habitat for aquatic organisms and amphibians.	de Groot 1992, Semlitsch 1998.
3. Production Functions			
3.1	Food	Conversion of solar energy into edible plants and animals.	Wilson et al. 2005, de Groot 1992.
3.2	Raw materials	Conversion of solar energy into biomass for human construction and other uses. Genetic materials.	Wilson et al. 2005, de Groot 1992.
4. Information Functions			
4.1	Aesthetic information	Attractive landscape features. Clear and clean water enhance sensory and recreational qualities.	Wilson et al. 2005, de Groot 1992.
4.2	Recreation	Water quality for recreation, boating, and swimming.	Wilson et al. 2005, de Groot 1992.
4.3	Science and education	Variety in nature with scientific and educational value.	Wilson et al. 2005, de Groot 1992.

Note: Compiled by Sethuram Soman.

and *informational functions* (the provision of opportunities for enrichment, cognitive development, and recreation afforded by natural ecosystems).

As a consequence of the ecosystem service richness of riparian areas, and the cost-effectiveness of investing limited taxpayer dollars in establishing riparian buffer areas or strips, in 1997 the U.S. Department of Agriculture (USDA) initiated the National Conservation Buffer Initiative (NCBI) and pledged to assist landowners in the installation of 2 million miles (up to 7 million acres) of conservation buffers by the year 2002 (Natural Resources Conservation Service 1998). The NCBI was joined by the National Conservation Buffer Council, a group of agribusinesses (e.g., Cargill, ConAgra, Farmland Industries, Monsanto, Novartis Crop Protection, Pioneer Hi-Bred International, and Terra Industries) as well as environmental groups (e.g., Trout Unlimited, The Nature

Conservancy, and Environmental Defense Fund) and trade associations (e.g., National Association of Conservation Districts, National Corn Growers Association, and American Farm Bureau Federation), all committed to protecting sensitive riparian areas (U.S. Environmental Protection Agency 1998). With limited dollars to invest in soil and water conservation practices, assisting landowners in the establishment of vegetated riparian buffer areas would be a cost-effective approach to maximizing on-farm conservation benefits as well as off-farm societal benefits of those limited funds. The National Conservation Buffer Council pledged an additional $1 million to aid the NCBI in meeting its goal of 2 million miles of riparian buffers by 2002.

Through the NCBI, the USDA would focus its varied programs to foster the accelerated installation of riparian buffers through agricultural areas in order to enhance water quality while providing the additional ecosystem services shown in Table 10.1 (see Natural Resources Conservation Service 1998). The USDA agencies would use the umbrella provided by the NCBI to marshal and funnel their limited resources to encourage landowners with riparian areas in nonconservation uses to shift these areas to vegetative covers and uses that would enhance their conservation value and their "production" of ecosystem services. To this end, through the NCBI, programs such as the Conservation Reserve Program, the Conservation Reserve Enhancement Program, the Wetland Reserve Program, and the various cost-share and technical assistant programs of the Natural Resources Conservation Service and its sister agencies were all used to facilitate the expanded installation of vegetated riparian filter strips throughout working agricultural landscapes. Subsequent research has shown that targeted installation of riparian buffers can be a very cost-effective method of providing water quality–based conservation (Yang et al. 2005).

As of 2004, approximately 1.55 million miles of riparian buffers had been installed in conjunction with the various programs participating in the NCBI (Natural Resources Conservation Service 2004). Figure 10.1 shows the contribution of these various programs in building the 1.5 million miles, 75 percent of the stated goal. With the shift from one presidential administration to another, and the accompanying change in priorities, the NCBI is no longer a program being used to focus national policy on the installation of riparian buffers for conservation purposes (Schnepf 2006). However, the NCBI does illustrate how the emerging understanding of the scientific significance of riparian areas to human health and welfare was combined with existing policy tools to develop a program that would stimulate enhanced production of ecosystem services from thin ribbons of land adjoining water bodies in agricultural landscapes.

Although the NCBI itself no longer appears to be a programmatic initiative, riparian buffers and their critical role in soil and water conservation con-

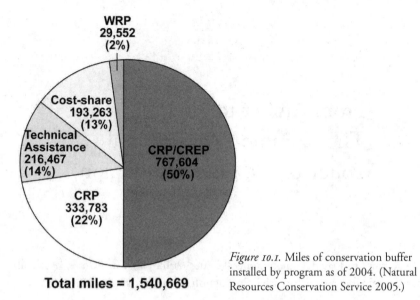

Figure 10.1. Miles of conservation buffer installed by program as of 2004. (Natural Resources Conservation Service 2005.)

tinue to be a focal point for national and state policy (see College of Agriculture and Life Sciences 2005; Natural Resources Conservation Service 2005, 2006). Through the continuous signup provisions of the Conservation Reserve Program, the USDA has retained its programmatic emphasis on riparian areas and the richness of their ecosystem services (Little and Knight 2004).

II   From Amber to Green:
The Common Agricultural
Policy of the European Union

The abstract to David Tilman and colleagues' 2002 paper in *Nature,* "Agricultural Sustainability and Intensive Production Practices," reads,

> A doubling in global food demand projected in the next 50 years poses huge challenges for the sustainability both of food production and of terrestrial and aquatic ecosystems and the services they provide to society. Agriculturalists are the principal managers of global usable lands and will shape, perhaps irreversibly, the surface of the Earth in the coming decades. New incentives and policies for ensuring the sustainability of agriculture and ecosystem services will be crucial if we are to meet the demands of improving yields without compromising environmental integrity or public health. (671)

Continued population growth, and the change in diets toward greater meat consumption that is accompanying increasing incomes in the developing world, present a great challenge to increase global food production. This expansion could result in the loss of already threatened natural capital and ecosystem services. How can an agricultural law and policy of ecosystem services navigate this twofold dilemma?

Futurologist Jeremy Rifkin's *The European Dream: How Europe's Vision of the Future Is Quietly Eclipsing the American Dream* (2005) contrasts American individualistic capitalism with European social democracy, finding the European model to be superior for addressing the global challenges of the twenty-first century such as those described by Tilman et al. (2002). In Chapter 15, we show how Europeans are driving forward on global carbon trading while the United States, which pioneered both the idea and the implementation of tradable pollution permits, has only begun to participate as a nonsignatory to the

Kyoto Protocol. Having examined in part the U.S. approach to the law and policy of ecosystem service provision from agricultural landscapes, how does the European Union (EU) system compare?

In addressing the issue of agriculture and the environment, Lowe and Baldock (2000) contrast an "impact" approach that views agriculture as a major source of pollution and a land use that inhibits ecosystem service provision, with a "public goods" approach that sees agriculture as a multifunctional enterprise that produces ecosystem services alongside food and fiber goods. These views, common to both sides of the Atlantic, are embedded in the structure of property rights to land and to the water that falls upon and runs through that land. They have considerable policy significance because the former suggests a regulatory, or "polluter pays," approach similar to environmental laws imposed on other industries that use public water and air resources for waste disposal, while the latter suggests a system of compensating farmers for the ecosystem services that they choose to produce from "their" land. Elements of both views lie embedded within policy initiatives in both the United States and the EU. Ongoing farm surpluses, described in Europe as "mountains" of food and "lakes" of drink, have also been a critical component driving policy developments, because, unlike in many Asian countries, ecosystem services can be gained from agricultural land in North America and Europe with little real trade-off in the form of marginally reduced food and fiber production. Also, agriculture figures prominently in ongoing World Trade Organization (WTO) negotiations where the common agricultural policy (CAP) has been accurately characterized over the years as even more protectionist than U.S. farm policies.

As discussed earlier, the Conservation Title of the 1985 U.S. Farm Bill set historic precedents in two ways. The first is through the initiation of environmentally targeted payments to farmers through the Conservation Reserve Program, partly motivated of course to stabilize Great Plains banks. Second, this was accomplished through forms of cross-compliance (sodbuster, swampbuster, and conservation compliance)—environmental prerequisites to maintain eligibility for farm subsidies. The 1980s also witnessed a Europeanization of environmental policy to match the CAP that has long coordinated European farm policy as well as a rhetorical shift in the EU toward integration of agricultural and environmental policy. In the 1992 "MacSharry reforms" (discussed later in this chapter), major changes in the CAP introduced an agri-environment program that embraced the concept of multifunctionality. In the 1996 Farm Bill, the United States implemented "decoupling" by shifting a considerable portion of agricultural subsidies from crop-specific price supports to "market transition payments," though the 2002 Farm Bill retreated from this policy. Decoupling and cross-compliance are key components of the 2003 EU reforms. On both sides of the Atlantic, farmers are subject to environmental regulations that establish a baseline of what farmers must and must not do and a system of cross-compliance that forms a quid pro quo of what farmers must or must not do to remain eligible for government subsidy programs. Through agri-environment and livestock extensification programs, European farmers, like U.S. farmers, have access to a system of government payments for specific voluntary actions.

Legg (2000), in addressing work from across the Organization for Economic Cooperation and Development, reaches the following conclusions about the interaction between agricultural and environmental policy: (1) decoupling benefits the environment by reducing the price incentives to crop marginal lands and use agrichemicals excessively; (2) targeted green payments provide farmers with needed incentives to provide ecosystem services, although (3) current policies are not cost-effective; (4) historic subsidies make it difficult to assess what ecosystem services would occur in their absence; (5) the higher environmental regulatory and cross-compliance standards are set, the smaller the investment in green payments that is required to achieve specific ecosystem service outcomes; and (6) the "polluter pay" policy should be applied to farmers, especially if they are eligible to receive green payments.

The argument for shifting the focus of agricultural subsidization from crop-based to ecosystem service–based subsidies that is beginning to be embraced by the EU is, therefore, a strong one and is a key component of the agricultural law and policy of ecosystem services. First, both U.S. and EU citizens are supportive of farmers and are willing to be taxed to support them. But what aspects of farming do they support, and what are they getting for their tax contributions? Crops are private commodities bought and sold in markets. Ecosystem services and rural landscapes are public goods. Subsidies are paid from public funds. It, therefore, follows that taxpayer-funded programs should support the public values of agriculture rather than the production of market commodities, especially when that production can undermine public good values of the agricultural landscape, including ecosystem services. Second, farmers respond to ecosystem service payments, or for that matter environmental regulations and cross-compliance, by reallocating their farm resources in a manner that marginally decreases crop production. The European Commission's Directorate General for Agriculture (2003) estimates that the 2003 EU reforms would decrease cereal production by 2 percent. Yet, the need for agricultural subsidies establishes that markets for most primary agricultural commodities suffer from surplus supply causing low prices that undermine farm profitability. Therefore, marginal reductions in the output of crops such as corn, wheat, soybeans, and the like actually help stabilize agricultural markets by tending to increase prices that farmers receive for these crops. Increased prices further reduce the need for crop-based subsidies; however, they also compete with ecosystem service payments for farm resources such as land. In this way, agricultural commodity markets and ecosystem service payment programs form an evolving dynamic that is expressed in rural landscape patterns and the ecosystem services they produce. Third, ecosystem service (green box) payments are regarded by the WTO as nondistorting in contrast to crop-based (amber box) subsidies that are regarded as trade distorting—that is, protectionist. Shifting subsidies from amber to green improves negotiating positions within the WTO, which regulates the labyrinth of national competition over trade.

Despite these similarities and the diffusion of policy concepts across the Atlantic, the effects of agricultural policies on U.S. and European ecosystem service provision have begun to diverge. Having examined the U.S. system, let's now turn our focus to the EU.

The 1992 MacSharry Reforms

In 1992, the EU began shifting subsidies away from crop-specific production toward decoupled payments to farmers and green payments for ecosystem services, known as "MacSharry reforms" after Ray MacSharry, the EU commissioner for agriculture at the time. The agri-environment program initiated with the 1992 reforms "provides for payments to farmers in return for a service" (European Commission, Directorate General for Agriculture and Rural Development 2005, 3) where farmers' conservation activities exceed the regional Code of Good Farming Practice—mandatory practices that farmers are expected to employ. Like the Conservation Reserve Program in the United States, farmers voluntarily sign a multiple-year contract on lands targeted for their environmental characteristics with payment rates set locally in accordance with competing uses of land. Spending for agri-environment grew from nothing in 1993 to around EUR 2 billion per year in 2000–2004, and these payments fit within the WTO definition of non-trade-distorting green payments.

EU and member states cost-share the agri-environment program, with EU contributions set higher for regionwide (Objective 1) environmental goals. Goals fall into two primary areas—reducing environmental risks, such as those associated with agrichemicals, and preserving nature and cultivated landscapes. Member states develop "schemes" for implementation of agri-environmental payments; "light green" schemes cover broad issues and areas, "dark green" schemes are more site-specific and may involve larger payments per farm. Outcomes are measured according to specific goals, including agrichemical input reduction, extensification of livestock (as opposed to large confinements), conversion of arable land to grassland, soil conservation measures (e.g., cover crops, tillage, and filter strips), biodiversity protection, rural landscape improvement, water conservation, and organic farming (European Commission—Agriculture 2005).

This last item, organic farming, illustrates a considerable divergence between U.S. and EU approaches. Organic farming is growing rapidly in the United States despite a general lack of governmental support in the form of policy or subsidization except for an organic certification process that establishes a separate market for organic food consumers (Duram 2005). In contrast, by 1996, all EU member states except tiny Luxembourg had implemented policies to accelerate the transition from agrichemical-based to organic farming, including the expenditure of EUR 260 million in 1997 (Lampkin et al. 1999). Organic methods were applied to 100,000 European hectares in 1985 (Lampkin et al. 1999), but this expanded more than fiftyfold to 5.8 million hectares by 2005, about 3.5 percent of EU farmland (Organic-Europe 2005). European Commission Regulation 2092/91 defined organic production and established policies to promote it on the basis of scientifically documented improvements in several ecosystem services, a reduction in agricultural surpluses, the emergence and growth of a separate, higher-priced organic food market that must be supplied, and the higher labor requirements of organic production that support rural employment goals. Transition payments to encourage conversion from agrichemical-based to organic production vary from EUR 181 for cereals to EUR 1,208 for fruit as per hectare EU averages.

A price support system does little to help organic farmers because they generally sell their products in higher-priced niche markets. The EU system of payments by hectare, rather than for crops grown, therefore encourages continued organic production after transition payments lapse. The expansion of agricultural research and extension services targeted toward organic production also fuels the conversion process because organic farming is more reliant than agrichemical-based farming on complex management regimes (Duram 2005).

The 2003 Reforms

Agenda 2000 constituted a major reassessment of CAP policy by the European Commission with eight years of experience under the MacSharry reforms. Floor Brouwer and Philip Lowe capture this reassessment in *CAP Regimes and the European Countryside* (2000). The major principles of this policy reform are (1) except under extraordinary circumstances, markets should determine the prices farmers receive for food and fiber commodities; (2) farmers should receive payments for their efforts in producing public goods in the form of ecosystem services; and (3) rural development is an important goal of CAP. From 1990 to 2002, the proportion of CAP payments targeted to crop-specific supports fell from more than 90 percent to about 30 percent. Support of rural development initiatives increased from nothing to 10 percent of the total over the same time frame and is scheduled to reach 25 percent by 2010. Nevertheless, in 2005 EUR 46 billion (US$55 billion) were paid in agricultural subsidies, compared to $12.5 billion in the United States in 2004. The cost of increased food prices is even higher—EUR 55 billion (BBC News 2005).

Based partly on Agenda 2000, on June 26, 2003, the EU made effective a sweeping reform of its CAP by decoupling most subsidies to farmers from crop production (sugar, wine, fruit and vegetables, and dairy remain to be reformed), by strengthening cross-compliance, and by increasing funding for its agri-environment program, thus shifting the focus of CAP spending from amber to green. These reforms include a single farm payment for EU farmers independent from production that is linked to environmental goals, including animal welfare, and a transfer of payments from large farms to rural development (European Commission–Agriculture 2005). The 51-article decoupling and the 81-article cross-compliance rules of the 2003 CAP reforms (Commission Regulations 795 and 796, 2004) establish that "direct payments to a farmer who does not comply with certain conditions in the areas of public, animal, and plant health, environment and animal welfare ('cross-compliance') shall be subject to reductions or exclusions." The rules establish that these are to be proportionate to the extent, severity, permanence, and repetition of violations due to negligence or intentional noncompliance. Member states must set up extraordinarily detailed monitoring systems that include sampling strategies and unannounced, on-the-spot checks. Remote sensing in the form of aerial photography and satellite imagery is normal practice and the use of geographic information systems is mandated for tasks as detailed as the counting of individual trees. The rules establish a computerized database for bovines based on ear-tag numbers as well as for slaughterhouses, and establish limits on the THC level of hemp grown for fiber (legal in the EU). Detailed rules are established governing farmers' single aid application and the transfer of payments in the event of land sale, farmer retirement, or death. These rules are embedded in tortuously detailed and legalistic documents that exceed even U.S. standards for bureaucracy. Many Americans, given the culture of private property rights and anti-government intervention, would find the degree of governmental micromanagement and bureaucratic control to be offensive. That very precision and tortuosity, however, reveal the thoroughness with which European farmers are reading the policies in order to play the rules in a way that maximizes their subsidies, whether they are comfortable with, or offended by, the bureaucratic structure administering them.

There is a strong emphasis in the 2003 agri-environment program on hedges, ditches, and walls, and on the maintenance of permanent pasture. Europeans have, in general, a rural landscape aesthetic that values visually delineated irregular fields emphasizing livestock pastures and a pacified landscape that Americans would find lacks wildness. The natural succession of forest on abandoned farmlands is strongly discouraged. In contrast, some 100 million acres in the eastern United States have returned to forest following farm abandonment, a land use trajectory that Americans view as favorable, both aesthetically and for ecosystem services.

With provisions of the 2003 reforms becoming effective in 2004 and 2005, it is too early to make an empirically based assessment of their economic and environmental effects. Nevertheless, the subsidy-shifting that was adopted by the EU in 2003, building upon the 1992 reforms, represents a progressive step, despite its bureaucritization. It helps build natural capital and supports ecosystem services in a number of ways, from control of polluted runoff to biodiversity to carbon sequestration as expressed in the concept of multifunctionality (Jervell and Jolly 2003).

Although the United States took the early lead in 1985 in developing a progressive agricultural law and policy of ecosystem services, the 1992 and 2003 reforms of the CAP have given the EU the lead. The EU now sets higher baseline standards through regulation and cross-compliance than does the United States, while a more aggressive approach to subsidy-shifting is also starting to succeed in providing farmers the rewards they need to increase their provision of ecosystem services. This divergence is especially evident with regard to organic farming, which epitomizes the notion of multifunctionality. Nevertheless, an EU-type policy approach—a transition from amber to green—could form an even more effective ecosystem service policy in the United States, given its larger land and agricultural resource base, enormous area of potentially restorable wetlands, and aesthetic for forests and wildness. Finally, this approach would help meet the challenge portrayed by Tilman et al. (2002)—to meet rising global food needs while protecting ecosystem services.

12 Ecosystem Services from an Agricultural Watershed: The Case of Big Creek

The Big Creek, a tributary of the Cache River in the southern tip of Illinois, is really quite ordinary. It's a rather muddy perennial stream that doesn't harbor many game fish or attract the attention of canoeing enthusiasts. Its watershed is a little hillier than the norm for Illinois and contains a mix of cropped fields, pasture, forest, and a few small towns. What can this unexceptional place possibly tell us about the law and policy of ecosystem services?

Watersheds as Geographical Units of Natural Capital

A watershed is a spatial unit of natural capital that usually has easily definable borders and also serves as a semiclosed system, at least with respect to surface water movement. If a construction site or a confined animal feedlot operation is placed outside the watershed boundary, for example, the pollution from these will show up in the watershed in which it occurs, not in Big Creek. Watersheds are also linked in a hierarchical arrangement. The 134 km² Big Creek watershed is a subset of the 1,944 km² Cache River watershed, which is a subset of the 528,000 km² Ohio River basin, which is a subset of the 4.7 million km² Mississippi River basin, the largest in North America and third largest in the world. This spatial hierarchy allows us to choose among different spatial scales in analyzing watersheds as "long-lived multi-product factories" harboring natural capital and producing ecosystem goods and services (after Gottfried 1992).

The value of the ecosystem services a watershed produces is also, as pointed out in chapter 2, tied to specific geographical relationships. Big Creek is the primary source of sediments to Buttonland Swamp on the Cache River, a unique and diverse wetland ecosystem that includes bald cypress and water tupelo swamps and more than a hundred Illinois state threatened and endangered species (U.S. Fish and Wildlife Service 1990). Buttonland Swamp, a wetland of international significance, forms the core of The Nature Conservancy Bioreserve and the Cypress Creek National Wildlife Refuge. Sediments have been found to greatly inhibit cypress tree regeneration (Middleton 1995), one of the most salient ecological restoration challenges in the Cache. Sediment retention is thus a critical ecosystem service within the Big Creek watershed because of this geographical relationship to a highly valued downstream resource. But the Big Creek watershed is also either a source or a sink for atmospheric carbon; a source or a filter for nitrogen, phosphorus, and other water pollutants; a habitat for various wildlife species; and a regulator of the hydrologic cycle, as well as a source of agricultural commodities. Different people have a vested interest in one or more of these individual ecological goods and services depending upon their livelihood, their values, and their location. They also have differing capacities to pursue these interests reflecting a constellation of social factors. In this way, the Big Creek shares with other watersheds the fundamental issues of scale and management of natural capital for the production of sets of ecosystem goods and services that society determines, through some political process, are desired.

Modeling Approaches and Big Creek

The primary reason we have something to learn from Big Creek is that some very good work has been done modeling the ecosystem services that it generates and could potentially generate. Applying a GIS-based linear programming model called GEOLP (Kraft and Toohill 1984) and the well-known watershed-based water quality model AGNPS (agricultural nonpoint source pollution simulator) (Young et al. 1989) to Big Creek, Lant et al. (2005) found a noteworthy interplay between agricultural and environmental policies, farmers' choice of land use, and the package of ecosystem goods and services that the Big Creek watershed can produce. For example, the Conservation Reserve Program (CRP), discussed in chapter 9, is a U.S. Department of Agriculture program that pays farmland owners for taking certain types of environmentally sensitive cropland out of agricultural production and replacing it with one of a number of forms of vegetative cover called conservation practices. T by 2000 is a State of Illinois program designed to induce farmers to incrementally reduce soil erosion to the "tolerance" level of T that soils can withstand without losing long-term productivity (explained in chapter 9). It is regulatory, unlike CRP, which is a positive economic incentive program.

What the analysis found, not surprisingly, is that T by 2000 reduces farm income and the CRP increases it. What is more interesting, however, is that farmer income with *neither* T by 2000 nor CRP is almost identical to farm income with *both* T by 2000 and CRP. However, with both programs in effect, soil erosion is only one-third as high as when neither is in effect, and sediment loads are little more than half (Figure 12.1). This important result is achieved through a complex process where each farmland owner or farm manager reevaluates his or her land use choices as the policy environment and economic opportunities shift. In this case, the acreage of land planted to corn and soybeans, the two most important crops in Illinois, falls by a third when T by 2000 and CRP are in effect, and the use of no-till, alfalfa hay, and CRP all dramatically increase, especially on the most highly erodible lands. Implementing T by 2000 in combination with CRP therefore maintains farm income while substantially improving the ecosystem service package from the watershed, even while crop production falls by about 25

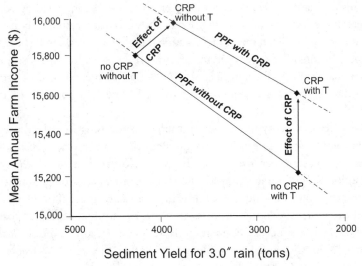

Figure 12.1. Sediment loads and farm income produced by Big Creek under four scenarios—with and without CRP, and with and without T by 2000. The results show that farm income and T by 2000, as a regulatory program, form a trade-off, but that the CRP, as a positive economic incentive program, increases the total delivery of both farm income and sediment retention from the watershed. This is shown as an expansion of the production possibilities frontier (PPF) from the origin. Farm income is modeled using GEOLP, sediment load using AGNPS. (Reprinted from: Lant, C. L., S. E. Kraft, J. Beaulieu, D. Bennett, T. Loftus, and J. Nicklow. 2005. Using ecological-economic modeling to evaluate policies affecting agricultural watersheds. *Ecological Economics* 55(4): 467–84, with permission from Elsevier.)

percent. However, in comparing these two scenarios with nearly equal average farm income, some farmers gain and some farmers lose income by as much as 10 percent.

Extrapolating from the Big Creek, if T by 2000 was in effect and CRP was expanded throughout the United States, crop production would marginally decrease, while ecosystem services from agricultural watersheds would substantially increase. Farm income would at least be maintained, but might increase because the reduction in crop supply could increase prices and thereby profits for farmers. Or these price increases could offset the need for crop-based subsidies, while at the same time making U.S. crop production less protectionist, improving international trade relations negotiated through the World Trade Organization (WTO). In their excellent overview of agricultural sustainability and intensive production practices published in *Nature*, Tilman et al. (2002) identify decreasing marginal returns to fertilization, shortages of available water for irrigation, ongoing difficulties in disease and pest control, and the potential loss of ecosystem services as central to the challenge of agricultural sustainability. They call for payments to farmers for providing ecosystem services as essential in changing the incentives farmers respond to in using land, water, and other farm inputs. National scale empirical studies confirm these conclusions. Ecosystem services derived from agricultural landscapes in Sweden are declining (Bjorklund et al. 1999). Environmental costs of agriculture in the United Kingdom amount to over $120 per acre per year (Pretty et al. 2000).

The analysis of ecosystem goods and service sets arising from land use patterns in Big Creek goes deeper yet. Because, as landscape ecology tells us, land uses affect one another over space in what Gottfried et al. (1996) term "economies of configuration," and due to the mathematics of permutations, the number of possible land use patterns in a watershed is equal to the number of land use units, such as fields, *raised to the power* of the number of possible uses of those fields, such as different crops, hay, pasture, forest, and so forth. If there are 1,000 fields and 10 possible uses for each, there are 10^{30} possible land use patterns. Which of these impossibly large number of land use patterns provides the best combination of ecosystem goods and services?

Lant et al. (2005) and Bennett et al. (2004) have explored this problem in Big Creek using genetic algorithms (GA). A GA is an algorithm metaphorically related to natural selection—which of the huge number of possible gene combinations in a species produces the organism most fit for survival and reproduction? The GA starts with a collection of random land use patterns and evaluates the ecosystem goods and services they would produce using economic, ecological, and hydrologic models. The best are maintained as the "parents" of the next generation that is produced by recombining elements from different parents through "crossover" and by "mutating" individual fields. Again, the best performers are maintained as parents for the next generation and so on for a few dozen generations until the performance stops improving and the ecological–economic "production possibilities frontier" (PPF) for the watershed has been discovered. Variations in the definition of "best performers," and how to implement the crossover and mutations make for a variety of different types of GAs for different types of problems. GAs do not always identify the optimal combination, but they always produce a family of diverse and high-performing combinations.

Figure 12.2 shows the trade-offs between gross marginal return (farm profits excluding land costs) and EBI scores. EBI, as described in chapter 9, is the Environmental Benefit Index used by the U.S. Department of Agriculture in evaluating CRP bids submitted by farmers and includes the characteristics of the land and the conservation practices to be used. Each dot represents the combined performance of a land use pattern or map. Land use maps from earlier generations lay closer to the origin. Over dozens of generations, both gross marginal revenue and EBI points grow until the PPF is clearly formed, showing that there is indeed a trade-off between these two components of the ecosystem goods and services set for the watershed. No land use patterns (i.e., dots) lay beyond the PPF, but land use patterns change considerably as we tour the PPF from the upper left (where gross marginal revenue is maximized) to the bottom right (where EBI points are maximized). As one would expect, row crops diminish and less intensive uses such as pasture, hay, and CRP increase. An interesting nuance is that the use made of an individual field can change many times in different positions along the PPF, demonstrating how precarious it can be to assign specific land uses to specific fields in pursuing optimality. Theoretically, a similar modeling approach could look for trade-offs and complementarities among individual ecosystem services such as carbon sequestration, water quality, flood control, and wildlife habitat along an N-dimensional PPF.[1]

What this advanced modeling exercise tells us is, first, that land use choices, and the land use patterns these create at larger spatial scales, are the driving factors in creating ecosystem services. Second, there is a limit to the package of ecosystem goods and services a watershed-scale landscape is capable of producing. Third, there can be trade-offs between ecosystem goods and ecosystem services, and society must choose what it wants from a set of possibilities. Fourth, current land use patterns may be suboptimal, inside rather than on the PPF, and so we may be able to improve multiple goals at the same time by closely examining these patterns.

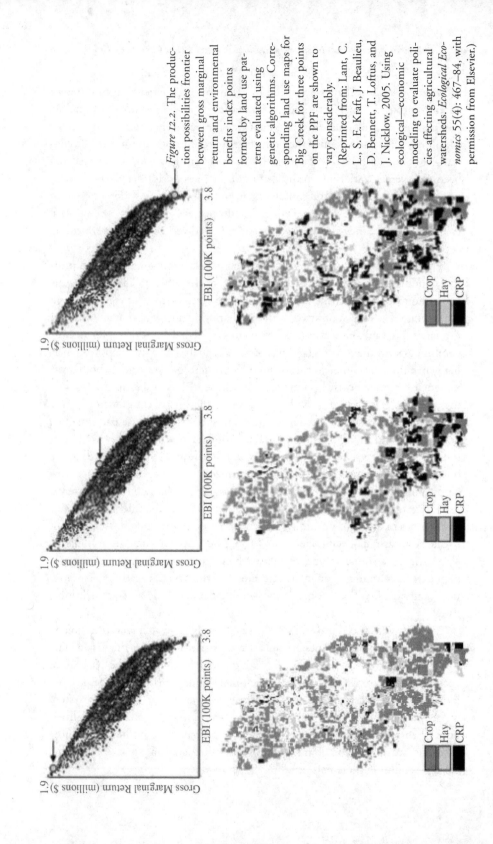

Figure 12.2. The production possibilities frontier between gross marginal return and environmental benefits index points formed by land use patterns evaluated using genetic algorithms. Corresponding land use maps for Big Creek for three points on the PPF are shown to vary considerably.
(Reprinted from: Lant, C. L., S. E. Kraft, J. Beaulieu, D. Bennett, T. Loftus, and J. Nicklow. 2005. Using ecological—economic modeling to evaluate policies affecting agricultural watersheds. *Ecological Economics* 55(4): 467–84, with permission from Elsevier.)

From Optimal Patterns to Individual Incentives

So what if the watershed-scale land use pattern is inside the PPF and therefore suboptimal? What does this mean to individual landowners or farm manager whose goals, be they financial or personal, most likely pertain only to their own land or farm? Where land is partly or wholly privately owned, as is predominant in the United States east of the Rocky Mountains and in many other countries, how is a land use pattern on the PPF to be put into effect? Moreover, different positions along the PPF reflect society's valuation of ecosystem services and farm income. Which position along the PPF is most desired? Even on large tracts of public land, finding society's most preferred land use pattern is a nearly impossible task. Unfortunately, individual land ownership leads to landscape fragmentation, not unlike suburban sprawl, where the benefits of landscape pattern, the economies of configuration, are not achieved. The simple answer is that optimal land use patterns cannot be imposed, they can only inform us of what is possible, that we can do better and by how much. Moreover, Gottfried et al. (1996) argue convincingly that even the best possible set of incentives cannot produce the "optimal" landscape; instead, each landowner would have to receive a specialized set of incentives, a proposition that can be rejected on political and practical grounds.

The problem then reverts back to the law and policy of ecosystem services. What incentives or decision-environments (sets of costs, prices, and policies) do landowners, farm managers, irrigators, and other users of natural capital face as they attempt to meet their own goals? What are those goals? Does the watershed itself have an institutional presence that affects landowners' goals or their decision-environment in pursuing their goals?

The Lant et al. (2005) analysis compared the real land use patterns in Big Creek in 2000 using remote sensing (41.4 percent hay grazing and meadow, 32.8 percent cropland, and 25.8 percent CRP enrollments) and the land use pattern that would have resulted under crop prices and policies that pertained in 2000 if all landowners were maximizing gross marginal revenue as determined by the GEOLP model (40.3 percent hay grazing and meadow, 37.2 percent cropland, and 22.4 percent CRP enrollments). This close correspondence validates the GEOLP model and tells us that the results from the T by 2000 and CRP scenarios are sound. It also tells us that farmers are managing their land with a great deal of economic rationality, and that the decision-environment of policies and economic opportunities and constraints therefore has such a strong influence on land use choices that we are tempted to mistakenly think that they completely determine them.

Lee (1992) in discussing "ecologically effective social organization" discusses information flow pathologies or ways in which ecological managers fail to respond constructively to the actual conditions of the ecosystems for which they are responsible. These include false analogy, insufficient detail, and a short observational series that does not capture the ecosystem's natural range of variability or disturbance cycles that drive the ecosystem. This can result in reacting out-of-phase with ongoing ecological changes. These pathologies also include institutional malfunctions, such as managerial detachment, ideological beliefs, or externalities and other market failures discussed in chapter 3. Large-scale, institutionalized land management has been shown to suffer particularly from information flow pathologies. Sociologist Walter Firey concluded as early as 1960 that future-oriented and group-oriented behavior toward the environment requires not only that individuals internalize these values but that they are reinforced both economically and in a sociopsychological sense: one's esteem within the group must be positively associated with sustainable behavior if people are to behave in a sustainable way. For example, planting trees and conserving soil are behaviors that depend on stable property rights so that the rewards of long-term investments of money or labor can be accrued. With stable property rights and positive social reinforcement, these behaviors can become habitual—part of the socially accepted, ethically supported, normal routine. In the United States, conservation tillage, recycling, and other environmentally beneficial behaviors have successfully crossed this threshold.

Put another way, there is a narrow set of social circumstances under which owners of natural capital will forgo current personal profit in order to improve long-term public assets by investing in natural capital or to shift current production to increase ecosystem services at the expense of ecosystem goods for which they currently receive market rewards. Finding that set of social circumstances is our challenge, and it is a difficult one.

13   Wetland Mitigation Banking: An Ecosystem Market without Ecosystem Services

As chapter 5 showed, conventional command-and-control regulation appears not to have made much headway toward accounting for natural capital and ecosystem service values, but what of the "reinvention" stage of environmental law that gained traction in the mid-1990s? The theme of regulatory reinvention has been to inject flexibility and efficiency into the environmental law system through three approaches: (1) government-stakeholder networks such as land conservation partnerships; (2) indirect governance mechanisms such as information disclosure requirements; and (3) market-based instruments such as pollutant trading programs (Hirsch 2001, 2004; Lazarus 2004; Ruhl 2004; Stewart 2001, 2003). Indeed, one of the original examples of regulatory reinvention was the federal wetlands protection program and its now decade-old practice of "wetland mitigation banking."

As explained in chapter 5, when a land development project involves filling of wetland areas regulated under the federal Clean Water Act (CWA) or similar state laws, one condition of the permit authorizing the activity is usually to require mitigation for the loss of wetland functions. Permit holders can accomplish this themselves directly through creation or enhancement of wetlands on the development site (on-site mitigation) or at an off-site location (off-site mitigation), or by paying a fee to fund wetland mitigation by a third party conservation entity in lieu of providing direct mitigation (in-lieu fee mitigation) (Environmental Law Institute 2002; Gardner 2005; Wilson and Thompson 2006). Wetland mitigation banking provides yet another means of satisfying mitigation requirements as a third-party variation on off-site mitigation (Environmental Law Institute 1993; Gardner 1993; Salzman and Ruhl 2000). This innovative market-based approach allows the developer to compensate for the resource loss by purchasing "credits" from another landowner—the wetland banker—who has created or enhanced wetland resources elsewhere. Indeed, wetlands mitigation banking today accounts for over thirty percent of all mitigated wetland acreage (Wilkinson and Thompson 2006).

The 1990 U.S. Army/EPA *Mitigation Guidance* explicitly endorsed mitigation banking as a form of compensatory mitigation and promised additional guidance on the subject. To fulfill that promise, in 1995 five federal agencies published a policy on mitigation banking, known as the *Mitigation Banking Guidance,* in order to detail the use and operation of mitigation banks (Army Corps of Engineers et al. 1995). The document's introduction declares that the "objective of a mitigation bank is to provide for the replacement of the chemical, physical, and biological functions of wetlands and other aquatic resources which are lost as a result of authorized impacts" (58606). This perspective is later broadened to acknowledge that "[t]he overall goal of a mitigation bank is to provide economically efficient and flexible mitigation opportunities, while fully compensating for wetland and other aquatic resource losses in a manner that contributes to the long-term ecological functioning of the watershed within which the bank is to be located" (58608). The *Mitigation Banking Guidance* thus qualifies the goal of replacing ecological functioning by acknowledging economic realities.

In the decade since this practice was put in use for purposes of satisfying mitigation requirements under the CWA, it has fueled an ongoing debate about its pros and cons (Sibbing 2005; Society of Wetland Scientists 2005). On the one hand, proponents of wetland mitigation banking claim it offers a number of significant advantages. Prior to the rise of wetland mitigation banking, the principal method for a land development project to satisfy regulatory wetland mitigation requirements was to compensate for resource losses through on-site creation, enhancement, or preservation of wetlands. The result of this practice, compounded over tens of thousands of land development projects, was an administrative nightmare for federal and state regulatory agencies administering wetland protection programs. Numerous retrospective studies have shown that individual project compensatory mitigation was usually poorly designed, inadequately implemented, and infrequently monitored (National Research Council 2001; Ruhl and Salzman 2006; Salzman and Ruhl 2000; Turner et al. 2001; United States Government Accountability Office 2005).

In a wetland mitigation banking program, by contrast, the banker is more easily subjected to permitting standards and close monitoring and has an economic incentive to produce and sustain the wetland values needed to generate credits to sell. Moreover, the product of wetland mitigation banking is large, contiguous wetland areas rather than a series of disconnected "postage stamp" mitigation sites. And along with these administrative and ecological features comes what appeals to land developers as well—a less time-consuming and more cost-efficient means of satisfying mitigation requirements. The Corps and EPA touted all of these attributes as benefits in their 1995 *Mitigation Banking Guidance* policy endorsing wetlands mitigation banking, and many wetland mitigation banking supporters continue to recite them (Mogenson 2006; Society of Wetland Scientists 2005).

On the other hand, while almost everyone acknowledges that wetland mitigation banking has practical advantages over individual on-site mitigation, wetland mitigation banking has attracted criticism on a number of grounds (Salzman and Ruhl 2000). One concern is that large, contiguous wetland tracts are not necessarily superior to smaller, separated tracts, as one large tract may be more prone to catastrophic degradation from invasive species, drought, and other natural disturbances (Semlitsch 2000). Also, given that wetland mitigation banks are in the business for profit, there is concern—and mounting evidence—that they will push permitting agencies for concessions that jeopardize the ecological performance of banks (Environmental Law Institute 1993). Wetland mitigation banking, in other words, is not universally regarded as an unmitigated ecological success story.

For the most part, however, this debate has focused on the relative advantages and disadvantages of banking programs in terms of administrative efficiency and ecological impact, with little attention being paid to the effects of wetland mitigation banking on *people*. As a convenient form of mitigation, wetland mitigation banking facilitates moving wetland resources from one location—the development project—to a potentially distant location—the bank site. It may well be that this provides, on balance, a net ecological advantage over on-site mitigation. Even assuming that is the case, however, it seems unlikely that the same human population will benefit from the ecosystem service values associated with the wetlands when wetlands mitigation banking is the mitigation method of choice. Simply put, if the wetlands move, their ecosystem services go with them (Brown and Lant 1999). Yet the debate over the *ecological* impacts of wetlands mitigation banking has thus far left out this potential *economic* impact as a relevant policy concern (Boyd and Wainger 2002b).

If environmental regulation broadly protects ecosystem services, one could reasonably expect evidence to that effect in the structure and performance of the wetlands banking program. In particular, a market-based program allowing what essentially amounts to trading of wetlands—exchanging acres destroyed in one location for acres created or improved elsewhere—ought to take into account the value and location of the services associated with the wetlands being traded. As with the general *Mitigation Guidance* discussed earlier, however, this has not been the case under the *Mitigation Banking Guidance*.

The *Mitigation Banking Guidance* describes the intricacies of creating a wetland mitigation bank but is vague on exactly what is being "banked." The document relies heavily on the term *function*. For example, the site selection criteria require agencies to give careful consideration to the ecological suitability of a site for achieving the goal and objectives of a bank—that is, it must possess the physical, chemical, and biological characteristics to support establishment of the desired aquatic resources and functions. Similarly, credit for wetland preservation is contingent upon the "functions" provided or augmented by the preserved land, and credit may be given for the inclusion of upland areas occurring within a bank only to the degree that such features increase the overall ecological functioning of the bank. Yet nowhere in the policy is "function" defined to include, or even refer to, ecosystem service values (Ruhl and Gregg 2001).

Federal and state wetland mitigation banking policies do employ some safeguards that might, whether intended or not, also sustain the delivery of ecosystem services to the particular human population situated around the wetlands being filled. The *Mitigation Banking Guidance* generally requires that the "swap" be for wetlands of similar kind and within a "service area" usually defined by relevant watershed boundaries. Some ecosystem services may thus be provided on the same basis to the human population within the service area regardless of where the development projects deplete the wetlands and the banks enhance them. But some of the ecosystem services flowing from wetlands are primarily local in terms of who benefits from them, or at least are more pronounced the closer to the wetland one is located. For example, research from Florida has shown that wetlands help regulate local moisture and temperature, which has proven to be of benefit to nearby agricultural lands (Marshall et al. 2003). Even small wetlands in urban areas have been shown to provide important pollutant control services to the local urban population (Keller 2005), and clusters of small isolated wetland areas provide important functions as a complex (Semlitsch 2000). Hence, moving wetland resources, even within a bank's defined service area, is likely to alter who benefits from the associated ecosystem services.

Indeed, there is good reason to believe that wetland mitigation banking, given its market incentive drivers, will systematically move wetland resources from urban areas to rural areas within a given bank's service area. Entrepreneurial bankers are in the business to make a profit, and are thus likely to seek the cheapest land that will produce the desired stream of credits for sale. Land developers are also in business to make a profit, and are likely to seek the cheapest land in the desired development market. It is highly unlikely, however, that bankers and developers will compete for land in the same market—bankers need large tracts capable of wetland restoration, which, if they do exist in a development market area, are likely to be too expensive for the banker to compete with the developers. One ought not be surprised, therefore, to find that development projects using wetlands mitigation banking to satisfy regulatory mitigation requirements are located in higher-priced urban markets and that banks are located in lower-priced rural markets. If so, wetland mitigation banking is likely to asymmetrically redistribute local ecosystem service values associated with wetlands between those two land markets.

Several limited empirical studies conducted early in the history of wetlands banking suggested that this concern was more than theoretical (Brown and Lant 1999; Jennings et al. 1999; King and Herbert 1997; Wainger et al. 2001). To test on a more comprehensive basis whether this effect is in fact experienced, particularly as the banking program has matured nationally, Ruhl and Salzman (2006) conducted an empirical study of the demographics of Florida's wetland mitigation banking program, one of the nation's largest.[1] The Florida program, which is an example of parallel federal and state authorities administering their respective wetlands protection authorities through coordinating implementation, has thirty banks actively selling credits, three that have sold all approved credits, and ten approved for operation but not yet selling credits. The permitted banks cover over 117,000 acres and have the potential, if they meet all permit conditions, to offer over 36,000 credits for sale within a combined service area that covers half the land mass of the state.

Taking the twenty-four banks for which adequate data were available, which represented over 95 percent of all credit sales completed through 2005, the study mapped each bank and its associated development projects for which reliable data were available and generated demographic data for the respective locations (Table 13.1).

The average distance from a bank to its associated project areas was considerable for many banks—over 10 miles for all but three of the twenty-four banks studied. Not surprisingly, the findings also confirm the hypothesized migration of wetland resources to less densely populated areas, which took place for nineteen of the twenty-four banks studied. For the banks exhibiting this urban-to-rural shift, the population density around the projects was on average over 900 people per square mile higher than for their associated banks.

The pattern for median income and minority population was less clear than for population density, but sharp differences prevailed. Project area median incomes were higher than bank area incomes for eleven banks, lower for eleven, and equal for two. Percentages of minority population were higher in project areas for fifteen banks, lower for seven, and within a percentage point for two. Nevertheless, although the directions were mixed, overall there were significant differences in median income and minority populations for project areas and banks. The average difference for median income was $11,750, and the average minority population difference was 13 percent. The majority of banks

TABLE 13.1. Demographic data on twenty-four Florida wetland mitigation banks. This table provides the following information for the twenty-four mitigation banks in Florida included in the study: (1) number of land development projects that have purchased credits from the bank; (2) total number of credits the bank has sold; (3) the population density of the local populations for the development projects and the bank; (4) the median income of the local populations for the development projects and the bank; (5) the percent minority of the local populations for the development projects and the bank; and (6) the average distance in miles from the bank to its development projects.

Bank	Permitted Projects	Credits Sold	Population Density (sq. mi.)		Median Income		Percent Minority		Average Distance to Projects (mi.)
			Projects	Bank	Projects	Bank	Projects	Bank	
Barberville	15	30	779	34	53,750	32,250	24	24	21
Big Cypress	20	126	553	4	50,500	31,250	17	70	35
Bluefield Ranch	24	85	748	66	35,000	29,000	17	40	17
Boran Ranch	44	74	413	35	31,250	37,500	18	10	28
CGW	14	40	425	1,975	42,000	35,250	20	29	4
East Central	46	144	2,349	39	43,500	37,750	31	12	16
Everglades	40	182	2,448	11	53,000	35,500	38	42	40
Farmton	136	404	789	486	48,250	53,750	21	11	20

Florida MB	93	588	1,024	1,246	41,750	64,250	37	39	9
Florida Wetlands	63	367	3,365	2,254	57,750	77,500	48	41	8
Lake Louisa	25	172	511	116	50,000	50,000	28	30	19
Lake Monroe	10	233	1,713	352	62,250	41,750	26	18	12
Little Pine	94	97	941	401	44,750	37,250	18	11	15
Loblolly	20	115	786	211	53,500	36,250	28	15	11
Loxahatchee	43	157	1,376	2,469	61,250	75,750	22	15	13
Mary A. Ranch	18	86	1,297	6	39,000	66,750	28	14	21
Northeast Florida	108	377	987	115	43,000	44,250	24	21	15
Panther Island	74	935	798	61	55,250	35,750	12	28	12
Reedy Creek	16	84	460	465	40,500	39,500	39	40	12
Split Oak	19	88	1,112	88	41,000	65,250	42	10	15
Sundew	13	67	348	31	32,500	36,500	24	2	18
TM-Econ	21	66	2,285	12	57,000	65,250	39	10	12
Tosohatchee	11	153	60	12	65,250	65,250	13	10	11
Tupelo	8	128	1,179	86	41,250	35,750	28	13	17
MEAN	41	200	1,114	441	47,635	47,052	27	23	17

Source: Ruhl and Salzman 2006.

exhibited higher incomes in whichever area had the lower minority population component.

When put together, the strong trend of shifting wetlands from urban to rural areas; the significant differences between bank areas and project areas for population density, median income, and percent minority; and the considerable distance between banks and their associated projects all point to the conclusion that completely different populations were winners and losers in terms of locally delivered wetland ecosystem service values. Hence, even assuming that wetland mitigation banking is administratively and ecologically superior to on-site mitigation, which may be generous assumptions, wetlands mitigation banking as implemented has unquestionably redistributed wetland ecosystem services from one set of human populations to another.

These findings raise more questions than they answer, simply because so little information is available about the economic effects of wetlands mitigation banking. It cannot be determined, for example, whether the effect of redistributing wetland ecosystem services is to increase or decrease overall social welfare. Ecosystem services are just one of the values associated with wetlands and land development, so we also cannot say whether any net loss of wetland ecosystem service values is offset by other considerations such as the economic impact of urban development facilitated by the wetlands banking program. Nor would either of those quantifications, if we could perform them, likely remain static. It is certainly possible, for example, that over time the population around wetland banks could grow, meaning that larger populations would enjoy their associated ecosystem services, and that the economic development in urban areas losing wetlands far outstrips the costs associated with the lost services. There is also the possibility that the services formerly provided in urban areas by wetlands, such as flood control, groundwater recharge, and sediment capture, are being replaced by services provided by technological structures such as storm water retention ponds and other measures required under state and local development regulations. One firm conclusion, however, is that wetlands mitigation banking does carry with it the significant potential for redistributing some wetland ecosystem services between human populations.

The wetland mitigation banking program thus has left the location of ecosystem services out of the calculus for evaluating bank credits and development project debits. In that sense, nobody can blame developers and bankers for not taking ecosystem service distribution into account, but neither can anyone reasonably claim that the "market" for credits produces the most efficient allocation of wetland resources. As long as federal and state wetlands regulation programs do not acknowledge the geographic distribution of ecosystem service values as a criterion for regulation and a factor in mitigation policy, the "market" for wetland mitigation credits will not do so either and one can expect only what has happened thus far—development projects in urban areas purchasing credits from banks located in distant rural areas.

The question, of course, is whether this should matter for wetland management policy. It is difficult to approach that question intelligently, however, given the data vacuum that exists about the scope and magnitude of the distributional effects. As Ruhl and Salzman (2006) report, wetlands mitigation banking procedures do not perform what would be necessary to test the policy implications of the phenomenon—that is, track the redistribution of wetlands, estimate the effects thereof on ecosystem service values, notify the affected public, and provide opportunity for public input. The "losers" in wetlands mitigation banking—the people in communities losing wetlands to the banking areas—do not even know that they are losing anything of economic value, much less what and by how much. And given that ecosystem services are economically valuable, one could reasonably expect the "losers" at least to be interested in knowing about their losses, so that they may make an informed decision about whether they care and whether any replacement of the services is adequate.

There is evidence that the Corps and EPA are cognizant of this concern. As mentioned in chapter 5, in March 2006 the Corps and EPA issued a proposed rule that would, if adopted, overhaul the wetland mitigation principles used under Section 404 (Department of Defense and Environmental Protection Agency 2006). The proposal does, for the first time in the program's history, expressly point to ecosystem services as relevant to decision making, but does so in a way that does not resolve the concerns Ruhl and Salzman 2006 identify. The proposal recognizes that compensatory mitigation might be sited away from the development project area, including in mitigation banks, but merely suggests that in such cases "consideration should also be given to functions, services, and values (e.g., water quality, flood control, shoreline protection) that will likely need to be addressed at or near the areas impacted by the permitted project" (15547). Hence, the proposal explicitly recognizes that populations around the development sites will lose wetland ecosystem services, yet nowhere in the extensive proposal does it elaborate on how the agencies will "address" those losses. This seems to advance the ball very little.

One can find it commendable that the Corps and EPA, like the USDA for national forests on public lands, have committed to "addressing" ecosystem services when regulating wetlands on private property. But the devil is in the details, and neither agency has said how it will manage the geographical redistribution of ecosystem services inherent in wetland mitigation banking.

14 Ecosystem Services and Pollution Trading I: A Sulfurous Success and a Nutritious Failure

Pollution is the antonym of *ecosystem services* with a chemical connotation; it implies that a substance poses either a risk to human health or a disruption of ecosystem function. Some manufactured substances, such as dioxins, PCBs, and most pesticides, are pollutants at any concentration. Regulations are, of course, the dominant form of environmental policy when dealing with such pollutants; for example, they have been instrumental in eliminating the use of DDT in the United States and in removing the lead from gasoline. Usually, however, whether a substance constitutes pollution is determined by its context and concentration. Ozone is a pollutant at ground level, a part of smog, and is associated with eye and lung irritation. In the stratosphere, however, it blocks ultraviolet light and thus provides one of our most essential ecosystem services, as we have all learned through the process of phasing out CFCs and other ozone-depleting substances through the Montreal Protocol, the most successful among this form of international agreements (Speth 2004). Nitrogen, usually in the form of nitrates, is essential to plant growth and aquatic ecosystems; but at high concentrations, nitrates cause eutrophication, a process wherein they stimulate plant growth, primarily of algae, which in turn dominates the aquatic ecosystem and results in oxygen depletion when they die and sink to the bottom of the water body—a very bad thing for fish and other aquatic fauna. In humans, drinking water high in nitrates is associated with methemoglobinemia, or "blue baby syndrome."

Similarly, carbon dioxide is also an essential part of the atmosphere. It fuels photosynthesis and helps maintain Earth's favorable temperature, and it was comparatively safe at its preindustrial level of 280 ppmv (part per million by volume). But as its current concentrations climb toward 400 ppmv, it has become the most important component of global warming. Sulfur is another naturally occurring and fairly common element essential to life, but it causes health problems and acid rain at high atmospheric concentrations. For pollutants that are harmful only in excess, and where complete elimination is extremely expensive or not called for on environmental grounds, economic incentives have substantial merit as a flexible form of environmental regulation.

Environmental economists have made a strong argument that emission fees and tradable pollution permits can be more cost-effective in attaining a pollution-control goal than can the "command and control" regulations that have dominated environmental law for decades. Although subtle, their arguments have become a core part of environmental economics. First, it is important to make a critical distinction between efficiency and cost-effectiveness. Efficiency in this context is associated with the paradoxical notion of "optimal pollution"—that is, where total benefits are maximized when pollution is controlled up to, and only up to, the point where marginal abatement costs equal marginal benefits of pollution reduction. Optimal pollution is therefore equivalent to potential Pareto efficiency and is, as described in chapter 3, dependent on the valid and accurate measurement of the economic value of pollution reductions—or ecosystem service improvements. The notion of optimal pollution has attracted many critiques on both philosophical and methodological grounds (see, for example, Sagoff 1988).

The case for emission fees and tradable pollution permits is not, however, that they are efficient in the optimal pollution sense. It is that they are cost-effective. That is, these policy mechanisms can achieve a specific, politically determined, pollution control or ecosystem service provision goal at less cost than some other forms of regulation. It is important to make this distinction because many of the arguments against a strict application of efficiency, as captured in potential Pareto optimality, optimal pollution, and cost–benefit analysis, do not generally apply when the criterion is cost-effectiveness. This is not to say that there are no other objections to using emission fees and tradable pollution permits, but that the philosophical arguments against optimal pollution do not transfer in any simple way to emission fees and marketable permits as economic incentives.

This two-part case study (chapters 14 and 15) explores tradable pollution permits from the perspective of the law and policy of ecosystem services. First, we review the basic argument of why they have the potential to be cost-effective. Second, we explore what is clearly the greatest success story to date in tradable pollution permitting—sulfur dioxide allowance trading initiated with the 1990 Clean Air Act Amendments. Third, we compare this with an equally disappointing failure of tradable pollution permits—nutrient trading between point and nonpoint sources of water pollution. Then, in chapter 15, we apply the lessons from these victories and defeats to the enormous challenge and intricate policy dilemma associated with global warming and the management of carbon, with special attention to carbon sequestration.

Why Marketable Pollution Permits and Emission Fees Are Cost-Effective in Theory

It is nearly always the case that firms in an industry or group of industries that emit a particular pollutant have different marginal abatement costs. In the case of sulfur in coal-fired electricity generation, for example, some firms have lower transport differentials between high-sulfur coal from the Ohio Valley and low-sulfur coal from Wyoming. Only a few coal-fired power plants have scrubbers (flue gas desulfurization), major investments made to comply with pre-1990 regulations. In the case of old plants, making a large investment in retrofitting scrubbers or fluidized bed combustion is not as sound an investment as it would be for a newer plant or, better yet, a plant still in the design phase. Because one plant might reduce sulfur emissions by, say, 50 percent at a greater cost than could another plant, both can gain if the high-abatement-cost plant pays the low-abatement-cost plant to reduce sulfur emissions by more than 50 percent while it abates by less than 50 percent. Within a group of plants, it has been shown that total abatement costs are minimized, not when each plant reduces emissions by the same amount, but when each has the same marginal abatement costs in dollars per ton of sulfur. This is the principle of equimarginality—abatement is cost-effective when all units abate up to the same level of marginal cost. Of course, the Coase theorem, as discussed in chapters 3 and 4, tells us that firms will gain by trading if the difference in marginal costs exceeds the transaction costs in trading pollution permits. The smaller the transaction costs, the closer we get to equimarginality in abatement costs among firms and the closer we get to cost-effectiveness in achieving a pollution-reduction goal.

Emission fees can also achieve equimarginality, and thus cost-effectiveness, in a slightly different way. If each firm has to pay a set fee for each ton of sulfur emitted, each will abate to a point where the marginal costs equal the fee—in this case, we don't even need the Coase theorem and trading. Moreover, both tradable permits and emission fees give firms an incentive to exceed regulatory standards and to develop technological or institutional means to control emissions at less cost, because by doing so they can sell permits or avoid buying permits or paying fees. However, absent knowledge of each firm's options and marginal-abatement-cost schedules, setting a fee that will result in a specific percentage reduction in emissions is a matter of trial and error; whereas in a cap-and-trade system, the number of pollution permits can deliberately be set by government (in this case, the U.S. EPA) at the level of reduction desired on environmental grounds (Tietenberg 2005). For tradable permits, however, uncertainty lies in the price of the permits and resulting abatement costs rather than in the pollution reduction achieved.

For these reasons, and because Americans have an ideological attachment to markets while they abhor taxes and fees, marketable pollution permits have been promoted by environmental economists. These policy mechanisms are also winning adherents within an environmental community that is increasingly realizing that environmental improvements must be made by changing private sector behavior within a capitalist framework and that we cannot regulate our way to sustainability. In practice, however, it gets tremendously more complex than even this subtle argument would suggest. But it is worth working through the complexities, because market-based policy mechanisms such as these are a critical and increasingly important component of the law and policy of ecosystem services.

A Sulfurous Success Story

Title IV of the 1990 Clean Air Act Amendments sets a cap on total sulfur dioxide emissions from coal-fired power plants at 8.95 million tons to be achieved by 2010 (a level roughly half of that which existed in 1980), distributes initial allowances at a rate of 2.5 pounds per million Btu (1.2 pounds after 2000 in Phase II), and allows firms to trade and bank these allowances. Thus allowances can be used to minimize abatement costs within a single coal-fired power plant over time (Ellerman et al. 2003) as well as among plants. As a political compromise to protect jobs in high-sulfur-coal-producing areas, 3.5 million bonus allowances were granted to utilities for installing scrubbers. The allowance trading zone is the forty-eight contiguous states. Phase I (1995–1999) applied to the dirtiest 261 electric-power-generating units, and Phase II (2000–2010) applies to most fossil-fuel units of 25 MW or greater.

Burtraw's (2000) analysis of the program for Resources for the Future, Inc., the most prominent natural resources and environmental economics think tank in the United States, describes it as "a noteworthy success from the standpoint of comparing benefits and costs" (3) and also an environmentally sound policy. There has been 100 percent compliance; in fact, affected facilities exceeded compliance in Phase I in order to bank 11 million tons of allowances for Phase II. Benefits of the program have exceeded costs by an order of magnitude, both because abatement costs have fallen from about $2 billion to about $1 billion and because subsequent research has pointed to the health benefits of reduced exposure to sulfates. Allowance prices fell from $132 per ton in 1995 to $68 in 1996, then rebounded to over $200 in 1999 on the eve of Phase II. These costs can be compared to EPA's 1990 prediction of 1997 marginal abatement costs of $235 per ton. In Phase I, sulfur emissions were reduced from 8.7 million tons to 4.4 million tons, with most of this reduction occurring in the first year that allowances went on sale (Arimura 2002). Similarly, emissions have been further reduced by 1.4 million tons in Phase II, 1.1 million of which occurred in the first year.

As shown in the case of wetland mitigation banking, the buying and selling of allowances necessarily changes the geographic distribution of emissions, and banking changes the temporal distribution. As a result, although Kentucky, Illinois, and Tennessee have increased their percentage of national sulfur dioxide emissions by buying allowances, this has not led to deterioration in air quality there or in the Northeast (where the effects of acid rain are greatest), because 75 percent of all abatement has occurred in the Midwest (Ellerman et al. 2003). Unlike acid rain, where emissions can affect ecosystem services far downwind, health benefits of sulfur emission reduction are more local. The changed geographic distribution of emissions has therefore had little effect on environmental benefits in this instance. The banking of credits has led to a more rapid reduction in emissions, but most banked allowances have now been used in Phase II (Ellerman et al. 2003).

The reductions in sulfur dioxide abatement costs attributable to the allowance trading system came not from major breakthroughs in scrubber technology, though there have been some improvements, but largely from utilities' agility in minimizing total abatement costs under the new, more flexible regulatory environment, largely through switching from high-sulfur coal to low-sulfur coal along with innovations in fuel blending. The costs of low-sulfur coal have declined as very large-scale production from surface mines in northeastern Wyoming and other areas has proceeded along with the railroad's need to maintain long-distance coal transportation as the backbone of its industry after the deregulation of the 1980s. The costs of mining low-sulfur eastern coal have also declined, though partly through the very damaging practice of mountaintop removal. This strategy has allowed electric utilities to avoid major capital investments in existing or new coal-fired generating plants at a time when there is great uncertainty about future regulatory regimes governing emissions of carbon dioxide—which coal-fired power plants produce in great abundance. Utilities have also maximized use of units already containing scrubbers in order to spread those substantial investments across more units of electricity production. There has been considerable loss of mining jobs in high-sulfur-coal states such as Illinois, Kentucky, and West Virginia. Arimura (2002) found that coal-fired power plants in those states have installed more scrubbers than would be economically optimal in an attempt to save local jobs.

Within utilities, responsibility for buying and selling allowances has shifted from engineers to financial officers responsible for fuel purchases. Initial transaction costs of 30 to 40 percent of the value of allowances have fallen to about 1 percent as participation in the program has been embraced and become routine (Burtraw 2000). Brokers and traders play an important role in facilitating trades, many of which involve swaps of vintage years rather than purchases of allowances (Ellerman et al. 2003).

Title IV of the Clean Air Act Amendments thus serves as the best model of successful real-world application of tradable pollution permits. It has allowed utilities to flexibly adapt to more stringent sulfur dioxide emission regulatory goals by comparing costs over time of installing scrubbers, fuel-switching, and purchasing, selling, or banking allowances. Transaction costs are low; the allowance trading market is large, both geographically and temporally, resulting in a fairly large number of potential traders. Ambitious environmental goals have been achieved, and costs of doing so have been nearly minimized. A ton of sulfur dioxide emissions is a highly fungible and measurable environmental good, and a regulation-induced market for this good has developed. Despite environmental impacts of low-sulfur-coal mining, loss of jobs in high-sulfur-coal mining, and delays in investing in modern coal-fired power plants, the program has been a sulfurous success.

A Nutritious Failure

Like the Clean Air Act, the Clean Water Act provides for pollution trading—not sulfur but the nutrients nitrogen (N) and phosphorus (P). N and P are leading sources of water pollution in the United States associated with eutrophication, including the hypoxic zone in the Gulf of Mexico. The cycling of P, and even more so of N, through ecosystems is complex and not the subject of this case study, but it is important to mention that both N and P are delivered to waterways from the runoff of fertilizers and that tertiary treatment is required to eliminate both from sewage. Atmospheric deposition is also a major source of N, which is soluble; detergents are an important secondary source of P, which is generally insoluble and tends to accumulate in sediments. Important waterborne fluxes of N and P occur naturally. Plants uptake N and P but also release it during decay or leaf fall. In anoxic environments, such as occur in waterlogged soils and wetlands, denitrifying bacteria transform nitrates (NO_3) into N_2 gas, which constitutes 78 percent of Earth's atmosphere, thus forming a benign storehouse of nitrogen that only becomes biologically available when it is fixed in the soluble forms of nitrates or ammonia. Vitousek et al. (1997)

estimate that deliberate human fixation of N, primarily in fertilizer manufacturing and fossil-fuel combustion, exceeds natural fixation, resulting in a surplus supply of N in many ecosystems.

The Clean Water Act distinguishes between "point source" pollution, which
is directly regulated under the National Pollution Discharge Elimination System (NPDES) system, and "nonpoint source" pollution, which is not, but is
subject to total maximum daily load requirements that are focused on ambient
water quality in watersheds rather than on discrete emitters. Nonpoint sources,
especially agricultural activities, increasingly constitute the majority of water
pollutants in the United States—76 percent of N and 56 percent of P reaching
waterways come from agriculture.

Forty-five of the fifty states have acquired authority over NPDES permits
within their boundaries. Of these forty-five states, eighteen have passed legislation allowing the formation of water pollution trading districts. It is widely
believed that nonpoint sources have lower marginal nutrient abatement costs
than do point sources. These circumstances have led to the notion that nutri-

TABLE 14.1. Water pollution trades occurring in the United States.

District Title	State	Year Adopted	Pollutant Traded	Actual Trades
Cherry Creek Basin Trading	CO	1985	P	3
Fox–Wolf Basin Trading Program	WI	1998	P	1
Lake Dillon Trading Program	CO	1984	P	2
Long Island Sound Trading Program	CT	1997	P	1
New York City Phosphorus Offset Program	NY	1997	P	1
Rahr Malting Permit	MN	1997	P	3
Red Cedar River Pilot Trading Program	WI	1994	P	1
Tar–Pamlico Nutrient Reduction Trading Program	NC	1990	P	1
TOTAL	8		P	13

Source: Adapted from King and Kuch 2003; Anebo 2005.

ent trading, especially with point sources as allowance buyers and nonpoint sources as allowance sellers, has the potential to achieve positive economic and environmental results such as those achieved with sulfur dioxide.

But that hasn't happened. While thirty-seven trading districts have been formed, only eight have conducted any trading, and the total number of trades in the United States is only thirteen—all involving P (Anebo 2005) (Table 14.1). Moreover, only one of these trades, occurring in Wisconsin, involves a nonpoint source. Trades that have occurred were approved by U.S. EPA on a case-by-case basis; no open-market trading has occurred. Why are nutrient trades, especially between point and nonpoint emitters, not occurring in a substantial and meaningful way?

King and Kuch (2003) provide a worthwhile analysis that helps us develop an answer to this question. First is an equity issue. Point source emitters, such as sewage treatment plants that are regulated under NPDES, see unfairness in a system where nonpoint emitters such as farms are also not regulated. They wince at the notion of paying farmers to reduce nutrient runoff when they are regulated to reduce nutrient emissions. In contrast, sulfur dioxide traders all fall within the same regulatory framework. On the other side of the coin, farmers do not want to take payments to reduce nutrient runoff because that would confirm that they are a source of pollution and set a dangerous political precedent threatening their unregulated status. Second, in many, if not most, instances, nonpoint sources deliver the vast majority of nutrient pollution. If this is the case, how can allowances purchased by point sources do more than scratch the surface of nonpoint runoff? Third, while point source emissions, like smokestack emissions of sulfur, can easily be measured on a common scale, nonpoint runoff of nutrients from a specific area of land is very difficult to measure with any accuracy and is dependent on a multitude of variables such as weather patterns, soil types, the location of drainage tiles, the juxtaposition of cropped fields, vegetative filter strips, surface water channels, and groundwater recharge zones. This means that nonpoint reductions must come in the form of surrogate land use changes that are inferred or estimated to cause nutrient runoff reductions, rather than direct abatement. Fungibility is compromised, a problem we revisit when considering carbon sequestration in chapter 15. Because of this uncertainty, trading ratios of 2:1, 3:1, or higher are introduced to make sure that a trade does not result in an increase in ambient nutrient concentrations. But with a ratio of 3:1, marginal costs of nonpoint reductions must be less than one-third as high as point source reductions to facilitate a win–win trade, even without considering transaction and information costs. While reducing fertilizer applications or planting streamside filter strips is likely less costly than tertiary sewage treatment on a per-pound-of-nutrient basis, is it three (or more) times less costly?

Fourth, land use changes that are used to create a nutrient reduction credit may have been undertaken anyway as part of a crop rotation, as an enrollment in the Conservation Reserve Program, because the farmland is no longer profitable for growing crops, as a transition to organic production or precision farming, because the farmer wants to retire, or for any number of other reasons. So there is a problem of "baseline" nutrient runoff to which must be applied the principle of "additionality"—what additional nutrient reductions can be attributed specifically to the land management changes associated with the allowance sold? Fifth, because EPA must approve each trade, because farmers and other nonpoint emitters do not normally participate in pollution trading, and because of information costs, transaction costs are extremely high. With this on top of trading ratios, the set of win–win trades approaches null.

Sixth, and finally, is the consideration of geography. While the location of sulfur emissions does matter at a regional scale, the specific location of nutrient runoff or emissions is absolutely critical. Trading nutrient pollution reductions in one watershed for increases in another is unsound because the ecological effects of the nutrients are specific to the location in which they occur. For this reason, the spatial extent of pollution trading must be defined by relatively small watersheds that generally do not contain a critical mass of potential traders. It has been estimated that nutrient trading could save $14 billion in the cost of reducing nitrogen emissions to the Gulf of Mexico from the Mississippi basin (King and Kuch 2003), but it is only at this very large scale that such savings could be realized. What does that mean for the quality of thousands of individual public water supply systems and thousands of unique streams, rivers, and lakes throughout that vast region? How does this affect the ecosystem service packages that inhabitants of specific locations within the Mississippi River basin enjoy?

For these reasons of regulatory inequity, compromised fungibility, and geographic specificity of pollution impacts, a nutritious failure has followed a sulfurous success. Interestingly, the location of greenhouse gas emissions, and even the location of carbon sequestration sites, makes no meaningful difference on the resulting concentration of greenhouse gases in the global atmosphere. With this note of encouragement, we turn in part II of this case study to consideration of carbon trading.

15 Ecosystem Services and Pollution Trading II: Carbon Trading to Ameliorate Global Warming

To a considerable extent, our chapter 14 discussion of trading with respect to atmospheric sulfur and waterborne nutrients is a prologue to a potentially much more important discussion about carbon. Global warming is, of course, a controversial subject; however, we do not here dig too deeply into these scientific issues. According to the Intergovernmental Panel on Climate Change (IPCC) (2007), representing an overwhelming majority of world scientific opinion, the burning of fossil fuels has increased the atmospheric concentration of carbon dioxide from a preindustrial level of 280 ppmv (parts per million by volume) to 379 ppmv in 2005, with a current rate of increase of about 2 ppmv per year. Methane concentrations (the second most important greenhouse gas) have also more than doubled from preindustrial levels of 715 ppbv (parts per billion by volume) to 1,774 ppbv in 2005. Nitrous oxides, ozone, CFCs, and other gases also make a contribution to global warming. These increases are due to the burning of fossil fuels as well as land use activities and have resulted in an increased average global temperature of 0.4 to 0.8°C (from about 59 to 60°F) over the 20th century. Temperature increases from the late 20th century to the last 21st centuries are expected to increase from 1.1 to 6.4°C, depending upon the climate model and the human response scenario used. These spatially and temporally variable increases in temperature have already had numerous effects on natural capital and ecosystem services in the form of sea level rise, changed precipitation patterns, glacial and sea ice melting, poleward migrations of species, and so forth that are described by the IPCC and a large volume of other scientific literature. High atmospheric carbon concentrations and many other global warming effects have considerable inertia, persisting for centuries after carbon emissions have been effectively abated. We take these conclusions of the IPCC and the global scientific community as our point of departure in discussing carbon trading as a key component of the law and policy of ecosystem services.

The Global Carbon Cycle

Carbon, along with hydrogen, oxygen, and nitrogen, is a foundational element of the biosphere and is contained in every living cell in every living organism on the planet. It is also found elsewhere, and its location in the biosphere is the critical factor in determining whether it is pollution or natural capital. Figure 15.1 diagrams Earth's carbon pools as rectangles and carbon fluxes as arrows, as they are currently understood by science. Carbonate rocks such as limestone store about 95 percent of the planet's carbon. Hydrocarbon fossil fuels—gas, oil, and especially coal—store most of the rest. Nevertheless, the oceans also store 39 trillion tons of carbon as dissolved carbon dioxide, soils store over 2 trillion tons, living organisms store 550 billion tons, and, critically, the atmosphere stores 750 billion tons. The quantity of carbon in the atmosphere in the form of carbon dioxide and methane is directly related to Earth's average temperature because these gases absorb long-wave heat radiation emitted by the earth and thereby raise the temperature of the atmosphere by reducing the amount of heat escaping into space. Given concerns about global warming, fluxes that deliver carbon to the atmosphere, especially in the form of methane (because it is 21 times as effective at absorbing heat as carbon dioxide), can be considered negatively. If the fluxes of carbon dioxide and methane are derived from human activities, they can be considered to be pollution in the sense discussed in chapter 14, even though neither is toxic and both are natural and necessary components of the atmosphere. Alternatively, reducing fluxes of carbon to the atmosphere, especially in the form of fossil fuel burning, can be considered positively as pollution prevention.

However, in examining Figure 15.1, we can also identify a number of fluxes of carbon from the atmosphere to other carbon pools. The oceans absorb 93 billion tons of atmospheric carbon per year and release 90 billion tons back into the atmosphere, with the difference absorbed by ocean biota such as shellfish, whose bodies ultimately form carbonate rocks. Over hundreds of millions of years, in fact, this process has removed many trillions of tons of carbon from the atmosphere to form the planet's primary carbon pool and to reduce the earth's greenhouse effect as the sun has increased its energy output. Photosynthesis not only provides free atmospheric oxygen as an ecosystem service of essentially infinite value; it also removes 102 billion tons of carbon per year from the atmosphere to form biomass. The land-based biomass carbon pool of 550 billion tons then represents only five to six years of photosynthetic activ-

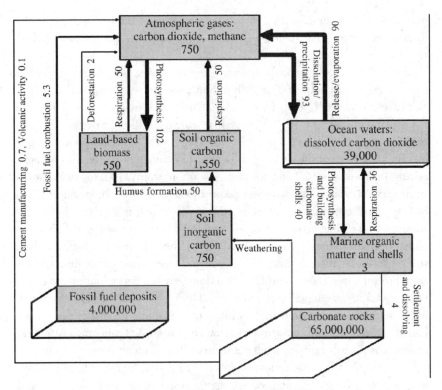

Figure 15.1. The global carbon cycle. Global carbon pools (millions of metric tons) and primary natural and human-induced fluxes (million metric tons per year). Based upon data drawn from Kyle 1993; Lal 2001; Lal et al. 1998; Mackenzie 1998, and Pickering and Owen 1994.

ity, and nearly one-seventh of all atmospheric carbon is removed by photosynthesis each year. Note, however, that respiration from organisms and soils and land use changes that reduce standing biomass, including fires, deliver this 102 billion tons back to the atmosphere in a gigantic global balancing act. Interestingly, these fluxes are profoundly unbalanced seasonally as photosynthesis greatly exceeds respiration in the Northern Hemisphere's spring and summer seasons while respiration is dominant in fall and winter. This seasonal biospheric inhaling and exhaling is clearly evident in the seasonal variations in atmospheric carbon dioxide concentrations alongside the longer-term increases. The decay of plants into soil organic matter is also an important flux of 50 billion tons per year. In examining these pools and fluxes of carbon, especially those that interface with the atmosphere, we learn that there are several opportunities for reducing the atmospheric pool of carbon in addition to reducing fossil fuel burning: increasing oceanic absorption, photosynthesis, or humus formation, or delivering the carbon content of fossil fuel burning to the lithosphere rather than to the atmosphere. These fluxes are termed "carbon sequestration."

From 1980 to 2000, global terrestrial ecosystems have likely been a minor net sink of carbon of about 200 to 700 million tons per year compared to global annual carbon dioxide emissions of 6,000 million tons (Cairns and Lasserre 2006). Houghton et al.'s (1999) article in *Science* quantifies the overall U.S. carbon sink and how it has changed over time (Figure 15.2). Land use change, primarily in the form of the clearing of forests and plowing of prairie soils for agriculture, released over 20 billion tons of carbon from 1700 to 1945, when cultivated soils lost an average of 25 percent of their stored carbon. Since 1945, however, land use changes have resulted in a substantial net carbon sink of 2.4 billion tons of carbon accumulating at 79 to 280 million tons per year. This reversal has primarily been due to the suppression of fire and reduced fuelwood harvest on forested lands, conservation tillage and the Conservation Reserve Program on agricultural lands, and farm abandonment and natural forest succession on marginal farmlands no longer cultivated. During the 1980s, this sink offset 10 to 30 percent of carbon dioxide emissions of 1.23 billion tons per year in the United States. By 1996, however, emissions had risen 220 million tons to 1.45 billion tons probably offsetting all net sequestration (Houghton et al. 1999).

The global carbon cycle illustrates well the relationships discussed in chapter 1 between natural capital, ecosystem processes, ecosystem functions, and ecosystem services. Carbon contained not only in fossil fuels but also in biomass or soils is natural capital. Carbon accumulations in the atmospheric pool, however, represent depreciation. The vast majority of carbon in carbonate rocks is relatively inert; what Erich Zimmerman, whose theories of the dynamism of natural resources were a breakthrough in the mid-twentieth century, termed "neutral stuff." Much of the cycle constitutes processes and functions that are

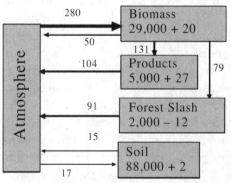

Figure 15.2. The U.S. terrestrial carbon pool and annual fluxes in 1990 (million metric tons). (Modified from Houghton et al. 1999.)

not ecosystem services, but components of the cycle that limit the atmospheric carbon pool are ecosystem services because they ameliorate global warming and its negative consequences—largely on other ecosystem services. Carbon trading would turn those components of the global carbon cycle that affect ecosystem services into a marketable commodity. This is a significant step in the relationship between humans and nature that recognizes the "human domination of Earth's ecosystems," as termed by Vitousek et al. (1997) in their article in *Science,* and the resulting need to manage overall human effects on planetary ecosystems. In fact, agreements such as the Kyoto Protocol create human institutions at the global scale to manage this global biogeochemical cycle, using the atmospheric concentration of carbon dioxide as the key indicator.

For the economic reasons discussed earlier, a cap-and-trade system for carbon emissions has the potential to achieve globally agreed-upon goals cost-effectively. However, we must keep in mind that it is the global atmospheric carbon pool at issue, and therefore a cost-effective system would either abate carbon emissions (with methane getting a 21:1 ratio compared to carbon dioxide) or sequester carbon anywhere on the planet up to the same marginal cost. Here we must keep in mind that perfect equimarginality is unobtainable, but a system that embraces an equimarginality approach, such as global carbon trading, has merit on theoretical grounds.

The Kyoto Protocol

The Kyoto Protocol on the United Nations Framework Convention on Climate Change was adopted in New York on May 9, 1992, with negotiations concluded in December 1997. The 2001 Marrakesh accords established rules for sequestration crediting through land use change and forestry. Kyoto came into force for all signatories in February 2005 when Russia ratified, thereby including 55 percent of all greenhouse gas emissions. As of this writing, the United States has not signed, and is not seriously considering signing, the Protocol. Article 3 of Kyoto is the heart of the complex, legalistic document. It reads, in part,

> The Parties included in Annex I (i.e., developed countries) shall, individually or jointly, ensure that their aggregate anthropogenic carbon dioxide equivalent emissions of the greenhouse gases listed in Annex A do not exceed their assigned amounts, calculated pursuant to their quantified emission limitation and reduction commitments inscribed in Annex I and in accordance with the provisions of this Article, with a view to reducing their overall emissions of such gases by at least 5 per cent below 1990 levels in the commitment period 2008 to 2012. . . . The net changes in greenhouse gas emissions by sources and removals by sinks resulting from direct human-induced land-use change and forestry activities, limited to afforestation, reforestation and deforesta-

tion since 1990, measured as verifiable changes in carbon stocks in each commitment period, shall be used to meet the commitments under this Article of each Party included in Annex I. The greenhouse gas emissions by sources and removals by sinks associated with those activities shall be reported in a transparent and verifiable manner. . . . The Conference of the Parties serving as the meeting of the Parties to this Protocol shall, at its first session or as soon as practicable thereafter, decide upon modalities, rules and guidelines as to how, and which, additional human-induced activities related to changes in greenhouse gas emissions by sources and removals by sinks in the agricultural soils and the land-use change and forestry categories shall be added to, or subtracted from, the assigned amounts for Parties included in Annex I, taking into account uncertainties, transparency in reporting, verifiability, the methodological work of the Intergovernmental Panel on Climate Change. . . . Any emission reduction units, or any part of an assigned amount, which a Party acquires from another Party in accordance with the provisions of Article 6 or of Article 17 shall be added to the assigned amount for the acquiring Party. . . . If the emissions of a Party included in Annex I in a commitment period are less than its assigned amount under this Article, this difference shall, on request of that Party, be added to the assigned amount for that Party for subsequent commitment periods.

In brief, the Kyoto Protocol calls for commitments to emission reductions in the 2008–2012 period from a 1990 base allowing for international carbon trading among Annex I countries that have ratified. It permits banking and limited carbon sequestration, but agricultural activities cannot be used in the 2008–2012 period. It empowers the IPCC to perform scientific functions such as establishing emission and sequestration estimates. It also includes a Clean Development Mechanism that provides credit for investing in emissions reductions or sequestration in developing countries. This has been used, for example, by European countries to pay China to build wind turbines in lieu of coal-fired power plants. The World Bank has also established a BioCarbon Fund to sponsor sequestration activities in developing countries that may also foster sustainable development (Antle and Young 2005).

The Prospects for Carbon Trading

Even if the United States does not ultimately sign the Kyoto Protocol, there are great possibilities for carbon trading, especially among the signatories. "It is, frankly, stunning the speed at which the international trading system [for carbon credits] has developed under the Kyoto Protocol" (Hayes 2005). However, Kyoto or some other national or international agreement that places regulations or costs on emissions of carbon to the atmosphere, or provides incentives for removal of carbon from the atmosphere, is essential in inducing trading. U.S. greenhouse gas emissions increased 14 percent in the 1990s. In a clear case of

Hardin's tragedy of the commons discussed in chapters 3 and 6, U.S. fossil fuel–based power plants, automobile drivers, and other fossil fuel users currently pay no penalties and have no restrictions on carbon emissions. The result is overuse of the atmosphere as a carbon sink, just as Hardin's pasture was overgrazed, even when the benefits of reductions in atmospheric carbon exceed the costs. Carbon emissions to the atmosphere are negative externalities; carbon sequestration is a positive externality. Why would a power plant, other carbon emitter, or even a national economy buy carbon dioxide emission permits when they can emit for free? Why would landowners invest in carbon sequestration, other than incidentally, if they are not paid for it?

This simple logic explains why large-scale carbon dioxide emission trading markets have not developed in the United States, although there has been some activity in anticipation of a future regulatory regime. Nine northeastern states have developed a coalition to reduce greenhouse gas emissions. PacificCorp has invested in forest preservation in Bolivia. Greenhouse Emissions Management Consortium has purchased soil carbon credits. Most interestingly, the Chicago Climate Exchange has facilitated numerous trades, with prices ranging from $3.19 to $6.92 per ton, in late 2004. These low prices, however, reflect the very low demand for carbon credits in the absence of legal restrictions on emissions. For comparison, the European Union market is selling carbon credits for $28 to $48 (Williams et al. 2005).

Beginning in 2005, the Annex I signatories to the Kyoto Protocol began trading carbon credits in a market that has grown rapidly as Europe, Japan, and Canada rush to overcome a 1.5 billion ton deficit in emission credits. The Clean Development Mechanism (CDM) has seen an "incredible avalanche" of applications for small to medium-size projects dealing with wind energy, small hydropower, control of methane from landfills, and reforestation, despite the tough bureaucratic environment for approval of CDM projects under U.N. auspices. Predictably, entrepreneurs are finding the "low-hanging fruit" of low-cost carbon reductions. The fast start to global carbon trading has generated a great deal of conversation in the U.S. Congress on what was a forbidden subject as late as 2005. U.S. companies are anxious, even as they become involved in carbon trading through subsidiaries or parent companies in Kyoto countries, and other U.S. companies are being advised by consultants to track carbon emissions (Hayes 2005).

The question then is, drawing from the experiences with sulfur and nutrients, if there were a regulatory, incentive-based, or penalty-based system put in place in the United States with respect to carbon dioxide emissions and/or sequestration, how well would emission trading work to meet environmental goals cost-effectively? In considering this question, it is fruitful to refer to the six issues raised earlier that help explain the failure of nutrient trading, though not necessarily in the same order. In doing this, we must also keep in mind some political realities. First, everyone who now emits carbon dioxide for free will fight very hard against having to pay for it; this is especially the case in considering whether and how permits would be allocated initially. Second, in a trading system every participant will play the rules of the game very hard and very smart, trying to minimize costs or maximize revenues associated with carbon even when these activities undermine the overall purpose of the trading system.

Geographic Considerations

Carbon dioxide is a uniformly dispersed pollutant; any emission of carbon to the atmosphere from any source on Earth will fairly rapidly mix with other atmospheric gases to marginally affect the global concentration rather than greatly affect the local concentration. This is also true of sequestration; on a per-ton basis, the regrowth of tropical forests in Brazil, the increase in organic content of soils in Ukraine, or the deep injection of carbon dioxide from a technologically advanced coal-fired power plant in the United States all have the same effect on the global concentration of carbon dioxide, and also have the same effect as emission reductions. Geography is not an issue—at least in this sense—opening up enormous possibilities for carbon trading on the grandest of spatial scales. As Thomas Friedman (2005) has told us with respect to global competition for jobs, the world is flat.

Ger Klaassen of Austria, Andries Nentjes of the Netherlands, and Mark Smith of the United States have collaborated on a set of experiments using college students and some modest but real incentives to gain a glimpse at how an international trading system in carbon might play out (Klaassen et al. 2005). Their experiments included the United States, the European Union (EU), Japan, Russia, Ukraine, and Central and Eastern Europe (CEE) in two trading systems employing bilateral sequential trading and a single bid auction with each region keeping secret its marginal abatement costs. Modeling established the theoretical least cost for meeting Kyoto Protocol goals for 2008–2012 as $7.65 billion with a marginal cost per ton of $38.70, a savings of 79 percent over Kyoto abatement costs of $36.05 billion where each region reduces an equal percentage of their 1990 emissions. With bilateral trading, the United States, the EU, and Japan made eighteen bilateral trades purchasing permits for 348 million tons of carbon from Russia, Ukraine, and CEE at an average price of $86 per ton. Total abatement costs were $8.78 billion; 96 percent of all possible savings were attained. With a single bid auction, bids fell from $62.50 in the first round to $37.50, slightly below the theoretical marginal costs, in the final round. Total abatement costs were $8.07 billion, nearly achieving the theoretical least-cost solution. Simple as they were, these experiments demonstrate the potential of international-scale trading in moving toward cost-effective solutions to atmospheric carbon reduction by taking advantage of differences in marginal abatement costs among nations and regions.

Emission and Sequestration Coverage

Sulfur trading has succeeded and nutrient trading has failed partly because the vast majority of sulfur dioxide emissions are covered under the Clean Air Act but only a minority (point sources) of nutrient emissions are directly and effectively regulated under the Clean Water Act. To be successful, trading must encompass most of the carbon fluxes into and out of the atmosphere that are affected by human activities. Primary fuels, mostly oil and gas, constitute about 65 percent of U.S. emissions, and electricity production, mostly coal, constitutes about 35 percent (Table 15.1). Carbon dioxide emissions are widely spread among the residential (20 percent), commercial (12 percent), industrial (34 percent), and transportation sectors (33 percent), illustrating how deeply

TABLE 15.1. Approximate percentages of U.S. carbon dioxide emissions from various sources.

Source	From Electricity	From Primary Fuel	Total
Residential			
Lighting, refrigeration, other appliances	8	0	8
Home heating	1	6	7
Water heating	1	2	3
Air conditioning	2	0	2
Commercial			
Commercial lighting	6	0	6
Cooling/ventilation	3	0	3
Commercial heating	1	2	3
Industrial			
Machine drive	8	0	8
Boiler fuels	0	7	7
Nonmanufacturing industrial	3	4	7
Process heat	0	6	6
Other manufacturing	2	4	6
Transportation			
Light-duty vehicles	0	20	20
Freight trucks	0	5	5
Other transportation	0	5	5
Air transport	0	3	3
TOTAL[a]	35	65	100

[a]Totals do not add to 100 due to rounding.

embedded carbon dioxide emissions are in the U.S. economy. Sequestration opportunities are also widespread. Lal et al. (1998) finds that U.S. agricultural lands could sequester 28 million tons of carbon per year through land conversion and restoration of degraded soils. Annual accumulation in agricultural soils could be equivalent to about 10 percent of Annex I carbon dioxide emissions. Canada could meet 10 percent of Kyoto Protocol requirements simply by not allowing postharvest forest slash to decompose or burn, presumably by burying it (Cairns and Lasserre 2006). Many other examples could illustrate that sequestration must be included if policies are to cover the great majority of relevant carbon fluxes. And let's not forget methane.

Equity among Emission and Sequestration Sources

Nutrient trading has failed partly because point sources face a stiff regulatory regime and nonpoint sources do not. In the United States, of course, there currently is equity in carbon emissions in that no one faces any regulations, but were this to change, how equally can all fluxes of carbon into and out of the atmosphere be treated? Emissions from coal and natural gas power plants can logically be subjected to a regulatory regime derived from the sulfur program—carbon dioxide in parallel with sulfur dioxide. But trading among tens of millions of gasoline users and homes with natural gas or oil furnaces can be dismissed as impractical. Holmes and Friedman (2000) suggest that allowances could be placed appurtenant to oil refineries, gas pipelines, and coal processing plants as the chokepoints in the fossil fuel production and distribution system, thus solving this problem. This would result in higher prices for gasoline, natural gas, and electricity acting as a carbon tax for end-use consumers. Extending a trading system on a per-ton-of-carbon basis in this way would place the vast majority of carbon dioxide emitters in the United States on an equal footing and appeals to the KISS principle (keep it simple, stupid). Nevertheless, two major issues remain: How are permits to be allocated initially? What about sequestration?

Holmes and Friedman (2000) and Cramton and Kerr (2002) identify the two obvious options: auctioning and allocating. The latter argue convincingly that auctioning all permits with none given away for free is the most equitable approach, especially if the permits can be banked for use in later years. It grants no rights to pollute simply on the basis that fossil fuel–based industries have always emitted carbon dioxide; it instead applies the polluter-pay principle throughout. It sets all sources of carbon dioxide emissions on an equal basis and in competition with one another per ton of carbon, driving the distribution of emissions among sources toward equimarginality and thus cost-effectiveness. Under this system, there would be about 1,700 U.S. traders, with 31 percent of current emissions coming from the oil industry, 25 percent from gas, and 44 percent from coal. An auction system would generate considerable revenue. U.S. emissions in 1990 of 1.411 billion tons would yield 1.246 billion tons of credits in 2008–2012 applying a 7 percent reduction as described in Kyoto. At $50 per ton, auctioned permits would net over $60 billion per year, about 5 percent of federal revenue that could be used to offset other taxes such as income taxes; this is called "tax shifting." Moreover, because this $60 billion would be a tax shift and not a tax increase, it would not slow down the economy as a whole while making it more carbon efficient (Cramton and Kerr 2002). Finally, it would avoid the political football of how to initially allocate permits. Most important from an equity standpoint, allocating or grandfathering permits would represent a windfall to energy companies in the form of rights to pollute at the expense of taxpayers. Here is where the question returns to the law and policy of ecosystem services. Do large-scale carbon dioxide emitters have the right to these emissions, which would be granted through the allocating or grandfathering of free permits, or does the nation as a whole have these rights, which it can choose to sell to the highest bidder through an auction with a capped number of permits? Cramton and Kerr (2002) conclude that "the arguments for auctions rather than grandfathering, on efficiency and distribution grounds, are overwhelming" (335). They recommend an ascending-clock auction where permits are initially offered at a low price, yielding a surplus of permit orders. The price is then raised until the number of permit orders equals the cap.

Unfortunately, it may be politically infeasible to use an auctioning system due to resistance from fossil fuel industries, fossil fuel–using industries, and folks like homeowners and drivers. Economists refer to the "energy paradox" in which households and even businesses do not make investments in energy conservation even though the annual return on those investments is as high as 20 percent. If that is the case, it would be time to play football. Remember that the Clean Air Act initially allocated sulfur permits on the basis of energy production, not past sulfur emissions. Many other allocation solutions reside in the entertaining political labyrinth of lobbying, horse trading, campaign finance contributions, and so forth, but none of them is as equitable or as efficient as a tax-shifting auction.

If it costs less to sequester a ton of carbon than to reduce emissions by a ton, and initial estimates show that this may well be the case, then sequestration must also be included in the trading system if the result is to be cost-effective. But sequestration gets tricky. Photosynthesis and dissolving carbon dioxide in the oceans occur without human action and are in rough annual balance as discussed earlier. Two broad examples of sequestration do require further analysis. First, technologies are now being developed that could result in a coal-fired power plant that emits no carbon dioxide to the atmosphere. FutureGen, a billion-dollar project proposed by the Bush administration, would develop a process where coal would be gasified into carbon dioxide and hydrogen, these gases would be separated, with the hydrogen being burned to produce electricity while releasing only water to the atmosphere. The carbon dioxide gas would be sequestered deep into the lithosphere in suitable geologic locations such as saline aquifers. Ongoing research is designed to determine how permanently such carbon will stay in the lithospheric pool and not be released to the atmospheric pool (Hepple and Benson 2002). If such research determines that it will stay sequestered, FutureGen would essentially redefine coal as a source of hydrogen, and power plants using the technology would not need to purchase permits since carbon dioxide would not be released to the atmosphere.

Second, carbon sequestration on forest and agricultural land has considerable potential but is fraught with complexities. Since sequestration is not being considered as a regulatory requirement, at this point we should consider only the notion of fossil fuel–based carbon dioxide emitters buying carbon sequestration credits from landowners or their representatives. For example, a coal-processing plant finds that reducing its carbon dioxide emissions has a marginal cost of $100/ton—all they can really do is sell less coal—whereas it would cost landholders in Brazil $30/ton of carbon sequestered to reforest cattle grazing lands cleared in the late 20th century. They make a deal with the coal plant, which pays some negotiated amount between $30 and $100 per ton to the Brazilian landowner to plant trees and sell off his cattle while the coal plant continues to sell coal for electricity production for the U.S. grid. The amount of atmosphere-derived carbon in the trees that grow on the Brazilian land is equivalent to the amount of carbon that would have been abated by the coal plant, and so the atmospheric concentration of carbon dioxide is the same with or without the deal. Is this equitable? One answer is, not only is it equitable but it also reverses the incentives that have led to tropical deforestation in the Amazon, with all of its consequent losses in biodiversity and other ecosystem services, provided the deal excludes clearing the land again and then claiming new credits to reforest. Here, however, is where carbon sequestration is different from reductions in emissions: sequestration is seldom permanent. For example, if the restored Brazilian rain forest burns, or subsequent generations clear it for a metropolis, the carbon returns to the atmosphere. For this reason, a better way of conceptualizing the ecosystem service provided is "supplemental carbon storage" rather than "carbon sequestration." Carbon stored even temporarily still has value, however, just as a house can be leased rather than purchased. Moreover, the terrestrial storehouse of carbon is a dynamic, shifting, natural capital asset that could potentially form a flexible approach to keeping large quantities of carbon out of the atmosphere.

The key points here are that, with due consideration of its permanence and other issues, sequestration should be part of a carbon trading system, both because this helps manage global atmospheric carbon dioxide concentrations cost-effectively and because sequestration of carbon on landscapes has the potential to provide ancillary ecosystem service benefits. Deforestation, desertification, soil erosion, and other forms of land degradation all involve a decarbonization of the landscape (Batjes and Sombroek 1997; Tiessen et al. 1998). Habitat loss, by transforming land to uses that contain lower carbon stocks, is the leading cause of biodiversity loss. Moreover, the purchase by fossil fuel emitters in developed countries of carbon sequestration credits in developing countries moves capital to poor countries as a form of trade-as-aid. Sequestration, therefore, has the potential to result in substantial investments in natural capital beyond its role in ameliorating global warming. Consider the prospect of oil refineries, gas pipelines, and coal processing plants earning carbon emission credits by making substantial investments in poor developing countries in the form of land restoration in heavily eroded countries such as Haiti, Ethiopia, or the Malagasy Republic, in tropical reforestation in Costa Rica, Congo, or Indonesia, or in rehabilitation of desertified lands in Uzbekistan, Chad, Niger, or Iraq. Conceptually, this has great appeal from a sustainable development perspective and as an integrated plan for investing in many forms of valuable natural capital around the globe, but, of course, many issues need to be resolved to make this possibility an actuality.

Fungibility

A ton of carbon—seems simple enough, and fungible. When applied to fossil fuel emissions at the power plant, refinery, or gas pipeline, with straightforward and well-known equivalencies for the carbon content per unit of energy produced from coal, oil, and gas, it is simple and fungible. With the appropriate formula for considering the greenhouse forcing per ton (21:1), methane, the most understudied component of global warming, could also be included. When applied to carbon sequestration, unfortunately, problems can arise similar to those for nonpoint source nutrient runoff reduction. It is difficult to measure the amount of carbon that has been sequestered through photosynthesis and resulting increases in biomass or soil carbon, and there are a large number of land uses that affect the terrestrial carbon pool (Table 15.2). Many of these activities result in 0.1 to 0.3 tons of carbon sequestered per hectare per year (Subak 2000). Simulations using the century model have shown that soil

Table 15.2. Sources and sinks of carbon from agricultural soils.

	Sources	Sinks
Transformations	Wetland drainage for cropland	Cropland set aside to grassland or woodland
	Grassland plowing for cropland	Wetland restoration on cropland
	Natural ecosystems for cropland	
Production	Lower residue yield	Higher residue yields
	Change to crop types with lower biomass	Change to crops with higher biomass
	Lower lignin content crops	Higher lignin content crops
	Longer fallow	Shorter fallow
Soil Conservation	Intensive till	No-till or minimum till
	Residue sales (e.g., straw)	Residue incorporation into soils
	Stubble burning	Cover crops
		Control of soil water
Other	Liming	Animal manure or sewage sludge storage

Source: Adapted from Subak 2000.

carbon sequestration projects require twenty to thirty years to reach their full potential (Antle and Young 2005). As discussed earlier, these pools are also in constant flux, raising issues regarding their permanence. Feng (2005) and Subak (2000) suggest that the fungible unit be not a ton of carbon but a ton-year of carbon storage, a system that would be similar to the vintage year of sulfur emissions and could potentially include banking. Because the biomass and soil carbon pools can be increased only to a certain ecological carrying capacity, after which they would either remain constant or decline, sequestration to expand carbon storage is an option that could be effective for only a few to several decades. Nevertheless, a few to several decades is also the amount of time it would take to develop an energy system free of carbon emissions using conservation, nuclear, wind, solar, hydrogen-based, and other technologies that are the ultimate solutions to the global warming problem. Even with current fossil fuel–based energy technologies, reductions in emissions are tied into the existing stock of manufactured capital—buildings, factories, automobiles on the road, even the spatial configuration of towns and cities and their sprawling suburbs, not to mention working power plants. Reducing emissions is therefore tied to the decades-long process of replacing this manufactured capital stock with one that releases less carbon to the atmosphere. Sequestration therefore buys time so long as progress is actually being made toward these long-term goals.

Transaction and Information Costs

Costs of carbon sequestration include land conversion, land treatment and maintenance, verification, opportunity costs of using the land in other ways, and option costs of forfeiting the ability to change land uses as new opportunities arise (King 2004). In addition to these, Susan Subak (2000) raises many issues regarding information costs in considering carbon sequestration in agricultural soils. She finds that a U.S. monitoring program that could establish the changes in soil carbon before and after the 2008–2012 Kyoto commitment period would cost about $1 billion compared to $1.5 billion in carbon offsets at $50/ton applied to 30 million tons of carbon sequestered. On this basis she suggests that carbon credits be applied to activities or Conservation Reserve Program–type "Conservation Practices," such as those listed in Table 15.2, rather than to direct measurements of carbon, and that losses of carbon as well as gains be included in the system. This approach, like nonpoint nutrient runoff, relies on well-known relationships between farming activities and their effects on soil carbon as well as an intensive monitoring program of those activities. Another approach would be to rely on less expensive remote sensing–based approaches rather than field-based approaches to measure soil organic carbon. With recent improvements in the spatial resolution (size of pixels), and especially the spectral resolution (number of wavelength bands sensed) of remote sensing satellites, such techniques may be able to derive fairly accurate measurements of soil organic carbon over broad geographic areas at much lower costs than those estimated by Subak (Jaber 2006).

However measured, the issue remains whether farmers should contract for expected carbon sequestration associated with a practice that is to be maintained over a period of years in a manner similar to the Conservation Reserve Program, or the carbon sequestration actually achieved over a period of several years based on before-and-after measurements or estimates. In either case, how does a farmer negotiate a carbon credit trade with a power plant, and what are the transaction costs? Again we see a parallel to the failure of nutrient trading between sewage treatment plants and farmers. Perhaps a better solution with lower transaction costs would be for the U.S. Department of Agriculture (USDA) to pay carbon credits to farmers alongside other subsidy programs such as the Conservation Reserve Program and have fossil fuel industries then buy credits from USDA based on their estimates of total carbon sequestered. The U.S. Forest Service could also act as a contract consolidator in this manner, trading with fossil fuel industries to lower transaction costs.

Additionality and Attributability

Because nearly every land use decision has an incidental effect on the accumulation or loss of the carbon content of biomass and soils, sequestration credits are potentially subject to game playing (i.e., finding loopholes and holes in the fence) that undermines the purpose of the system. This is especially true if activities serve as substitutes for measurements. Regulators would have to "score" sequestration trades using long and complicated formulas, rules, and regulations—a difficult business that drives up transaction costs. Credibility of such trades requires that strictness and possibly ratios be applied, yet these have greatly contributed to the failure of nutrient trading. Success of the market, in terms of an abundance of trades that actually reduce abatement costs, requires leniency and flexibility but can undermine credibility. Unlike in markets for commodities, buyers and sellers of carbon sequestration credits have a common interest in making unsubstantiated claims about the amount and permanency of sequestered carbon that only smart and persistent regulation of the system can contain.

Dennis King of the University of Maryland Center for Environmental Science has developed a "universal carbon sequestration credit scoring equation" that takes into account not only the amount of carbon removed from the atmospheric pool but also time and risk. Risks include performance risk (expected sequestration may not occur), durability risk (sequestered carbon may later be released to the atmosphere), baseline risk (credited sequestration may have occurred anyway), and displacement risk (reforestation associated with a sequestration project, for example, may cause deforestation elsewhere). Finally, future sequestration associated with, for example, a reforestation project that maximizes the rate of biomass accumulation after forty years must be discounted relative to present emission reductions. Establishing carbon storage credits that are year-specific is helpful in resolving this issue as well as in facilitating banking. Consider applying these forms of risk to a large carbon trade where an energy company paid about $11 million to reforest about 100,000 acres of publicly owned land at a proclaimed cost of about $1 per ton. King's formula applied to a similar hypothetical scenario raises the cost per ton sequestered from $1.74 per ton to $56.67 per ton, demonstrating the importance of how sequestration scoring rules take into consideration additionality, attributability, risk, and time.

Lessons Learned

So, where do we take the possibilities for this largest of markets in ecosystem services? Clearly, the answer is that this opportunity must be vigorously pursued because an active, well-designed ecosystem service market is a quantum leap forward from the tragedy of ecosystem services that obtained worldwide prior to 2005 and still obtains in the United States. It is a set of circumstances that closely resembles a classic tragedy of the commons, but one that can be overcome by requiring that users of the commons purchase the right to use it within an agreed-upon greenhouse gas carrying capacity constraint through a cap-and-trade system. Nevertheless, the primary ingredient needed to properly set the table of incentives remains lacking—a regulatory requirement in the United States to reduce net greenhouse gas emissions. With no cap, there's almost no real trading because there's no demand for credits.

Including carbon sequestration in a trading system is a more difficult issue for all of the reasons discussed above. Does this mean that it should not be pursued? We think, rather, that it should be pursued for two reasons. First, sequestration presents opportunities to reduce greenhouse gas concentrations at lower marginal costs than are available for emission reductions. Second, it provides great opportunities to improve a variety of other ecosystem services that can potentially come along with a recarbonized landscape. Yet, actualizing these possibilities, while foreclosing risky and inappropriate but inexpensive approaches to carbon sequestration, is indeed a tremendous challenge, one of the biggest in environmental policy design. Perhaps it is for this reason, rather than despite it, that carbon sequestration trading policies should be designed in concert with carbon emission trading policies to form a centerpiece of the law and policy of ecosystem services.

Part IV Designing New Law and Policy for Ecosystem Services

Based on the foundational chapters in parts I and II and the lessons learned from the case studies in part III, part IV forges an approach for the *design* of new law and policy for ecosystem services, working from the current baseline and taking into account the inherent limitations their ecological, geographic, and economic contexts present. The progression of the topics follows the choices that law and policy will have to make to put such an approach into action. First, it is essential to identify the important drivers of the existing status of natural capital and ecosystem services and to develop models of how they can be moved and the likely consequences of doing so (chapter 16). Policy choices then must confront the reality that taking more account of natural capital and ecosystem service values in natural resource decision making will not necessarily be a "win–win" for all stakeholders. Trade-offs are inevitable, and some people will be "winners" and others "losers" in the transition (chapter 17). Once policy is set, the appropriate instruments and institutions must be identified for policy implementation (chapter 18). In this sense, ecosystem services are likely to encounter the same tensions that environmental law in general has experienced as federal, state, and local governments, the courts, and interest groups jockey for position and authority. Only when all these choices are made in a cohesive, cogent institutional framework will the law and policy of ecosystem services have "arrived" and begin to fuse ecosystem services with resource commodities, manufactured products, and human-supplied services into a fully integrated decision-making framework for natural resources, one in which everything that matters is counted.

16   Drivers and Models

In the year since Hurricane Katrina focused the nation's mind on the importance of coastal wetlands for protection against storm surges, the secretary of agriculture announced that the agency will "broaden the use of markets for ecosystem services" in its administration of the national forest system, the Army Corps of Engineers and Environmental Protection Agency proposed that decisions about mitigation for development of wetlands will consider the "services and values . . . that will likely need to be addressed at or near areas impacted by the permitted project," a court in Rhode Island held that a landowner's plan to build a subdivision in a marsh would constitute a public nuisance because it would interfere with how the natural area "actually filters and cleans runoff," and the Environmental Protection Agency announced an agency-wide strategy to enhance its ability to identify, quantify, and value ecosystem services (U.S. Environmental Protection Agency 2006). These are not random, unconnected events. Although they stand against the great weight of evidence that law and policy have, heretofore, been largely ignorant of the value of natural capital and ecosystem services, they stand just as surely as evidence that a turning point is on the horizon. Indeed, the disciplines of ecology, geography, and economics leave law and policy little choice but to eventually incorporate natural capital and ecosystem service values into the substantive decision making of natural resources management. The question is how to do so.

Using the proposed wetland mitigation rule discussed in chapters 5 and 13 as its primary reference point, part IV examines the design issues law and policy must confront if natural capital and ecosystem services are to become part of the natural resources decision-making calculus. The proposal truly sets the stage for the question. On the one hand, the proposal defines ecosystem services and their values and advises that "compensatory mitigation should be located . . . where it is most likely to successfully replace lost functions, services, and values" (Department of Defense and Environmental Protection Agency 2006, 15536). Beyond that, on the other hand, ecosystem services are at best an ephemeral component of the proposed rule. Mentioned several times in the preambulatory justification of the proposed regulatory text, "services" and "values" fail to appear in the proposed *rules* tethered to any substantive or procedural standards. Provisions requiring permit applicants to prepare "baseline information" about the impact and mitigation sites omit all mention of services and values, as do provisions relating to amount of mitigation, public review and comment, site selection, ecological performance standards, monitoring, and adaptive management. The Corps and EPA have, commendably, recognized the problem Ruhl and Salzman (2006) identified—that mitigation as administered historically has redistributed wetlands, and thus the services associated with them, far from the development areas—but left a blank slate for how to solve it.

The three chapters in part IV suggest a three-stage process for designing such solutions. This chapter addresses the critical first stage of developing a firm understanding of the driving forces behind the status of natural capital and ecosystem services in discrete policy settings and a model of how different policy options could affect the delivery of their values. No policy decisions are inherent in this model-building stage; rather, it is designed to inform intelligent policy choices. Chapter 17 then deals with the thorny problems of policy choice—the trade-offs inherent in any particular option and the impacts to different stakeholders in the transition from the prior policy regime to the new policy regime. Once these policy choices are hashed out—a process that can often be as much about money and politics as about science and common sense—chapter 18 turns to the matters of instrument choice and institutional design. What legal mechanisms will deliver the chosen policy, and what institutions will administer them?

These are the hard decisions the Corps and EPA left out of the proposed rule, but that they will have to make eventually if they hope to deliver on the promise of ensuring that compensatory mitigation policy accounts for the ecosystem service values lost in connection with the development of a wetland area. Indeed, these are the questions the law and policy of ecosystem services must answer across the board.

Drivers

Drivers are the "natural or human-induced factors that directly or indirectly cause a change in an ecosystem" (Millennium Ecosystem Assessment 2005, 64). A driver can affect ecosystems either directly, as in land development, or indirectly by influencing the operation of direct drivers. Because drivers operate at different spatial and temporal scales, they are difficult to assess and more difficult to manage. The Millennium Ecosystem Assessment (2005), for example, identifies five indirect drivers operating at the global scale: (1) demographic drivers such as population growth, (2) economic drivers such as rising per capita income, (3) sociopolitical drivers such as levels of education and democracy, (4) cultural and religious drivers such as beliefs about the environment, and (5) science and technology drivers such as advances in food production capacity (64–67). In addition to global climate change, which has ubiquitous direct effects on ecosystems globally, the most influential direct drivers at the global level are identified based on ecosystem type:

- Terrestrial ecosystems: land cover change and overexploitation
- Marine ecosystems: fishing
- Freshwater ecosystems: modification of water regimes, invasive species, and pollution, particularly from nutrient loading (67–70)

Assessments of drivers at this global scale are, without question, important for the formulation of policy. Nevertheless, agencies such as the Corps of Engineers are neither formulating nor implementing *global* policy. They must be attuned to national policy objectives, such as the current "no net loss" policy goal for wetlands, but their work takes place primarily in discrete local settings in which the agency issues permits, imposes mitigation conditions, and enforces regulations. The critical first step in policy selection, therefore, is translating the general description of direct and indirect drivers employed in the Millennium Ecosystem Assessment into context. In their model of Brazilian Amazonia deforestation, for example, Portela and Rademacher (2001) open with an examination of regional drivers, finding that clearing for new ranches has been driven by speculative investment forces, whereas clearing for new farms has been driven by "the shifting nature of cultivation and political and economic conditions that drive population influx into the Brazilian Amazon" (118). Only with this understanding of drivers operating at the appropriate scale can the policymaker begin to explore and test different approaches for influencing the management of natural capital and ecosystem services.

As discussed in chapter 5, in the case of wetland mitigation policy, many of the human system drivers are well established (Nagle and Ruhl 2006; Ruhl and Salzman 2006; Salzman and Ruhl 2000). The background indirect drivers present a mixed bag:

- Demographic: population trends in the United States evidence increasing population growth in coastal areas where wetland resources exist. Over 50 percent of the American population now resides in coastal counties.
- Economic: developable coastal property has skyrocketed in value as a result of the population growth, making every square foot dear to land developers

hoping to extract a return on residential, commercial, and vacation land development projects.

- Sociopolitical: a complex web of local, state, and federal land use and environmental regulations govern land development in coastal areas, allowing a multitude of interests to jockey for or against additional development and the type of development.
- Cultural and religious: wetlands are no longer considered wastelands useful only if drained, but it is less certain how local populations value wetlands versus alternative land uses.
- Science and technology: improved engineering and construction methods may allow replacement of significant wetland service components through compact, affordable technological alternatives.

Within this background context, the intense demand for development in wetlands and for the opportunity to accomplish mitigation off-site has been fueled by direct drivers associated with terrestrial ecosystems, principally urban growth and road construction. The Corps' initial policy favoring on-site mitigation proved impractical in both settings. Land developers in expensive urban land markets viewed on-site mitigation projects as significant opportunity costs, and the narrow linear footprint of state and local road projects made on-site mitigation an expensive addition to right-of-way acquisition. For its part, the Corps also found administering a vast array of "postage stamp" mitigation sites a daunting drain on resources. Studies documenting the failure of compensatory mitigation under these conditions (National Research Council 2001; United States Government Accountability Office 2005) were a surprise only in the degree to which they revealed the program fell short of the goal of full compensation of wetland functions. Pressure to facilitate mitigation banking thus was strong from private and public sectors and seemingly well justified as a policy matter.

Yet mitigation banking has its own set of drivers that, as Ruhl and Salzman (2006) reveal, have systematically redistributed ecosystem services associated with urban wetlands to distant rural areas. If banks can locate in rural, low-priced land markets and sell credits in urban, high-priced land markets, they will. And if land developers can fill wetlands in high-priced urban land markets and satisfy regulatory mitigation requirements by purchasing credits from banks located in rural land markets, they will. The only startling revelation of Ruhl and Salzman's study is not that this does happen but that it happens so profoundly, with the vast majority of banks located on average more than 10 miles from development projects and in areas with one-tenth the population densities.

This much, at least, the Corps and EPA appear to have recognized. The proposed mitigation banking rule does, for the first time in the program's history, expressly point to ecosystem services as relevant to decision making and recognizes that when compensatory mitigation is sited away from the development project area, including in mitigation banks, "consideration should also be given to functions, services, and values (e.g., water quality, flood control, shoreline protection) that will likely need to be addressed at or near the areas impacted by the permitted project" (Department of Defense and Environmental Protection Agency 2006, 15547). Yet knowing that a phenomenon occurs, and even what drivers are behind it, does not establish the understanding of causal relationships necessary for deciding how the phenomenon can be effectively "addressed."

Models

Resource management policy will be most effective when the resource stakeholders "share an image of how the resource system operates and how their actions affect each other and the resource" (Ostrom et al. 1999, 281). A vast literature exists about developing models of how resource systems operate and how human systems operate, but there is a much smaller set of integrated models combining descriptions of resource systems with human systems (Millennium Ecosystem Assessment 2003, 162–165). In other words, we need integrated models of how the drivers operate and interact, not just an appreciation that they are in operation, in order to evaluate how to most effectively influence them (Villa et al. 2002).

As with drivers, consequently, so too must models be conceived on multiple spatial and temporal scales. Consistent with its global focus, for example, the Millennium Ecosystem Assessment (2005, 71–73) describes what it refers to as plausible global scenarios of the future, within which the identified direct and indirect global drivers are shaped, or reshaped, according to different assumptions about economic, political, and social structures (see also Butler and Oluoch-Kosura 2006). In far more detailed terms, Boumans et al. (2002) used multiple, intricate, nested STELLA-based submodels to build a global unified metamodel of the biosphere (GUMBO) representing "the dynamic feedbacks among human technology, economic production and welfare, and ecosystem goods and services within the dynamic earth system" (529). Similarly, Portela and Rademacher (2001) used their study of Brazilian Amazonia deforestation drivers to construct a detailed regional model integrating land use, ecosystem services, and ecosystem valuation.

All such models, to be useful to policy development, necessarily must simplify their subject matter to answer the following set of policy-relevant questions (King 1997, 8–9):

1. What functions are provided by this ecosystem?
2. What services, products, and amenities do these ecosystem functions generate?
3. How much value, at least in relative terms, do people place on them?
4. Could they be provided just as well by other nearby or distant ecosystems?
5. Are there manufactured substitutes that exist or could be developed?
6. What determines an ecosystem's ability to generate certain services and values?
7. With what reliability, precision, and frequency should ecosystem changes be measured?
8. Do changes in characteristics at one level in an ecological hierarchy reflect changes at other levels (e.g., forage or food fish)?
9. Do changes in an ecosystem at one location (e.g., a single wetland within a watershed) reflect changes at other locations (e.g., all similar wetlands within a watershed)?
10. How can normal fluctuations and cycles in the mix of ecosystem features and resulting services and values be distinguished from significant trends?
11. How reversible are ecosystem changes naturally or through technology?
12. Are biophysical relationships within the ecosystem linear, or are there important threshold points beyond which there are abrupt shifts in the mix of services and values provided?
13. How can and do people adapt to not having certain ecosystem services?
14. How do economic conditions and specific policies affect the production of ecosystem services?
15. And, most important, if waiting to measure actual ecosystem services is impractical or dangerous, what are useful "leading indicators" of them?

These questions are directly relevant to the Corps' and EPA's professed intention to address the loss of services at wetland development sites. The focus of compensatory mitigation on wetland *functions* has, over time, produced considerable advancement of wetland resource system models (Ruhl and Gregg 2001). To construct an integrated model, however, the Corps and EPA must at the most fundamental level also develop and implement a method for assessing the services and values being provided by a wetland prior to conversion, the extent of their loss through site conversion, and the natural or technological means by which, and extent to which, they will be replaced for the relevant human population. Unfortunately, neither existing agency policy nor the proposed rule establishes the framework for building and employing such a model. Ruhl and Salzman (2006) report having devoted nearly a year to assembling their data merely on the *location* of project sites and bank sites in Florida, notwithstanding that such data are required by law to be maintained and made publicly available by the mitigation bank operator, state agencies, and the Corps. No data are so much as collected, much less made publicly available, in connection with wetland permit application decisions regarding the services and values provided by preconversion wetlands, the extent of their losses, or of the natural and technological means of replacing those values.

In essence, the Corps and EPA have promised to "address" a well-recognized problem equipped with absolutely no operational model of how the human system component of the problem can be addressed and no policy for systematic data collection through which to construct such an integrated model. As chapter 17 explores, this does not bode well for successful implementation of the newly pronounced policy goal, not simply because the agencies may fall short of the goal, but because they may make decisions in the hope of meeting the goal that are wildly misinformed about the trade-offs and transitions that are inherent in any policy decision.

17 Trade-Offs and Transitions

The purpose of building an integrated model of the type presented in chapter 12 and envisioned in chapter 16 is to evaluate different policy options. The Corps of Engineers, for example, has several rather obvious policy options for meeting its goal of addressing lost service values at wetland development sites: it can either (1) approve less development in wetlands, (2) require that compensatory mitigation be accomplished closer to or on the development site, or (3) require the developer to replace the lost services through technological structures. Either of these options involves a change in current policy. A reliable integrated model, therefore, would allow the Corps to evaluate how implementing these options or combinations thereof would affect the relevant resource and human systems. Clearly, different options are likely to have different outcomes over time in both systems in terms of total costs and benefits and the distribution of costs and benefits. There are, in other words, trade-offs and transitions inherent in every policy option relevant to every other policy option.

Trade-Offs

Chapter 1 discussed how the complexity of ecosystem functions—the interlinked nature of inputs and outputs between ecosystems however we draw their boundaries—means that managing any particular ecosystem to yield a preferred service regime will have consequences for the flow of services from that ecosystem and from others. These feedback and feedforward effects will transpire at different spatial and temporal scales as well. There are, in other words, ecological trade-offs to be faced within and between ecosystems as a consequence of decisions about how to manage a service regime at a particular place and time, and it is critical that any working model take these into account (Barbier and Heal 2006; Heal et al. 2001).

Human systems are no less complex than resource systems, and thus face similar trade-off issues. As chapters 3 and 4 explained, to the extent that improved property rights and information mean that natural capital and ecosystem service values are more fully integrated into our market economy, overall social welfare cannot help but rise, as resource owners and resource users would make more informed decisions about what is the most economically efficient investment when the values of natural capital and ecosystem services are included. And chapter 5 explored how regulatory interventions to manage services that behave as public goods could also enhance overall social welfare. It is highly unlikely, however, that everyone will share equally in the net gain to social welfare, or even that everyone will experience a gain. In particular, people who used ecosystem services as if they were free would find themselves paying for at least some of their use, either through prices charged in the market or through the cost of complying with regulatory prescriptions. Moreover, people who might have enjoyed opportunities were natural capital to be overdeveloped—say, the people who would have been employed at a farm or construction site that has instead been devoted to providing water quality services—may find fewer such opportunities. On the other hand, new and more profitable opportunities might open up for businesses and landowners and others as a result of more efficient decisions about natural capital and ecosystem services (Athanas et al. 2006). Recall, for example, from chapter 12 how the Conservation Reserve Program maintained farm income in Big Creek watershed, at the same time generating winners and losers, while substantially improving its ecosystem service provision.

In short, any time a market defect is corrected through improved property rights and information, and any time a public policy intervenes to alter economic opportunities associated with public goods, there will be some winners and some losers. An aggregate benefit–cost analysis of the new set of conditions will demonstrate the overall rise in social welfare, but it would not reveal the *distributional* effects throughout the economy among the winners and the losers. Those on the losing end, particularly if their losses are significant, are unlikely to take much solace in the fact that overall social welfare has gained. They are more likely to perceive the new market or regulatory conditions as ushering in a wealth transfer from them to the winners.

These kinds of trade-offs within human systems, of course, are what make some people advocate market or policy change and others resist it. There is strong evidence, for example, that environmental policy in general has favored higher-income classes with disproportionately high benefits (Baumol and Oates 1988, 235–256; Johnson 2005, 361–369). Hence, as Adams et al. (2003) observe with respect to trade-off conflicts over common pool resources,

> One cannot . . . simply analyze the economic interests of different claimants to rights over a defined resource. Different people will see dif-

> ferent resources in a landscape. They will perceive different procedures appropriate for reconciling conflict. Moreover, perceptions will change, because different elements within a landscape will become "resources." For example, a market may develop for something previously regarded locally as useless or destructive of value, such as wildlife tourism. In these situations, the realm of conflict between beneficiaries and others will be both cognitive and material. (1916)

What makes the trade-off problem devilishly hard in the case of ecosystem services is that solutions designed to minimize the trade-offs and soften the blow to the losers are limited by the complexity of the resource systems. The stakeholders in one ecosystem of defined scale may come to agreement over the most efficient allocation of resources, but their decision about how to manage that ecosystem will have ecological effects, and thus economic effects, in other ecosystems. This is why models employed for purposes of decision making about natural capital and ecosystem services must be integrated between resource and human systems and between spatial and temporal scales.

As noted earlier, for example, the Corps could achieve its new wetlands policy goal of maintaining ecosystem service values at development project sites through some combination of approving less development of wetlands in the first place, requiring compensatory mitigation at or near the development site, and mandating technological measures. Any combination thereof would represent a change in policy with trade-off consequences in the relevant resource and human systems. In the resource system, as chapter 5 explained, the very purpose of moving toward wetland mitigation banking as a policy option was to improve the ecological performance of mitigation sites by consolidating them into large, contiguous parcels. And this also had the advantage in the human system of reducing the Corps' administrative costs and making mitigation less expensive for developers. As Ruhl and Salzman (2006) have shown, however, those advantages incurred a trade-off as well by favoring rural areas for wetland gain and urban areas for wetland loss, which in turn has opened up the concern about whether and how the lost urban wetland services are being replaced.

Consider the different approaches the Corps might use to address that concern. If the Corps were to adopt policies that make it more difficult to develop in urban wetlands, or that require compensatory mitigation closer to the development site, wetland banking could become less attractive and its purported advantages less realized. On the other hand, if the Corps decided to promote technological fixes as the means of replacing lost urban wetland services, wetland banking might become even more attractive than it already is. A myriad of interests are already established with stakes in the current wetland mitigation policy—for example, landowners and local governments in urban and rural areas, potential developers in urban markets, potential wetland banking entrepreneurs, and citizens and businesses in urban and rural areas—and each is likely to perceive any such change in Corps policy with a different eye from the others. Hence, assuming there are several policy combinations that meet the goal of replacing lost services (or avoiding losing them in the first place), it would behoove the Corps to have developed an integrated, multiscale model for evaluating how each combination might play out in terms of trade-offs in the relevant resource and human systems.

In particular, it would be important for the Corps to understand whether the distributional effects of the proposed policy option will impose significantly disproportionate wealth effects for different stakeholder groups. For example, if the Corps' shift in policy were to reduce demand for purchase of wetland bank credits, existing wetland banks that have already invested in providing mitigation values on the prospect of a higher demand could find themselves facing financial ruin. Or, if the Corps were to approve less development of urban wetlands, developers already invested in urban markets may find their expectations sharply undercut. Groups put in that extreme loss position are likely to vociferously oppose the new policy. By contrast, the people gaining as a result of the new policy are likely to be urban dwellers whose delivery of services is maintained, but each of whom realizes only a small increment of gain as a result. People put in that small incremental gain position are less likely to mobilize in support of the new policy. The result could be that the Corps finds only significant opposition to its policy choice even though it can demonstrate a net overall gain in social welfare.

It is no revelation that these kinds of interest group battles happen—their prevalence and potential for inefficient policy outcomes are the basic subject matter of public choice theory, the seminal works of which date to the early 1960s (Buchanan and Tullock 1962; Olson 1965). The point here is not to settle the debate over the merits of public choice theory versus counterexplanations for public behavior (Pierce et al. 1999), but rather to illustrate that even efficient solutions made possible through improved operation of markets or regulatory intervention may face stiff resistance if the consequences are to put a significant interest group at a substantial disadvantage financially or politically. And environmental policy has by no means been a stranger to interest group politics (Lazarus 2004; Zywicki 2002). In such cases, therefore, the advocates and supporters of the change must have a plan for navigating the *transition* to the new set of rules.

Transitions

Given their substantial but unrecognized values in many settings, trade-offs inherent in changes to market and regulatory conditions governing natural capital and ecosystem services are likely to alter settled expectations for many interest groups (Banner 2002; Salzman 2005; Wyman 2005). It may very well be that once the new conditions have been in place for a time, there is likely to be general and widespread contentment with their operation. The difficulty is moving from the current position to the new position when a powerful set of interests believe their new circumstances will be substantially less advantageous than the status quo. Particularly in the western United States, where environmental and recreational interests are displacing agricultural and extractive industry uses and communities, transition has been plagued by social, political, and economic friction (Laitos and Carr 1999; Rasband 2001). There are several means of dealing with this kind of transition problem.

One method is to make any changes prospective and to "grandfather" all prior vested interests, as is frequently done in the land use context when local zoning law amendments change the status of prior legal uses. The so-called nonconforming uses are allowed to remain indefinitely or for a substantial amortization period so as to avoid suffering acute economic losses (Callies et al. 2004). Similarly, if the Corps were to adopt more restrictive permitting or mitigation policies regarding development in urban wetlands, it could exempt current owners of development or mitigation property and allow them to operate under the prior set of policies.

One obvious disadvantage of doing so is that the goals of the new policy are compromised—ecosystem service values remain unrecognized within the subset of land uses exempt from the new policy. Moreover, where the grandfathered interests compete with other interests, as would be the case in urban development and mitigation markets, their exempt status is likely to put them at a significant competitive advantage. Indeed, the prospect of securing a competitive advantage through grandfathered status may lead to a race to "vest" grandfathered rights when a policy change is suspected (Dana 1995). Where the potential size of the exempt segment of land users is significant, therefore, grandfathering is likely to be an ineffective, if not counterproductive, means of managing transition.

At the opposite extreme, another transition strategy is to simply ignore the disadvantaged interests and weather the battle they might initiate. Even more aggressive, opponents to the new policy could be painted as self-interested and undermining worthy social goals. Indeed, as the wave of environmental legislation in the 1970s shows, legal change is often the agent of change in social norms (Salzman 2005), and it could be the case that opponents to the new policy become increasingly isolated and marginalized.

Concerns of fairness, one would hope, put a limit on this transition approach, the formal legal representation of which is the Fifth Amendment's prohibition against uncompensated taking of private property and the Supreme

Court's jurisprudence of regulatory takings (Doremus 2003b). Landowners whose property rights are severely impaired through a change in regulation may be able to establish the right to compensation. This protection, however, is quite limited in scope. A landowner must establish a nearly complete loss of economic opportunity, and even then is not protected against changes in common law property rights or against changes in regulation that merely codify restrictions already imposed by common law property rights. Indeed, this has largely been the story of regulatory takings law in the wetlands context. For example, the *Palazzolo* decision discussed in chapter 5 is the most recent installment in a long chain of litigation over the landowner's claim that a series of changes in wetland regulation policy so severely undercut his settled expectations as to constitute an unconstitutional taking of property. The Rhode Island court's ruling that the land development Mr. Palazzolo had in mind would have constituted a public nuisance insulated the regulatory changes from such a challenge. Almost all other regulatory takings cases involving wetlands regulation have reached similar conclusions—either the landowner's losses were not sufficient to rise to a taking, or the regulation simply implemented existing common law restrictions (Meltz 2005). The result is that many landowners feel unfairly treated, but the law provides little or no safe harbor.

Grandfathering may be an appropriate transition mechanism in some settings, and regulatory takings law may protect landowners in limited circumstances, but most transitions associated with changes in market or policy conditions related to natural capital and ecosystem services are likely to fall between those two extremes—that is, grandfathering would too substantially undermine the new policy goal, regulatory takings law provides no protection, and thus a substantial set of interests suffers a disproportionate loss as a result of the change in conditions. In such cases it may be effective to use "transition payments" as a means of facilitating approval of the new conditions and satisfying fairness concerns (Salzman 2005, 2006). Transition payments are in effect a legislated extension of the constitutional backstop of regulatory takings law, providing compensation to the otherwise disadvantaged interests in order to ameliorate their economic losses and remove their opposition to the policy change. The compensation may take different forms—cash subsidies, tax credits, or transferable development rights—and may not fully compensate for the expected loss. It may also be subject to conditions and time limits. But it is widely believed that "short-term transitional payments act as 'circuit breakers,' easing the internalization of and transition to a higher duty of care. Such short-term and conditional payments help retain support of the political and local communities as contested or uncertain property rights are redefined" (Salzman 2005, 951). They are frequently used in land use contexts, where local authorities provide transferable development rights to ease the sting of more restrictive zoning and growth control regulations (Callies et al. 2004), and Salzman (2005) believes transition payments could be employed effectively in the context of policy development for natural capital and ecosystem services.

Questions about the appropriate use of grandfathering, regulatory takings, and payments in transition contexts are a subset of the larger question of implementation. Assuming that a policymaker such as the Corps has constructed an integrated, multiscale model for evaluating trade-offs and acute transition problems under different policy choice scenarios, implementation choices extend far beyond the transition stage. In short, any proponent of a change in policy must identify the methods and agents of implementation. Chapter 18 turns to these questions of instrument design and institutional structure.

18   Instruments and Institutions

Models and an understanding of their underlying drivers will carry natural capital and ecosystem services only so far. Solutions for improved accounting of their values must be integrated with, and if necessary alter, an existing mosaic of instruments and institutions governing property rights, regulation, and social norms. This chapter explores how to accomplish this task.

Any such endeavor, however, must proceed with caution. No problem of the environment has been solved easily or quickly. There is no silver bullet instrument—no elegant doctrine of property law or innovative regulation—that will solve the Tragedy of Ecosystem Services in one fell swoop. Nor is there a superagency that can marshal all the power of property law and regulation under one roof and administer the law and policy of natural capital and ecosystem services from above. Rather, accounting for natural capital and ecosystem values will happen, if at all, only incrementally, through a combination of instruments, and with the concerted effort of a wide variety of institutions. With that caveat firmly in mind, therefore, this chapter surveys the range of instruments that can be employed in the project and explores some of the institutional configurations that show promise for implementing them.

Instruments

A number of impressively long and detailed treatises survey, evaluate, and compare the various tools that can be used to shape economic and social behavior consistent with chosen policy goals, both as a general matter (Salamon 2002) and in the context of environmental policy (Baumol and Oates 1988; Johnson 2004). The objective here is not to repeat that work but to demonstrate how better accounting of natural capital and ecosystem service values can be effectuated in property rights, regulation, and social norms through a variety of instruments.

Property Rights

In the modern context of highly articulated, nearly pervasive environmental regulation, it will be important not to bypass property rights in the conversation about ecosystem services. Indeed, to move directly to regulatory instruments without first considering how property rights can be reconfigured to overcome the wall of anti-ecosystem bias would be potentially a huge mistake, as property rights can shape the norms that provide the foundation for regulatory initiatives. Hence, it is fitting in this chapter to begin the exploration of instruments with an examination of the oldest resource management instrument of all.

Private Property Doctrines

As discussed in chapter 4, one impediment to more effective representation of natural capital and ecosystem services in law has been a systematic and long-bred bias of common law private property doctrines in favor of developing land. But the common law is profoundly adaptive. As Justice Scalia acknowledged in the *Lucas* case covered in chapter 4, "changed circumstances *or new knowledge* may make what was previously permissible [under common law] no longer so."[1] Hence, there is no reason why the common law cannot make an adaptive move to account better for natural capital and ecosystem service values in the law of property rights.

Yet it is too easy to propose that the common law do so by simply reversing direction and placing a holistic "green thumb on the scales of justice" in favor of protecting ecosystems in general (Sprankling 1996, 587–589). There has to be a concrete theme to motivate the interest and action of private litigants and the courts, and that theme must have dimensions fitting within the basic contours of common law doctrine and institutions. This includes articulating a coherent statement of rights and liabilities that are susceptible to analysis through commonly understood and applied principles of proof of breach, injury, and causation, as well as a remedial system that provides efficient and equitable outcomes. In other words, the approach has to be legally practical.

Valuation of natural capital and ecosystem services can provide the foundation for such an evolution of the common law. Rather than imposing an exogenous policy demand on the common law, valuation of natural capital and ecosystem services can work from within, factoring into the existing utilitarian calculus as Klass (2006) suggests could be the case for the doctrine of adverse possession:

> In order to justify a judicial departure from many years of disparate treatment for adverse possession of wild lands, it may be helpful to base such a development not only on changes in public sentiment regarding

conservation, but also on an ecological and economic framework that attempts to quantify those values. That framework is the growing field of "ecosystem services." . . . [T]he growing body of scientific, policy and legal thought and data surrounding ecosystem services presents a framework for valuing undeveloped land in its own right (i.e., apart from its development potential), thus upsetting the calculus that was relied upon to create special rules for adverse possession of wild lands. Once "value" is placed on undeveloped land in its own right, it becomes an easier case to justify abandoning the special standard for adverse possession of wild lands. (330–332)

Moreover, this shift in outcome under the common law would not require exact quantification of the ecosystem service values any more than courts two centuries ago demanded precise quantification of land development values. Klass continues,

The ability of ecosystem services to actually influence policy and transform markets in a way that provides greater protection for natural resources certainly faces many hurdles because of the difficulty in valuing and creating market exchanges for nonfungible resources such as wetlands and species diversity. However, such difficulties do not impede the use of ecosystem services in the adverse possession context. The lesser standard for adverse possession of wild lands is based on the idea that undeveloped lands have little or even negative value. Whether or not the field of ecosystem services can capture the precise market value of any particular piece of undeveloped land is beside the point. Because ecosystem services provides a theoretical and economic framework for justifying that such lands have some value (and in many cases, significant, quantifiable value for clean air, clean water, open space, aquifer recharge and wildlife habitat), it provides a concrete basis for abandoning an outdated legal standard for adverse possession of wild lands. In this way, the science of ecosystem services can work in tandem with modern-day public sentiments regarding protection of undeveloped land to justify a modification of current doctrine. (332)

Although the process of change in the common law can be slow, there is no reason why any doctrine of property law that suffers from the anti-ecosystem bias built up over the past three centuries could not transform from within as Klass describes.

PRIVATE AND PUBLIC NUISANCE

Indeed, of all the property doctrines ripe for such an evolutionary move, nuisance law is unsurpassed. As Ruhl (2005a) suggests, it is remarkable how straightforward an exercise it is to outline a set of common law rights and liabilities that put ecosystem services into play as the essential fabric of a new stage in the development of the common law of nuisance.[2] Every law student learns the black letter doctrine of nuisance: one commits a nuisance when his or her use of land unreasonably interferes with another person's reasonable use and enjoyment of his or her interest in land. Lawyers through the ages have had no problem agreeing that odors from a pigsty, or fumes from a copper smelting plant, or chemical pollution of a lake or stream are within the ballpark of nuisance so defined. Why should matters be any different when one person's use of land severs the flow of economically valuable ecosystem services to another person's use of land?

A thought exercise drawing from the pollination example can help illustrate the spectrum of possibilities suggested: A commercial apple orchard is situated between an industrial facility on one side and a forested tract on the other. The owner of the apple orchard has suffered a substantial decline in commercially marketable apple production and can prove both the cause and the economic damage. The alternative causes to consider are these:

- Emissions from the industrial facility drifting into the orchard are damaging the bark of the trees, causing them to decline in productivity.
- Emissions from the industrial facility drifting into the orchard are blemishing the skin of a substantial percentage of the unripe apples, causing them to be unmarketable.
- Emissions from the industrial facility drifting into the orchard leave a residue on the apple tree leaves that interferes with photosynthesis, causing the trees to decline in productivity.
- Emissions from the industrial facility drifting into the orchard are deterring visits from wild pollinators residing in the forest tract habitat, thus causing a reduction in successful fruit production.
- The owner of the forest tract cuts down all the trees to build a shopping mall, eliminating that source of wild pollinator visits and thus causing a reduction in successful fruit production.

The first two of these scenarios are classic fodder for nuisance claims. To be sure, there may be much to resolve about questions of liability and remedy, but these cases are squarely within the tradition of nuisance law. The next two scenarios involve land uses that sever the flow of ecosystem services to the orchard by interrupting the delivery of the service, photosynthesis in one case and pollination in the other. If these causal connections are proven, it is not clear why the common law would fail to recognize them as cognizable causes of action in nuisance if it does recognize the first two scenarios as such. Indeed, while the causes in the first two scenarios are described in familiar terms—damage to tree bark and blemishes on apple skins—in fact the causal chain in those cases is the interruption of ecosystem functions that support the trees and their production of unblemished fruit. Why should it matter that the cause of the reduced fruit production was the chemical reaction of the pollutant on tree bark or apple skin in the first two cases versus its effect on sunlight or bees in the next two? Why treat any of the first four scenarios differently?

The more difficult case is the fifth scenario, because the flow of ecosystem services is severed at the source property through destruction of the natural capital—the forest supporting the pollinators—rather than at the benefited property through interruption of the service at the point of delivery. But the end result is the same—the orchard produces less fruit. If the orchard owner can prove that the reduced fruit production is the loss of pollinators that once resided on the shopping mall tract, why would that not be cognizable in nuisance?

The quick response might be that the conversion of the source property from forest to shopping mall is not unreasonable, whereas pollution drifting in from the industrial facility is. But that does not answer the question, which was whether the orchard owner's case is *actionable* in nuisance, not whether it would prevail. The termination of pollination is, after all, interfering with the orchard owner's use and enjoyment of the property. That opens the door to a nuisance claim, with the central question being, as it is in most nuisance cases, whether the allegedly wrongful behavior was unreasonable. Nuisance law is quite a thicket on the question of what is unreasonable, but that is both the beauty and the frustration of the common law. It is made for this kind of balancing inquiry, which Justice Scalia described in *Lucas* as an "analysis of, among other things, the degree of harm to public lands and resources, or adjacent private property, posed by the [landowner's] proposed activities, the social value of the [landowner's] activities and their suitability to the locality in question, and the relative ease with which the alleged harm can be avoided through measures taken by the [landowner] and the government (or adjacent landowners) alike."[3]

To be sure, it is not expected that every loss of natural capital should be or would be branded unreasonable under this test. Some natural capital is more critical than other natural capital, in that its degradation or destruction leads to significant economic injury on other lands. But given that we increasingly know where natural capital is located, where the ecosystem services it produces flow, and the value of those services at benefited properties, there is no reason why nuisance law in both its public and private stripes could not sort through questions about whether the destruction of natural capital in discrete cases is reasonable or not.

For example, one can see quite palpable evidence of the importance of coastal dunes to the mitigation of hurricane storm surge damage at inland locations throughout the Gulf Coast. There is a staggering difference in outcome between inland areas shielded by intact dunes and those inland of coastal development that did not retain dunes. Under Justice Scalia's version of the nuisance balancing test, the harm to the public resources and private property resulting from the impaired dune and wetland systems unquestionably was severe, likely far outweighing the social utility of development that destroyed the resources, and the owners of prior intact dune areas were in the best position to avoid the harm. Were those resources thus critical natural capital, the destruction of which was unreasonable in relation to the expectations of inland property owners whose homes and businesses are now in splinters?

The common law is designed and equipped to answer that question and others like it. The fact that it has not until now attempted to do so does not mean that it cannot, or will not have the opportunity, or simply is against all notion of it. The only missing ingredient until now has been the storehouse of knowledge ecologists and economists are building about the value of ecosystem services. This is precisely the kind of new knowledge Justice Scalia confirmed in *Lucas* can transform the common law and "make what was previously permissible no longer so."[4] As sovereigns and landowners become aware of this new knowledge and begin to appreciate the cost imposed to them when others sever the flow of ecosystem services to their lands, they *will* sue in public and private nuisance actions. Indeed, such a claim was initiated in federal court with respect to the losses suffered in Katrina, alleging that those responsible for the disruption of wetland processes are also responsible in tort for the economic losses that followed.[5] And when lawyers and experts use this new knowledge to demonstrate to courts the cause of the injury and the value of the services lost, the courts *will* award damages, injunctions, and other relief. And that will be a natural evolution of the common law of nuisance—one motivated by the common law's central objective of promoting efficient use of land. On Dry Hill, to draw from the first case study in part III of this book, a strengthened approach to nuisance law could have been a tool of local residents in opposing the construction of a pumped storage facility along Lake Michigan's dunes, had the conservation efforts with Grand Traverse failed. More generally, modifying land uses in ways that substantially affect evapotranspiration and stream flow, as discussed in chapter 8, or water quality, as discussed in chapters 9–13, could, under a modified nuisance law, be subject to injunction when those changes injure downstream parties. Cloud seeding is an even clearer example where downwind parties could employ nuisance law to prevent a diminishment of valuable ecosystem service flows (i.e., precipitation stealing).

Public Trust Doctrine

As chapter 4 explained, the common law also extends its reach to one important doctrine of *public* property rights—the public trust doctrine—which appears already to have begun moving toward greater recognition of environmental values in contexts such as reserving instream flow in western states. As in the case of private property doctrines, valuation of natural capital and ecosystem services can work from within the public trust doctrine to strengthen that trend. When public trust lands provide valuable services in their undeveloped state, they are in a very real economic sense "working" for the public welfare and enjoyment, which the public trust doctrine is designed to promote. While this would not extend the public trust doctrine to resources other than tidal and submerged lands, it would provide grounds for using the doctrine to protect in situ resources, such as supporting greater instream flow demands, and for curtailing extractive uses and other development of public trust lands. And as the public trust doctrine has never demanded precise quantification of the values of different public and private uses competing for access to public trust land, neither should it demand precise quantification of ecosystem service values to make this evolutionary move.

For example, when the Rhode Island Superior Court in *Palazzolo v. Rhode Island*[6] found that the state's denial of a permit to fill 18 acres of salt marsh was not a taking, the court described the tidal pond adjoining the salt marsh as a "particularly fragile ecosystem," and the salt marsh itself as a "valuable filtering system" for runoff and pollutants.[7] In addition to finding that the plaintiff's proposal to fill the salt marsh would constitute a public nuisance (see chapter 4), the court also found approval of the project would have violated the public trust doctrine.[8] This ruling did not require any unanticipated or anomalous extension of the public trust doctrine, nor did the court demand an exact quantification of the value of the "valuable filtering system."

PUBLIC LAND MANAGEMENT

Whereas the common law public trust doctrine operates in the hybrid property context where public trust values limit private property rights, publicly owned resources are for the most part governed by statutory and regulatory assignment of rights. As chapter 4 explained, although significant public resources have been reserved for wilderness and other low-intensity access, vast expanses of the public domain are subject to a multiple use/sustained yield (MUSY) mandate that purports to balance a number of inherently competing uses in the interests of promoting overall social enjoyment of and utility from our public resources. As with the common law doctrines, however, recognition of natural capital and ecosystem service values in the public resources context does not require any fundamental alteration of the MUSY mandate, but rather can work from within.

Given its open-ended, multifactored character, the MUSY mandate is inherently flexible. For example, when the Forest Service and other federal land management agencies in the 1990s began incorporating principles of ecological sustainability into programs subject to the MUSY mandate, many interest groups associated with grazing, timber, mining, and other extractive industries objected on the ground that doing so illegally tilted the MUSY balance (Thomson 1995). The Forest Service's retort (Department of Agriculture 2000, 67521) illustrated the malleability of the MUSY mandate:

> The proposed rule's focus on sustaining ecosystems is fully compatible with the Forest Service's underlying statutes. In order to ensure that the multiple-uses can be sustained in perpetuity, decisions must be made with sustainability as the overall guiding principle. Ecological sustainability lays a necessary foundation for national forests and grasslands to contribute to the economic and social needs of citizens.

There would be no less basis for incorporating more systematic use of natural capital and ecosystem service values into MUSY-based decision making. Indeed, it is difficult to envision how ecological sustainability could "contribute to the economic and social needs of citizens" without more explicit and concerted effort by the public resource management agency to account for natural capital and ecosystem service values. The decision to do so, of course, is a policy that federal resource management agencies would need to implement through assessment of natural capital and ecosystem service values and appropriate regulation of public access to and use of the public resources. Hence, whether in private or public land contexts, relying on more than the common law for improved accounting of natural capital and ecosystem service values will require selecting the most effective tools of regulation.

Regulation

Regulation has an undeservedly bad reputation. As chapter 5 explained, regulation, even the conventional "command-and-control" variety of prescriptive regulation, can provide effective and efficient solutions for management of public goods. It can also be disastrously ineffective and inefficient. To avoid the latter, the "reinvention" movement discussed in chapter 13 has sought to expand the regulatory toolbox beyond prescriptive regulation to encompass a spectrum including reflexive instruments such as information disclosure and economic instruments such as pollutant trading programs (Johnson 2005). The Corps of Engineers' proposal to begin taking the distribution of ecosystem service values into its accounting of its wetland regulation program, discussed in detail in chapter 13, thus provides a useful setting for exploring the expanded set of options. They are arranged in this section according to a loose approximation of the extent of coercion exercised on the regulated party.

INFORMATION DISCLOSURE AND MANAGEMENT

The starting point for any such instrument analysis should be a focus on information, where natural capital and ecosystem services face a bit of a catch-22 dilemma: because current law and policy do not adequately account for natural capital and ecosystem service values, little reliable information is available about them, but law and policy cannot intelligently account for them through regulatory prescriptions without such information. A way out of this predicament that has proven remarkably successful in other contexts is to require disclosure of information without attaching regulatory consequences directly thereto. For example, the Toxic Release Inventory, which requires specified industrial categories to disclose annual pollutant emissions, is credited with enriching the information base available to the public and regulators about pollution in general, as well as with motivating many industrial facilities to reduce emissions even though not required to do so (Johnson 2005; Karkkainen 2001; Schroeder 2000). Consumer product certification programs, such as sustainable timber certification, also provide information upon which the public can make better-informed choices (Millennium Ecosystem Assessment 2005, 96). Human systems rely on flows of information, and measures that enrich the information base can build into existing feedback mechanisms to illuminate reflexive responses by regulators, the regulated, and the public with respect to natural resources decision making (Karkkainen 2001; Orts 1995; Stewart 2001). Better decisions are possible not only for regulation but also in the market.

Indeed, as Ruhl and Salzman (2006) discuss, one of the most disturbing aspects of the wetlands mitigation program is the lack of organized data about mitigation in general and its effects on ecosystem services in particular. The mitigation banking program, touted by the Corps and EPA as a win–win–win for developers, the agencies, and the environment, has operated in a virtual data Dark Ages, making the claims of its success little more than speculation. It is no wonder that the so-called market in wetlands mitigation has had asymmetric effects outside its narrow scope of selling and buying wetland function "credits."

Unfortunately, in their proposed wetland mitigation rule the Corps and EPA have missed a golden opportunity to correct this defect. Although the rule requires developers intending to provide compensatory mitigation to prepare a mitigation plan including "baseline information" about the development site and the mitigation site, the specified information does not include an evaluation of pre- and postmitigation ecosystem service values at either site (Department of Defense and Environmental Protection Agency 2006, 15539). Even if no regulatory consequences were attached to the results of such an evaluation, at the very least the agency ought to wish to know the effects of its program on ecosystem services, and it unquestionably has the authority to require permit applicants to disclose such information.

Of course, collecting information from resource users does little good if the information is not effectively managed. All of the data Ruhl and Salzman collected in their study of wetlands mitigation banking, for example, was obtained by government agencies through mandatory disclosure requirements. From there, however, the agencies did nothing to manage or distribute the information in a way that would have been useful to policy development of any kind, much less for the purpose of addressing ecosystem services distribution. By contrast, the success of the Toxic Release Inventory lies in the ease with which members of the public, regulated entities, and government agencies can access the data in user-friendly interfaces and metadata formats. Similarly, information gathered about ecosystem services through mandatory disclosure programs such as the one suggested above for wetlands mitigation could be aggregated regionally and nationally, provided to the public in formats that can be appreciated by lay persons and accessed by researchers, and used by all to learn about the local, regional, and national effects of wetland mitigation on natural capital and ecosystem service values.

In addition to improving the public's assess to information relevant to private market decisions, information management can also be focused on public decisions such as how to manage publicly owned lands, how to design and apply other regulatory instruments, and basic fiscal policy. For example, economists have identified many gaps in our national income accounts, including omitting childcare values provided by parents and, pertinent to this study, the economic role of natural resources and the environment. Ironically, as Repetto et al. (1989) observed, the nation "could exhaust its mineral resources, cut down its forests, erode its soils, pollute its aquifers, and hunt its wildlife and fisheries to extinction, but measured income would not be affected as these assets disappeared" (3). In response to this defect, the U.S. Department of Commerce began work in 1992 to develop shadow accounts designed to track natural resources through conventional capital stock and depletion accounting. Congress halted work on that project pending completion of an external report on the concept, which was provided two years later by the National Academy of Science's *Nature's Numbers* report (National Research Council 1994). Directly on point with the suggestion made above, the academy concluded that "development of environmental and natural-resource accounts is an essential investment for the nation. It would be even more valuable to develop a comprehensive set of environmental and nonmarket accounts" (9).

Even if this step is not taken at the national accounts level, it seems appropriate to compile macroaccounting that includes natural capital and ecosystem service values for discrete public resource ownership regimes such as the national forests and regulatory programs such as wetlands mitigation. This is not to suggest that the national forests, wetland mitigation, or other public decisions should be held to a strict cost–benefit accounting standard, as that methodology is fraught with its own set of concerns and controversies (Driesen 2006), but if costs and benefits are to matter at all, natural capital and ecosystem service values ought to be included in the inventory.

Planning and Assessment Requirements

Moving beyond mere provision and management of information, the reflexive effect of information enrichment can be incorporated on an immediate basis by mandating additional planning and assessment steps. The most notable of such measures is the National Environmental Policy Act (NEPA), which requires all federal agencies planning to carry out, fund, or authorize major actions or decisions to prepare a statement evaluating the environmental impact of the proposed action and its alternatives.[9]

NEPA has no regulatory consequences—it is purely procedural, and the action agency may (subject to other legal constraints) choose the alternative that has the worst net effect on the environment, even if it is a significantly adverse effect. Yet by forcing an agency to go though the process of comparing alternatives, and by allowing the public to review and comment on the agency's evaluations, NEPA has had a profound reflexive effect on agency decision making (Karkkainen 2002).

Regulations implementing NEPA, however, do not explicitly require agencies to include evaluation of impacts to natural capital and ecosystem service values, nor must the analysis of alternatives consider the relative merits of maintaining service values through natural capital versus technological means (Fischman 2001). Both such requirements would be consistent with the scope of NEPA, and both would improve not only the information available about how to value ecosystem services, but also how to design land uses with them in mind. Indeed, mandating such planning and assessment steps through natural resource management programs would assist in improving the knowledge base relevant to natural capital and ecosystem services, with no direct regulatory consequences at stake.

For example, the Corps' and EPA's proposed mitigation rule could go further than requiring an assessment of ecosystem service values at development and mitigation sites by requiring permit applicants to evaluate project design alternatives for addressing the potential loss of service values at development sites. What would be the effect on the development of retaining more in situ natural capital versus employing technological solutions? Would moving the mitigation site closer to the development site retain service values for the relevant population? Without answers to these and related planning and assessment questions, it will be difficult for the agencies to fulfill the promise of addressing potential loss of ecosystem service values at development sites in a manner that is both effective and efficient.

Subsidies

Although information disclosure and planning and assessment requirements are not voluntary or costless to the regulated parties subjected to them, they rely on the reflexive mechanism to alter behavior. No positive or negative incentives are directly attached to them. A more direct means of altering behavior commonly used in environmental policy involves using positive incentives in the form of subsidies—economic payments in the form of cash or avoided costs such as reduced fees or taxes (Baumol and Oates 1988, 211–234; Johnson 2005, 93–124; Roodman 1998).

In essence, subsidies pay someone to do what is right, or what Baumol and Oates (1988) famously described as "analogous to the case of a holdup man who appeals to his victims to finance the costs of going straight" (211). As such, they carry the risk of diluting norms chosen policy seeks to reinforce, and they can lead to moral hazard problems as potential beneficiaries may have the incentive to do what is "wrong" in order to set up the case for receiving the subsidy (Salzman 2005). Nevertheless, if the objective is to implement policy in the most effective and efficient manner, the argument·on behalf of subsidies can be compelling in some contexts. In the agricultural context, as shown in chapters 9–13, after decades of doling out income support subsidies that actually promoted inefficiently harmful use of the environment (Millennium Ecosystem Assessment 2005, 95), our farm policy is shifting toward greater use of "green" subsidies to induce farmers to take high-value natural capital out of production (Swinton et al. 2006). Given the political power of agriculture in our society, as well as the American enchantment with an idealized (and fictional) picture of agricultural life (Ruhl 2000; Salzman 2005), few commentators have suggested that command-and-control regulation could more efficiently produce the same results and thus attention has turned to how better to use subsidies to increase the flow of ecosystem services from agricultural lands (Antle and Stoorvogel 2006; Smith 2006). In the case of carbon sequestration, as seen in chapter 15, there is no good argument that farmers and other rural landowners have an *obligation* to withdraw carbon from the global atmosphere and store it on their land. Rather, subsidies, perhaps funded by carbon emitters, are a perfectly appropriate mechanism to compensate them for what has been a positive externality since atmospheric carbon concentrations began to climb several decades ago—so long as these are paid for "additional," not long-standing or incidental, carbon storage.

Green subsidies have been largely confined to the agricultural sector, where international trade policy suggests that the trend away from income support subsidies will continue. But there is no inherent reason why green subsidies could not be used in other sectors to effectuate chosen policy regarding management of natural capital and ecosystem services (Millennium Ecosystem Assessment 2005, 96). In the wetlands context, for example, urban landowners could be paid to conserve urban wetlands, and mitigation projects could receive subsidies for locating closer to areas where development has depleted urban wetland resources. Subsidies could be targeted using criteria designed to identify the most valuable natural capital and ecosystem services (knowledge of which could be gained through the information disclosure and planning and assessment provisions discussed earlier), and could be efficiently distributed using reverse auction techniques through which willing participants bid payments they will accept and subsidies are awarded based on lowest bids (Salzman 2005).

Perhaps the public status of developers and mitigation bankers is not as amenable to this form of treatment as it is for farmers, but as Salzman suggests, "why not . . . simply recognize this situation for what it is—the provision of valuable services to consumers—and realize that through an explicit arrangement of payments for services rendered?" (888). In other words, to the extent natural capital and ecosystem services are economically valuable, society ought to approach their management through the eyes of a business. If property rights would allow a landowner to deplete wetland resources that it would otherwise be more efficient in terms of overall social welfare to conserve, it should not matter to society whether the person to whom it is paying the subsidy is a farmer or a shopping mall developer.

TAXES AND FEES

Baumol and Oates (1988, 14–56) showed that, in theory, properly designed taxes and fees assessed on externality-producing behavior can provide negative incentives that will sustain optimal levels of such activities in a competitive system. Taxes and fees are, in other words, the sticks to subsidies' carrots, and there has been considerable debate over which instrument is most effective and efficient—the carrot, the stick, or a combination of both. In the wetlands context, for example, would a tax on development of urban wetlands serve as a perfect substitute in place of a subsidy payment for conservation of urban wetlands? Baumol and Oates (211–234) find the general comparison between subsidies and taxes a mixed bag, with symmetry between the two instruments in some circumstances and significantly different effects on externality-producing behavior in others.

The question is particularly complex in the context of natural capital and ecosystem services, because both positive and negative externalities are potentially in play given the status of property rights outlined in chapter 4. Even if the common law were to develop as proposed earlier by correcting the anti-ecosystem bias of property doctrines and formulating principles for application of natural capital and ecosystem service values in nuisance law and the public trust doctrine, not all natural capital resources will be removed from potential development. Landowners will still be generating positive externalities by leaving intact natural capital they otherwise could develop without violating the evolved common law doctrines. It may be effective, therefore, to consider using subsidies to compensate landowners for continued supply of ecosystem services from natural capital that falls clearly outside the reach of the common law prohibitions, using regulatory prohibition to deter development of natural capital that falls clearly within the reach of common law prohibitions but which may be more efficiently controlled through regulatory instruments, and using taxes to deter development of natural capital that falls in neither of those categories. Clearly, different configurations of subsidies and taxes, in combination with yet other instruments, can be tested in different contexts to gain an understanding of which is most effective.

Market-Based Systems

Subsidies, taxes, and fees build on economic incentives to derive the desired optimum level of externality-producing behavior. They have the disadvantage, however, of relying on a regulatory authority to set the level of carrots and sticks, which in turn depends on reliable models of the relevant resource and human systems (Malloy 2002; Parkhurst and Shogren 2003). As discussed in chapter 14, market-based systems, such as the tradable emissions allowances used in the Clean Air Act for sulfur dioxide emissions from power plants, have the advantage of using market pricing forces to find the optimum distribution of effects (Baumol and Oates 1988, 177–189; Johnson 2005, 125–187). Although not in every case a success story, under the right conditions economists find tradable permits and other market-based programs have been shown to provide effective rationing of access to common-pool resources (Ali and Yano 2004; Tietenberg 2006; Freeman and Kolstad 2007).

Their success in practice has made market-based solutions attractive in theory to commentators proposing more complete representation of natural capital and ecosystem service values in resource decision making (Millennium Ecosystem Assessment 2005, 96; Salzman 2005). For example, Thompson (2000) proposes that government agencies enter the market directly, acting as "brokers" of resources acquired by purchase or through compensatory mitigation requirements, buying and selling natural capital as a full market participant.

As with all the other instruments surveyed in this section, however, market-based systems are only as effective as their design permits. Wetlands mitigation banking, for example, is a market-based solution to the pitfalls of project-specific compensatory mitigation, allowing market pricing rather than the Corps of Engineers to dictate the clearing price of mitigation credits. But the banking market exists only because the regulatory program created it, meaning it is susceptible to structural obstacles to efficient operation if the agency does not set up the market correctly (Robertson 2006; Ruhl and Salzman 2006). As with other credit-based and cap-and-trade programs, the Corps has set a maximum on wetland losses through its "no net loss" policy and determines the quantity of wetland credits a bank may sell and debits a development project must offset, leaving it to the wetland banking program to decide where the losses and compensating gains are experienced within a bank's marketing service area and at what prices. As Ruhl and Salzman's study (2006) shows, this kind of indifference to geographic distribution can lead to "hot spots" of negative environmental effects. Because "no net loss" measures only wetland functions, and because it has been blind to geographic distribution within each bank's marketing area, it has failed to prevent, or has even facilitated, asymmetric redistribution of ecosystem services associated with wetland resources.

The Corps could address the "hot spots" effect by integrating wetland service values into its credit and debit calculations, to which the banking market would respond through market incentives. If higher loss of ecosystem service values resulted in higher debit loads for developers, and if higher provision of ecosystem services yielded higher credit profiles for banks, developers would have less incentive on the margin to develop in urban wetlands, and bankers would have more incentive on the margin to locate banks near urban areas. Because the Corps and EPA have failed to implement that approach in their proposed wetland mitigation rule, the Corps is therefore left in the position of having to "address" the potential loss of services at development sites through alternative, as yet unspecified, devices.

In the case of carbon emissions, however, we have seen that the spatial considerations that have undermined nutrient trading in a water quality context and have created inequities in wetland mitigation banking are not evident because it is a "global" atmosphere with uniform carbon concentrations. In this context, the commodification of carbon fluxes into and out of the atmosphere through the development of carbon trading represents a progressive policy that resolves critical components of the Tragedy of Ecosystem Services as it pertains to global climate change by applying the polluter-pay principle to those who use the atmosphere as a carbon sink and by transforming carbon sequestration from a positive externality to a compensated ecosystem service.

PRESCRIPTIVE REGULATION

It may be that the Corps (or another authority) simply falls back on plain vanilla prescriptive regulation to address the ecosystem service distribution effects of the wetland mitigation program. It might, for example, take some areas out of the mitigation markets by prohibiting development of important urban wetlands, or mandate tighter marketing areas for mitigation banks in order to keep ecosystem services within more locally drawn human communities. State or local authorities might accomplish the same through environmental and land use controls.

Prescriptive regulation may no longer be the default rule for environmental policy, but it is by no means off the table as an alternative. As chapter 5 observed, when transactions costs associated with enforcing common law property rights are high, regulation can provide a superior solution. Regulation is also less likely to induce the moral hazard problem associated with subsidies and the "hot spots" effect associated with markets. Most proposals for comprehensive management of natural capital and management solutions that take ecosystem service values into account thus include prescriptive regulation in the mix of instruments likely to be employed (Heal at al. 2001; Salzman 2005).

Social Norms

As much as law is on precarious ground when it is too out of step with social norms, both the common law and regulation have the capacity to shape social norms. It seems unlikely that legal and policy developments designed to take better account of socially valuable natural capital and ecosystem services would shock the normative conscience of American society. On the other hand, it is not clear, despite recent unfortunate events such as Hurricane Katrina, that the American public understands the concepts of natural capital and ecosystem services at all, no less appreciates them in their complete ecological, geographic, and economic terms.

To be sure, through increased information and through more deliberate analysis in court opinions and agency promulgations, it is far more likely that the instruments discussed above will improve the visibility of natural capital and ecosystem services in public discourse. But law and policy need not be passive in this regard. Concerted education and support for social institutions can facilitate the refinement of social norms. As discussed in chapter 6, for example, Lavigne (2004) found little official support for many of the watershed councils he studied, rendering them largely ineffective as a means of forming and implementing social norms for watershed management. It will be important, therefore, for government to "get the word out" about natural capital and ecosystem service values, and to promote social institutions that coalesce around the theme.

One difficult aspect of such a campaign, however, is the degree to which formal public participation rights are included in the regulatory instruments outlined above. Public participation rights are a hallmark of American environmental law and have the capacity to reinforce the law's reflexive qualities, but they are not an unmitigated positive (Rossi 1997; Seidenfeld 2000). At some level the direct involvement of the public in decision making can bog down deliberation and defeat any efficiency qualities the decision making may have otherwise enjoyed. Hence, it will be important, as the process of testing and implementing regulatory instruments progresses, to balance among education, support, and participation as vehicles for the legal and policy developments to shape social norms.

Institutions

Settling upon a set of instruments does not settle the question of what set of institutions will implement them and under what structural conditions. The common law instruments are, of course, the domain of the courts, though legislatures have great leeway to influence how they are implemented, such as by designating certain uses a nuisance per se, and even to preempt them, such as by protecting uses from claims of nuisance (Dukeminier et al. 2006). For the most part, therefore, questions of institutional design focus primarily on (1) whether the common law or regulation is the more effective institutional context (Cole and Grossman 1999; Hylton 2002), a topic explored in detail in chapter 5, and (2) where regulation is part of the mix, how to design regulatory institutions that are effective at implementing the regulatory instruments but also exhibit the key attributes of legitimacy—transparency, accountability, and efficiency (Heal et al. 2001; Millennium Ecosystem Assessment 2005). The latter of these questions is the subject of this section.

Environmental Federalism

By no means has the challenge of institutional design in this sense been for want of options—environmental regulation alone is administered by a myriad of federal, state, and local agencies of all sizes and shapes. Rather, the problem has been more a function of the abundance of agencies and how to coordinate them. In particular, our federalist tradition makes it difficult for any level of government—federal, state, or local—to decide how to design and coordinate a comprehensive institutional framework for implementing environmental regulation. Any solution for improved accounting of natural capital and ecosystem services must therefore mesh within a complex federalist structure that, because of its constitutional origins, is not optional.

As national policy with respect to environmental goals is formulated, the federal government must consider implementation options that range quite broadly with respect to federal–state relations (Buzbee 2005; Fischman 2005; Schroeder 2005). In some fields, the federal government may have no power at all, failing to match the policy with an enumerated authority over federal public lands (Appel 2001), interstate commerce (Klein 2003), spending (Binder 2001), or other federal domains. At the other extreme, federal power is clearly present and pervasive, and the national interest so overriding of particular state concerns, as to lead the federal government to preempt all state action. Options between these two extremes represent increasing dominance of policy outcomes by the federal government. So, for example, in some contexts the federal government would have authority to act but instead relies on state action to address the issue. Or, in the once-prevailing model known as "dual federalism," the federal government might take action to regulate the field but take no position with respect to the exercise of concurrent authority by the states other than to preempt state actions repugnant to the federal program. Under the model prevailing since the New Deal era, known as "cooperative federalism," the federal government uses its authority to act in a particular field and its power to preempt state authority as leverage to enlist states to implement state programs consistent with the federal objectives without impermissibly commandeering them.

The shift from dual federalism to cooperative federalism tracks the rise of the modern regulatory state and the increasing proliferation of social issues that transcend any conception of precise lines between federal and state interests. In the environmental policy context, where both phenomena are keenly represented, the federal government has given life to cooperative federalism through a wide spectrum of tools. At one extreme, the federal government can provide a strong impetus to states to pursue national goals by agreeing to subordinate federal policy to state decisions, such as by incorporating state law on a particular matter as federal law or policy or by binding federal agencies to act consistent with state program requirements. A less constricting approach places federal agencies in an advice-seeking role by requiring them to consider state positions as an important factor in making federal action decisions, or at least to provide states heightened opportunities to receive notice of and provide comments about proposed federal actions. Incentive-based approaches to cooperative federalism include the provision of federal financial and staff support of state planning efforts designed to initiate state programs that satisfy national goals, and broader unconditional federal financial and staff support for implementing state programs that facilitate national policies. The federal government can pursue more persuasive strategies by making financial and staff support contingent on meeting federally prescribed criteria, offering states a package of benefits if they agree to opt in to a program the federal government implements within the state, or allowing a state to work as a partner with the federal government in implementation decisions, albeit in a junior partner status. Under the most coercive of the cooperative environmental federalism options, known as the "delegation" model, the federal government threatens to enforce a comprehensive federal regulatory program in a state unless the state uses its authority to regulate the field in a manner that satisfies federal program criteria.

Overall, then, the federal government's decision to employ cooperative federalism uses the presence of pervasive federal authority as the leverage to influence states to act in line with federal policy choices, but the tools of cooperative federalism range from "soft" strategies that place the state in the driver's seat for policy development, to "firm" arrangements under which the federal government prescribes national goals but enters partnership relationships with states for implementation, to "hard" prescriptive approaches that dictate regulatory details and leave the states little latitude. Combinations of these different approaches also can be used within one comprehensive federal program, such as control of water pollution, making cooperative federalism potentially quite complex as a system for implementing national environmental policy objectives.

Cooperative environmental federalism has been both a blessing and a curse. As a way of mobilizing state and local entities to support and implement national environmental policy, it has been successful in addressing numerous environmental problems it appears the states were unlikely to manage effectively on their own (Buzbee 2005; Fischman 2006; Schroeder 2005), although the degree to which states would otherwise have lagged in environmental policy without federal intervention is hotly debated (Engel 1997; Revesz 2001). On the other hand, by its nature cooperative federalism proliferates agencies and programs in an often uncoordinated and ineffective manner, with the end result not always providing an effective distribution of authority for achieving stated policy goals (Ruhl 2005b). The Corps of Engineers and EPA, for example, have proposed to "address" the loss and replacement of ecosystem services at locations where the Corps approves development of wetlands, but state and local agencies are primarily responsible for approval of site and engineering plans that would dictate how technological solutions might satisfy that goal. The Corps is not authorized to direct state and local agencies in their decisions in that regard. In many other cases the decision-making authority needed to effectively manage natural capital and ecosystem service values is also not under the same roof.

Toward an Integrated Institutional Structure

Although the Supreme Court has in recent years displayed some interest in adjusting the constitutional edges of federal authority relative to the environment (McAllister and Glicksman 2000), for the foreseeable future environmental federalism is a given. The question, therefore, is how should the relevant institutions be organized? The Millennium Ecosystem Assessment (2005) has articulated several criteria for that purpose, aimed mainly at the international level but relevant here as well:

- Integration of ecosystem management goals within other sectors and within broader development planning frameworks
- Increased coordination among multilateral environmental agreements and between environmental agreements and other international economic and social institutions
- Increased transparency and accountability of government and private-sector performance in decisions that affect ecosystems, including through greater involvement of concerned stakeholders in decision making
- Development of institutions that devolve (or centralize) decision making to meet management needs while ensuring effective coordination across scales
- Development of institutions to regulate interactions between markets and ecosystems
- Development of institutional frameworks that promote a shift from highly sectoral resource management approaches to more integrated approaches (93–94)

The few studies that have applied these criteria to management of ecosystem services in the United States focus on watershed-based institutions at state and local scales (Heal et al. 2001; Ruhl et al. 2003). In particular, based on a study we and our colleagues conducted of watershed resource management in the Cache River watershed in Illinois—a large, working landscape comprised of multiple private and public landowners and subject to the jurisdiction of numerous government agencies—Ruhl et al. (2003)[10] propose several overarching institutional design goals:

1. The institutional structure for watershed management must enjoy the type of power and authority generally associated with centralized administrative governments, such as the federal or state governments, but must also be capable of establishing democratically based legitimacy at regional and local levels where many regulatory actions are implemented. This requires going beyond federal or state laws enabling local districts to take action. Rather, much like watersheds themselves, a nested hierarchy of interrelated federal, state, and local governmental authorities will be necessary.

2. The institutional structure must have the authority and the responsibility to manage watershed issues "holistically" on a system level. This requires, at a minimum, some form and level of authority over surface and ground water, over water quality and water quantity, and over key physical and biological effects on aquatic ecosystems such as flood control, soil conservation, wetlands conservation, fisheries, recreation, stream entrenchment, dams, reservoirs, pollutant sources, and land uses with significant watershed impacts.

3. The institutional structure must rely on more than voluntary governance and voluntary compliance with specified standards and goals. In particular, where implementation relies on local units of governance, accountability must be lodged at the local level. The full range of financing mechanisms should be made available (e.g., taxes, fees, surcharges, bond) and the full range of compliance instruments should be capable of being used effectively as appropriate (regulatory, market-based, incentives, reporting and information requirements, planning requirements, voluntary).

4. The institutional structure must have the capacity—the budget, staff, and expertise—to carry out complex scientific, economic, and social analysis functions, and the responsibility to make policy and regulatory decisions through public, transparent procedures and based on the record of best available evidence it generates through its capacities.

5. The institutional structure should be generalizable across watershed types, scales, and political units, and the information-gathering capacity and protocols should be standardized so as to allow sharing of information vertically (e.g., within a state from local to higher levels) and horizontally (e.g., between local districts and between states). (934–935)

A critical premise of this approach is that putting these design parameters into play requires a comprehensive and coordinated effort led by the states and implemented at several levels of governance within each state. Several factors strongly suggest that a comprehensive federal regulatory law is very likely not the most effective or efficient vehicle for carrying out the necessary policy challenges posed by improved watershed management. First, watersheds vary across many dimensions throughout the national landscape and respond primarily to local land use and resource use actions. Also, support for centralized regulation of natural resources in general eroded during the 1990s as the desire for more state and local control of key land use decisions intensified (Esty 1996; Nolon 2002).

We do not suggest, on the other hand, that the federal government remove itself entirely from the objective of influencing state and local ecosystem service initiatives. As the discussion of the Coastal Zone Management Act (CZMA) in chapter 5 illustrates, federal law can be useful as a motivator for state action without intruding on basic design choices. There undoubtedly are some national objectives, which, while not lending themselves to nationally uniform standards, may nonetheless justify federal support for states that satisfy the national concerns as they become increasingly and more formally involved in ecosystem service management within their boundaries. Like the CZMA, a federal ecosystem service initiative could express broad national goals and standards and establish a mechanism for states to submit their respective watershed management programs for federal approval, offering in return federal financial support for design and implementation as well as the commitment that federal agencies will not carry out, fund, or authorize actions inconsistent with the state plan. At the very least, the federal government can and ought to maintain an important role as a source of scientific data and research that has broad usefulness to state-based institutions and as an environmental engineering contractor, such as through the Corps of Engineers. At the most, however, the federal government might consider ways to influence state policy through a statute, like the CZMA, that provides cooperative support for state action. Full-blown command-and-control federal regulation imposing ecosystem service policy is ill advised.

For different but equally as compelling reasons, however, Ruhl et al. (2003) argue that effective management regimes cannot rely exclusively on the initiative of local governance, particularly if channeled through conventional local political entities. Even putting aside the lack of match between conventional local political boundaries and watersheds, local governments face several constraints to effective watershed management. First, while most state political systems allow considerable local authority—certainly enough to establish watershed ordinances—management of transboundary effects is often outside their authority or able to be accomplished only through burdensome interlocal coordination procedures (Tarlock 2002). Second, many watershed management issues will present difficult political choices with potentially significant economic consequences, and local governments, particularly those constituted by popular election, may be reluctant to make economic sacrifices not being made by others. Finally, even with most local governments committed to watershed management, it is doubtful that all could afford the intensive scientific, social, and economic data gathering and analysis necessary to carry it out effectively. It is not surprising, therefore, that soil and water conservation districts, which in many states are elected and have political boundaries corresponding to county borders, have generally failed to live up to their promise of comprehensively managing soil and water quality issues (Davidson 2002). The emerging generation of "place-based" resource management proposals, while stressing local autonomy, should strive to avoid repeating that history.

Hence, on the one hand, there is good reason to believe that the federal government should not attempt to initiate a sweeping federal regulatory scheme for national management of ecosystem services. On the other hand, ecosystem service management cannot effectively rely exclusively on the initiative and authority of local governance. States, therefore, will likely have to carry the primary burden of designing and empowering the institutional structure. Nevertheless, several of the design parameters also suggest that states should design their internal political frameworks around a hierarchy of physical units, most likely watersheds, and should consider ways to achieve interstate coordination of their respective management efforts. To be sure, one advantage of initiating ecosystem service management at the state level is to accommodate policy diversity across states and within states, but by sharing the same basic governance framework states can more freely exchange data and experience and thus work in a more coordinated and efficient pattern to solve both intrastate and interstate problems. Historical experience also indicates that state-based institutions can effectively exercise "horizontal federalism" in managing interstate transboundary resource management issues on an ecosystem-level scale, such as the Great Lakes states have recently accomplished through state compact (Hall 2006). Accordingly, Ruhl et al. (2003, 939) concur in the National Research Council's finding that

> [o]rganizations for watershed management are most likely to be effective if their structure matches the scale of the problem. Individual local issues related to site planning, for example, should be the purview of local self-organized watershed councils, while larger organizations should deal with broader issues. These larger organizations, however, must include the nested smaller watershed groups within their area of interest, and must account for downstream interests. (1999, 15)

Hence, borrowing, combining, and enhancing a number of features from a variety of watershed management examples, we and our colleagues designed a multitiered approach for watershed management that distributes funding, authority, and other resources in a way that can address many of the design parameters discussed above. In particular, this approach aims to establish legitimacy for watershed management at the local level while not sacrificing broader state and regional concerns. The framework relies on creating and coordinating institutions at three levels of government: (1) the state watershed management agency, (2) appointed regional watershed coordination districts, and (3) elected local watershed management councils. Each level of government must prepare a watershed management plan for its respective scale of focus. In the case of the regional and local entities, the plan must be consistent with the plan that is vertically above it in the tiered system. The state agency would continue to direct policy for matters of statewide concern, including developing a state watershed management plan, but would delegate most watershed management policy development, implementation, and enforcement authority to the regional districts. The regional districts would develop regional plans to implement the state plan and would be the locus of most planning and policy expertise. They would have staffs including engineers, biologists, economists, hydrologic modelers, information specialists, conservation experts, and lawyers. Yet the regional districts would rely in large part on the elected local boards for final policy development, implementation, and enforcement.

To fulfill this role, the local councils must be not only more than mere "special districts," lest they wither the way many other special district initiatives have, but also more than conventional local governments. The local councils would be organized around watershed-based boundaries and held accountable to state and regional interests through the requirement that their local plans be consistent with the regional (and thus state) plans. Perhaps even more important, local councils would coordinate the review of all land use decisions by other existing state and local authorities, such as state highway agencies and municipal and county zoning authorities, for consistency with the state, regional, and local watershed management plans, thus extending the policy reach of watershed planning beyond the direct management of water resources.

This framework allows an institutional structure that matches the physical realities of watershed resources in both vertical and horizontal dimensions. The vertical integration of local, regional, and state planning and regulatory authority matches the nested hierarchies of watershed scales. The ability at each level of this structure horizontally to examine the impacts that decisions of other governmental authorities have on watershed resources matches the dynamics of watershed processes at each physical scale. Accounting for each of these dimensions in the institutional design is necessary to successful implementation of watershed management, but neither is sufficient alone.

Building on this three-tiered approach for watershed management, one can easily envision a legislative initiative expanding the scope beyond water resources to establish an institutional structure for comprehensive natural capital and ecosystem service management. It would look something like the following in terms of structural integration of authority and responsibility:

State Ecosystem Resources Management Agency

All states already have a state agency, if not multiple agencies, responsible for developing law and policy for environmental protection and the management of natural resources. All such responsibilities would be consolidated into a single state agency or division referred to perhaps as the State Ecosystem Resources Management Agency. This agency would continue to serve as the original authority for statewide environmental regulation, including implementation of federal laws such as the Clean Water Act. Under the state ecosystem resources law, however, the agency would also be required to

1. Prepare a State Ecosystem Resources Management Plan specifying the goals for ecosystem resources management in the state, including the conservation of natural resources based on their capacity to supply ecosystem service values.
2. Specify the format for all state, regional, and local agencies to evaluate the potential impacts of proposed actions and decisions on the quality, quantity, and distribution of ecosystem services.
3. Delegate to Regional Ecosystem Resources Coordination Agencies responsibility for implementing programs that affect "matters primarily of regional or local ecosystem resources significance." Matters of primarily regional or local significance would be defined in the statute to include (1) rules and decisions specified in the statute and (2) any other types of rules or decisions that the State Ecosystem Resources Management Agency prescribes by rule.
4. Include in its deliberations on statewide decisions and rules within its authority any information and comments supplied by the Regional Ecosystem Resources Coordination Agencies.
5. Review the Regional Ecosystem Resources Management Plans for compliance with the State Ecosystem Resources Management Plan and provide corrective elements in the case of a plan that is deficient.
6. Review and comment on the actions of all other state and regional agencies and, to the extent permitted, of federal agencies that are deemed to have "substantial ecosystem resources effects." Substantial ecosystem resources effects are any effects the State Ecosystem Resources Management Agency concludes could substantially interfere with the State Ecosystem Resources Management Plan, any Regional Ecosystem Resources Management Plan, or any Local Ecosystem Resources Management Plan.

REGIONAL ECOSYSTEM RESOURCES COORDINATION AGENCIES

The Regional Ecosystem Resources Coordination Agencies would be organized to the extent practicable along watershed boundaries, such as the 222 subregional hydrological units the U.S. Geological Survey (2006) has defined for the nation, as constrained by state boundaries. The Regional Ecosystem Resources Coordination Agencies would be appointed boards with significant staff and budgets. Because they would take over many functions previously managed by the state agency, their budgets would be state appropriated. Each Regional Ecosystem Resources Coordination Agency would do the following:

1. Establish the Local Ecosystem Resources Management Council boundaries as it deems appropriate, but to the maximum extent practicable according to the 2,150 U.S. Geological Survey ecosystem resources cataloging units. (See United States Geological Survey [2006].)

2. Establish a Regional Ecosystem Resources Management Plan demonstrating how it will satisfy compliance with (1) all federal and state laws governing the quality and quantity of ecosystem resources, and (2) the State Ecosystem Resources Management Plan.

3. Decide all the matters of primarily regional ecosystem resources significance that are prescribed in the statute or by the State Ecosystem Resources Management Agency.

4. Review Local Ecosystem Resources Management Plans and develop one for any Local Ecosystem Resources Management Council that fails to meet the State Ecosystem Resources Management Plan and Regional Ecosystem Resources Management Plan criteria.

5. Identify Special Ecosystem Resources Areas, which are areas of natural resources that produce ecosystem services of exceptional value.

6. Define the criteria for land use and water project developments to be classified as a Development of Regional Ecosystem Resources Impact.

7. Review local government land use and water project decisions that are either (1) in Special Ecosystem Resources Areas or (2) for a Development of Regional Ecosystem Resources Impact, and impose the conditions it deems necessary to ensure compliance with the Regional Ecosystem Resources Management Plan.

8. Hear appeals from local governments and citizens of Local Ecosystem Resources Management Council decisions on local government development matters, including whether a project is in a Special Ecosystem Resources Area or is a Development of Regional Ecosystem Resources Impact.

9. Provide the scientific, economic, and social data gathering and analysis capacity for implementation of the Regional Ecosystem Resources Management Plan and the various Local Ecosystem Resources Management Plans within its jurisdiction. This will include geographic databases of all Special Ecosystem Resources Areas as well as other information on ecosystem resources and ecosystem services in the region.

10. Notify the State Ecosystem Resources Management Agency of any state agency or regional agency action it believes may substantially interfere with the Regional Ecosystem Resources Management Plan.

11. Maintain an ongoing partnership with federal agencies, such as the Army Corps of Engineers, with respect to past, present, and future public and private environmental development projects that may have a substantive

impact on the achievement of the Regional Ecosystem Resources Management Plan.

LOCAL ECOSYSTEM RESOURCES MANAGEMENT COUNCILS

The Local Ecosystem Resources Management Councils would be generally elected local governmental bodies. They would have the following authorities and responsibilities:

1. Prepare a Local Ecosystem Resources Management Plan demonstrating how it will achieve compliance with the Regional Ecosystem Resources Management Plan.
2. Review all land and resource project development applications affecting ecosystem resources within its jurisdiction. State, regional, and local governments would be required to provide advance notice to the Local Ecosystem Resources Management Council of such proposed actions and decisions and an evaluation of the impact of their proposed actions and decisions to the quality, quantity, and distribution of ecosystem services in the format specified by the State Ecosystem Resources Management Agency. The Local Ecosystem Resources Management Council then would either (1) find the matter has no significant impacts to local ecosystem resources, regional ecosystem resources, or Special Ecosystem Resources Area, in which case the agency will take no action; or (2) for those matters the Local Ecosystem Resources Management Council deems to have the potential for significant local ecosystem resources impacts, provide the local government the conditions the Local Ecosystem Resources Management Council deems necessary to ensure compliance with the Local Ecosystem Resources Management Plan; or (3) for those matters the Local Ecosystem Resources Management Council deems to be located in Special Ecosystem Resources Areas or to constitute a Development of Regional Ecosystem Resources Impact, refer the review to the Regional Ecosystem Resources Management District.
3. To acquire (including by eminent domain) and manage lands it deems of importance to local ecosystem resources management and fulfillment of the Local Ecosystem Resources Management Plan.
4. To establish incentive and support programs for conservation of natural resources based on their capacity to supply ecosystem services.
5. To finance its operations and implement its Local Ecosystem Resources Management Plan through appropriate rules, subsidies, tradable rights, property taxes, recreational user fees, water utility fees, and development permit fees, including fees levied as a surcharge portion of local government

property taxes, user fees, water utility charges, and development permits, and through bonds.

6. To notify the State Ecosystem Resources Management Agency of any state agency or regional agency action it believes may substantially interfere with the Local Ecosystem Resources Management Plan.

Neither this proposal, nor the ecosystem service districts proposed by Heal et al. (2001), nor the environmental broker agencies suggested by Thompson (2000) is likely to materialize in full form without a careful exploration of drivers, a methodical construction of robust models, a keen appreciation of trade-offs inherent in different policy options, a coherent plan for dealing with transition, and a willingness to test different combinations of instruments ranging from the common law, to regulation, to purely economic measures. With those predicates in place, however, institutions such as these could mesh effectively into political, economic, and social contexts to produce a fuller accounting of natural capital and ecosystem service values in resource decision making.

Conclusion

Ecosystem services are complex ecologically, geographically, and economically. So are invasive species, climate change, nanotechnology, poverty, and a host of other "wicked" issues that challenge law and policy. The difference is that all those problems have found the attention of policymakers and have been addressed, albeit with varying success, in tangible ways through law and policy, whereas ecosystem services have been largely ignored.

The central purpose of this book is to focus debate on the essential need to construct a law and policy of ecosystem services and how it can be configured. Part I emphasized that ecosystem services are complex in all their dimensions, but that the disciplines of ecology, geography, and economics are making tremendous strides in forming qualitative and quantitative understandings of the value of natural capital and the ecosystem services it provides at various scales. Part II thoroughly surveyed the relevant sources of policy at society's disposal for managing natural capital and ecosystem service values—property rights, regulation, and social norms—showing each to be woefully insufficient in their present configurations. In short, parts I and II showed we know with certainty that natural capital and ecosystem services are valuable and of critical importance to the continuance of modern society, but also that law and policy do not adequately take those values into account. Ecological economics professor Robert Costanza has defined such failures as *social traps*, situations "in which the short-run, local reinforcements guiding individual behavior are inconsistent with the long-run, global best interest of the individual and society" (1987, 408). This book argues that ecosystem services are caught in such a social trap.

Parts I and II thus present a diagnosis of this social trap we call the Tragedy of Ecosystem Services. Like Garrett Hardin's (1968) famous tragedy of the commons, the Tragedy of Ecosystem Services lies in the "remorseless working of things," in the systematic failure of institutions to take important values into account and to provide an informational feedback to managers of critical resources so that they respond to these values in allocating resources. But the Tragedy of Ecosystem Services is a different story than the tragedy of the commons.

In Hardin's commons (more accurately an open access regime) the road to folly lies in *over-consumption*—too many cattle graze the pasture, too many fish are caught from the ocean, too many greenhouse gases are released to the atmosphere, too much fertilizer runs off to the streams and rivers—because the benefits of an additional cow, fish caught, mile driven, or bushel produced accrue directly to the owner while the marginal costs of over using the commons are distributed widely to the larger community as *negative* externalities. The Tragedy of Ecosystem Services, by contrast, is a case of *under-production* that happens because mechanisms are missing for rewarding investments in natural capital that produce ecosystem services. No one devotes resources to improving the pasture or habitat for fish in the ocean or sequestering carbon or restoring wetlands that filter water pollutants because they would not receive a resulting revenue stream to finance the investments made. Rather the ecosystem services made possible by these investments are public goods, they are common pool benefits that accrue over time to a population within a geographical area affected by the improved ecosystem service flows. Once produced and made available to the larger community as *positive* externalities, no one in this geographical area can be denied the benefits, and the owner of the natural capital therefore lacks a means to charge the beneficiaries, even if they would be willing to pay the cost of ecosystem service provision and even if the generation of ecosystem services is the highest and best use of natural capital resources such as land. For beneficiaries of ecosystem services, there is great temptation to be a free rider, to let others pay for ecosystem service provision while still enjoying them. For owners of natural capital, the ecosystem services they happen to produce are *positive* externalities, and rarely does anyone have a right to their continuance, no matter how valuable they have become.

How has this happened? We have argued that property, the foundation of much of our common law, has evolved over the centuries to facilitate the conversion of natural capital, nearly always in surplus, into economic goods to serve the driving need of the socioeconomic system to increase manufactured capital as the infrastructure of an industrial society and to increase consumable goods as an investment in human capital. The resulting economic growth has, of course, delivered the very positive development of a middle class society. American common law has in fact carefully constructed a wall, preventing consideration of ecosystem services from interfering with this process of converting natural capital into other more valuable economic goods. But natural capital is no longer generally in surplus, and so its conversion into other goods has a rising ecological opportunity cost, as well as an implicit economic opportunity cost, in the form of increasingly scarce ecosystem services lost in the conversion.

Reflecting this ecological opportunity cost in the definition of property, and thereby signaling its value in decisions governing natural capital, is a central challenge for the evolving law and policy of ecosystem services. Consider, for example, how intellectual property rights have rapidly evolved as an economy based on knowledge has developed over recent decades, with intellectual capital rapidly eclipsing manufactured capital as the most important input to economic production. So we do have evidence of the resilience and adaptability of the common law to meet the challenges posed by a rapidly changing world. The question is how to motivate the same development in the law of natural capital.

Part III presented nine empirical case studies demonstrating how the themes elucidated in parts I and II apply to specific parcels of land and to the hydrologic cycle. It then explored the realm of agricultural land use and watershed management starting with the Conservation Reserve Program and National Conservation Buffer Initiative as important domestic ecosystem service subsidy programs, then shifting across the Atlantic to present the recently more determined shift from crop-based (amber) subsides to ecosystem service-based (green) subsidies in the European Union. These policies were then analyzed for a typical watershed using place-based process models. Part III then investigated the successes, failures, and potential of market-based instruments for encouraging pollution reductions, investments in natural capital, and the consequent delivery of ecosystem services in the arenas of wetland mitigation banking and tradable pollution permits for sulfur emissions, nutrient discharges, and carbon fluxes.

There are no simple solutions to the Tragedy of Ecosystem Services and we have not attempted to provide a silver bullet. Nevertheless, with a proper diagnosis we can point to potential cures along three avenues that will guide the transition from the status quo to more desirable conditions: (1) changes in the common law of property as described above, (2) readjusting the economic playing field into an ecological-economic playing field by signalling the value of ecosystem services in decisions over the allocation of natural capital, and (3) the development of geographically defined governmental institutions for the regulation of natural capital and the provision of ecosystem services as public goods. In this last regard, part IV surveyed a toolbox of instruments to be used by a proposed institutional structure built around a three-tiered array of state, regional, and local agencies. The discussion was necessarily exploratory in this respect, raising many more questions than it answered. But the exploration is not of a blank slate, as experience with other environmental issues lends itself to the consideration of different options. While not a panacea for our environmental challenges, there is good reason to believe that, were the institutional structure proposed in part IV or some variation thereof to be put into place, it would substantially contribute to a more complete accounting of natural capital and ecosystem service values in our economy.

Just as there is no silver bullet to bring ecosystem services fully into the fold of law and policy, neither are ecosystem services the silver bullet that will deliver environmental law and policy from the problems of our present and future. They are simply one facet to be considered, and in that respect imply a purely utilitarian calculus. For reasons having nothing to do with efficient, sustainable use of land and resources, society may wish to exploit more or less natural capital than its economic value suggests. And to be candid, even if efficiency were all that mattered, we will never know everything needed to make perfectly efficient decisions over the long run and into future generations. It is beyond debate, however, that we could be making far better decisions, throughout government and the economy, than we are today when it comes to natural capital and ecosystem services. This book is devoted to that end.

Endnotes

Introduction

1. A wealth of physical data about the ACF River basin is available in the Draft Environmental Impact Statement (EIS) the United States Army Corps of Engineers (1998) prepared in connection with its ongoing management of reservoirs and navigation on the river system. Much of the cultural background presented here is based on the Draft EIS and a series of probing articles journalist Bruce Ritchie (2001a–2001d) wrote for the *Tallahassee Democrat* during November 2001.
2. Moore (1999) and Stephenson (2000) provide more details in their comprehensive histories of the early stages of the tristate water dispute.
3. The congressional allocation and interstate compact mechanisms for interstate water apportionment are constitutionally enabled means of resolving such matters. The Supreme Court first announced its power to apportion interstate waters in *Kansas v. Colorado*, 206 U.S. 46 (1907).
4. The Court provided its most complete exposition on the equitable apportionment doctrine in *Nebraska v. Wyoming*, 325 U.S. 599 (1945).
5. As Moore (1999, 67) observes, "the 'natural flow regime' approach to allocation proposed by Florida elevates environmental concerns to a new level in water quantity disputes."
6. *Nebraska v. Wyoming*, 515 U.S. 1, 12–13 (1995).
7. The Court extended the doctrine from water to fish and other resources in *Idaho v. Oregon*, 462 U.S. 1017 (1983). Ruhl (2003) has suggested that this case may be the critical piece of the Court's jurisprudence for integrating a broader vision of ecosystem services into interstate water dispute resolution.
8. Although references to a duty to conserve could be found in earlier cases, the first instance in which the Court clearly articulated and applied the duty in an interstate water dispute was in the relatively recent case of *Colorado v. New Mexico*, 459 U.S. 176 (1982).
9. This high standard of injury was first articulated by the Court at the beginning of

the 20th century in *Missouri v. Illinois,* 200 U.S. 496 (1906), in which Missouri alleged that Illinois had fouled the Mississippi River by reversing the flow of the Chicago River to lead to the Illinois River and dumping Chicago's sewage into it rather than into Lake Michigan. The Court was skeptical of Missouri's scientific theory of waterborne disease and convinced by Illinois' argument that Chicago's sewage would be diluted by water from Lake Michigan.

Chapter 1

1. 16 United States Code § 1271.
2. 16 United States Code § 1531(a)(3).
3. 43 United States Code § 1701(a)(8).

Chapter 3

1. The assumptions for the perfect competitive economy include (1) a large number of small-sized firms or economic agents relative to the whole market; (2) sellers are price takers; there are no barriers to market entry; (3) products of a similar type are homogeneous; (4) all resources are perfectly mobile and can readily move in and out of markets in response to changes in prices; (5) participants in the markets have perfect information; and (6) there is a lack of governmental involvement. Not only do we have the assumptions for the perfectly competitive economy, we also have a set of requirements related to an efficient structure of property rights so that there can be a well-functioning marketing economy. Generally these requirements include (1) universality—all resources are owned and entitlements are completely specified, (2) exclusivity—all benefits and costs of using/owning a resource accrue to the owner/user, (3) transferability—all property rights are transferable through voluntary exchange, and (4) enforceability—property rights are secure from involuntary seizure or encroachment by others (Tietenberg 2006).
2. Or, what economists frequently refer to as comparing marginal cost (MC) to marginal revenue (MR).
3. Other factors influencing consumer demand include taste and demographics, and prices of substitutes and complements.
4. Named after Vilfredo Pareto (1848–1923), the person who first suggested it.
5. The discount rate for the individual will reflect her cost for capital, her risk preferences, and inflationary expectations.
6. *Private* in this sense should not necessarily be confused with the form of property ownership commonly referred to as "private property." Other forms of property ownership, covered in chapter 4, include open access, common property, and state property (Burger and Gochfeld 1998). As chapter 4 explains, each form of ownership can give rise to its own set of specific problems when combined with different forms of natural capital and the ecosystem services they provide.
7. While air is characterized as a public good, from the perspective of a regime of property ownership it is frequently referred to as an open-access resource—one for

which there is no effective owner determining who has access, when, and how. However, increasingly through legislation such as the Clean Air Act and its amendments, the public is beginning to exercise some control over these issues of access.

8. There are a number of reasons (Bator 1958) for which market failure can develop, and we will be examining some of them as they relate to ecosystem services.

9. Externalities result from the requirement of exclusivity referred to in note 1 being violated. That is, as a consequence of production and/or consumption activities, there are spillover effects that negatively or positively impact the consumptive or productive activities of others.

10. A third was mentioned as well but not developed—open-access resources, such as air.

11. Another one of the regimes of property ownership is state property, which includes ownership by local, state, and federal units of government. However, even with this form of ownership there can be significant disputes about how the resources are going to be used and who is going to have access to units. All of which can have implication for the diversity, quality, and quantity of the associated ecosystem services—for example, use of federal lands for grazing and use of the U.S. Forest Service lands. These concerns are covered in more detail in chapter 4.

12. While in this example we will be considering one ecosystem service and its potential value, the reader is encouraged to remember that a parcel of natural capital such as described in the example could well be providing a suite of ecosystem services, each one of which is facing unique challenges of identification and valuation for the purposes of resource allocation.

13. This also assumes there are not any additional ecosystem services of interest derived from the same unit(s) of natural capital.

14. These positive and negative economic effects resulting through these chains of connected ecosystems are the same types of positive and negative externalities discussed earlier if they are not captured and reflected by existing market forces. Given our discussion so far in chapters 1, 2, and 3, we would expect that most of these effects would be some type of externality.

15. Opportunity costs refer to what a resource could earn in its next best alternative use. In the context of the contingent valuation survey, since the respondent really does not have to spend her money as she is asked to in the survey, it is easy for her to say she would be willing to pay more than she would if she actually had to sit down and write a check for it or if there was a tax levied on her purchases to pay for the desired service.

16. As defined earlier, an open-access resource is one that violates the requirement of universality for an efficient system of property rights for a market-based economy.

17. For many ecosystem services for which there are no markets, a critical challenge is determining the relevant frames of reference for developing values: geographical and temporal scales. If a wetland's flood mitigation services are valued only before a storm, a different value will be obtained than if they are valued during a drought. Market prices for many commodities and services are reported by economists at a number of spatial and temporal scales designed for different uses. Our challenge is to match the ecosystem services to the relevant scales for the appropriate uses.

Chapter 4

1. A classic example of this judicial behavior is *Madison v. Ducktown Sulphur Copper & Iron Co.,* 113 Tenn. 331, 83 S.W. 658 (1904), in which the Tennessee Supreme Court found that, despite overwhelming evidence that fumes from several copper smelting plants in eastern Tennessee had caused substantial injury to nearby agricultural lands, the plants were not committing a private nuisance. Later, in *Georgia v Tennessee Copper Co.,* 206 U.S. 230 (1907), the United States Supreme Court found that fumes from the plants had caused a public nuisance to citizens in Georgia and required the plants to reduce emissions.

2. The landmark case for this development, a staple of law school property classes, is *Boomer v. Atlantic Cement Co.,* 26 N.Y.2d 219, 257 N.E. 2d 219 (1970), in which New York's highest court abandoned the rule that the exclusive remedy for private nuisance was injunctive relief, shifting instead to a compensatory damages approach.

3. For example, see Dukeminier et al. (2006), the leading property law text for first-year law students, which covers the case in the first chapter.

4. *Van Ness v. Pacard,* 27 U.S. (2 Pet) 137, 145 (1829).

5. *McNeal v. Assiscunk Creek Meadow Co.,* 37 N.J. Eq. 204, 209 (1883).

6. *Kerwhacker v. Cleveland, Columbus & Cincinnati Railroad Co.,* 3 Ohio St 172, 179 (1854).

7. *United States v. Riverside Bayview Homes, Inc.,* 474 U.S. 121 (1985). Other cases in which courts assign similar value to wetlands are discussed in Blumm and Ritchie (2005, 337).

8. 505 U.S. 1003 (1992).

9. 505 U.S. at 1029.

10. 505 U.S. at 1035 (emphasis added).

11. 505 U.S. at 1031 (citing *Curtin v. Benson,* 222 U.S. 78, 86 [1911]).

12. 201 N.W.2d 761 (Wis. 1972).

13. For a famous example, see *Fontainebleu Hotel Corp. v. Forty-Five Twenty-Five, Inc.,* 114 So.2d 357, 359 (Fla. Ct. App. 1959) (noting that the English doctrine "has been unanimously repudiated in this country").

14. *Southwest Weather Research, Inc. v. Rounsaville,* 320 S.W.2d 211 (Tex. Civ. App. 1958).

15. *Prah v. Maretti,* 321 N.W.2d 182, 189 (Wis. 1982)

16. 505 U.S. at 1031 (emphasis added).

17. *Hastings v. Crukleton,* 3 Yeates 261, 262 (Pa. 1801).

18. The Property Clause of the United States Constitution provides that "Congress shall have the power to dispose of and make all needful Rules and Regulations respecting the Territory or other Property belonging to the United States." U.S. Const. Art. IV, § 3, cl. 2.

19. 16 United States Code § 531.

20. *The Clinch Coalition v. Damon,* 2004 U.S. Dist. LEXIS 7933 (W.D. Va. May 6, 2004).

21. *Public Lands Council v. Babbitt,* 529 U.S. 728 (2000) (quoting language from the Public Rangelands Improvement Act, 43 U.S.C. § 1901(a)(1)).

22. USDA Press Office, Transcript Release No. 0335.05, available through the USDA

home page, http://www.usda.gov/wps/portal, under Newsroom/Transcripts and Speeches.

23. 16 United States Code § 1531(c)(2).
24. *New Mexico v. General Electric Company,* 335 F.Supp.2d 1185, 1202 (D.N.M. 2004)
25. Nevada is the clearest example. See, for example, *Nevada v. Morros,* 766 P.2d 263 (Nev. 1988).
26. *Lake Shore Duck Club v. Lake View Duck Club,* 50 Utah 76 (1917).
27. 146 U.S. 387 (1892). ·
28. Portions of the history of the public trust doctrine appearing in this section are reproduced from Ruhl (2005a) with permission.
29. *National Audubon Society v. Superior Court of Alpine County,* 658 P.2d 709, 728 (Cal. 1983).
30. *Just v. Marinette County,* 201 N.W. 2d 761 (Wis. 1972).
31. Prominent examples include *Friends of Van Cortland Park v. City of New York,* 750 N.E.2d 1050 (N.Y. 2001) (doctrine covers public parks); In re Water Permit Applications, 9 P.3d 409 (Haw. 2000) (doctrine covers groundwater because absolute private ownership of groundwater never existed in Hawaii property law); *Weden v. San Juan County,* 958 P.2d 273, 284 (Wash. 1998) (doctrine may be used to regulate personal watercraft use on state waters to protect water resources and wildlife); *Pullen v. Ulmer,* 923 P.2d 54 (Alaska 1996) (fish occurring in their natural state are covered under the doctrine); *Vander Bloemen v. Wisconsin Department of Natural Resources,* 1996 WL 346266 (Wis. Ct. App. 1996) (doctrine extends to protection of lakeside ecology); *Selkirk-Priest Basin Association, Inc. v. Idaho ex rel. Andrus,* 899 P.2d 949 (Idaho 1995) (doctrine allows environmental group standing to challenge timber sales on ground that sedimentation could injure fish spawning grounds); and *Marks v. Whitney,* 491 P.2d 374, 380 (Cal. 1971) (doctrine includes "preservation of those [trust] lands in their natural state").
32. See *Allegretti & Co. v. County of Imperial,* 138 Cal.App.4th 1261 (2006).
33. 505 U.S. 1003 (1992).
34. 505 U.S. at 1029.
35. 505 U.S. at 1031.

Chapter 5

1. Portions of Ruhl 2005a are reproduced in this section with permission.
2. 206 U.S. 230 (1907).
3. *Id.* at 238–239.
4. 237 U.S. 474 (1915).
5. *Attorney General v. Sheffield Gas Consumers Co.,* 3 De G.M. & G. 304, 320 (1853).
6. *City of Chicago v. Commonwealth Edison Co.,* 24 Ill. App. 3d 624, 632 (1974).
7. *New Mexico v. General Electric Co.,* 335 F. Supp. 2d 1185 (D.N.M. 2004).
8. *Palazzolo v. Rhode Island,* 2005 WL 1645974 (R.I. Superior Ct. 2005) (emphasis added).
9. 26 N.Y.2d 219, 257 N.E.2d 870, 871 (1970).

10. *Connecticut v. American Electric Power Co., Inc.*, No. 04 Civ. 5669 (LAP) (S.D.N.Y., opinion and order of Sept. 15, 2005).
11. 42 United States Code § 1344(a).
12. *Solid Waste Agency of Northern Cook County v. United States Army Corps of Engineers*, 531 U.S. 159 (2001).
13. 42 United States Code § 1344(b)(1).
14. Portions of Ruhl and Gregg (2001) are reproduced in this section with permission.
15. *Id.* at 9,212 ("In determining compensatory mitigation, the *functional values* lost by the resource to be impacted must be considered.") (emphasis added).
16. Va. Code Annotated § 62.1–44.15:5.
17. Neb. Administrative Code Ch. 7, § 101.
18. Conn. General Statutes Annotated § 22a–28.
19. Conn. Agencies Regulations § 22a–39–6.1.
20. 16 United States Code §§ 1451–1465.
21. 16 United States Code § 1451(i).
22. 16 United States Code § 1455(d)(2).
23. 16 United States Code § 1455(d)(8).
24. 16 United States Code § 1455(d)(9).
25. 31 Tex. Administrative Code § 501.24.
26. N.J. Statutes Annotated § 13:9A-1.a.
27. Fla. Administrative Code Annotated §§ 62B-33.005(4)-(5).
28. Mass. Regulations Code title 301, § 21.98(6).
29. Portions of Nagle and Ruhl (2002) are reproduced in this section with permission.
30. 1885 N.Y. Laws ch. 283, p. 482.
31. Act of March 3, 1891, ch. 561, 26 Stat. 1095, 1103.
32. 30 Stat. 11, 34-35.
33. 15 United States Code § 473.
34. 16 United States Code § 528.
35. *Parker v. United States*, 307 F. Supp. 685 (D.Colo.1969).
36. *Sierra Club v. Hardin*, 325 F. Supp. 99, 123 (D. Alaska 1971).
37. 522 F.2d 945 (4th Cir.1975).
38. Public Law No. 94-588, 90 Stat. 2949, codified at 16 United States Code §§ 1601-14.
39. 16 United States Code § 1604(g)(3)(B).
40. 36 Code of Federal Regulations part 219.
41. S.C. Code Annotated title 48, chs. 23–36.
42. Minn. Statutes Annotated § 89.001
43. Oregon Administrative Regulations chapter 629.\

Chapter 7

1. The actual year in which Burnham ceases to exist is difficult to determine since the people gradually moved away and the buildings were either torn down or moved south to Arcadia.

Chapter 9

1. Public Law 95-192.
2. Public Law 95-192 § 5(a)1.
3. T is the amount of soil measured in tons per acre per year that an acre of land can lose without impairing its long-term productive capacity. For example, for a parcel of midwestern soil with a T level equal to 5 tons per acre to be eligible in the first CRP sign-up, it would have to have had an average soil loss of 10 or 15 tons per acre per year depending on its Land Capability Class.
4. Conservation Compliance was another provision of Title XII of the 1985 Food Security Act that required producers with highly erodible land who wanted to remain eligible for USDA program benefits to obtain an approved conservation plan from SCS by January 1, 1990 and to have the plan fully implemented by January 1, 1995.

Chapter 12

1. This is currently being pursued by Lant et al. in a project funded by the National Science Foundation Biocomplexity in the Environment program. Entitled "Virtual Watershed: Agricultural Landscape Evolution in an Adaptive Management Framework," the project is investigating the relationship between economic circumstances and policies on the one hand and ecological-economic outcomes for the watershed on the other.

Chapter 13

1. Portions of Ruhl and Salzman (2006) are reproduced in this section with permission.

Chapter 18

1. 505 U.S. at 1031 (emphasis added).
2. Portions of Ruhl (2005a) are reproduced in this section with permission.
3. 505 U.S. at 1031 (citations omitted).
4. 505 U.S. at 1031.
5. *Barasich v. Columbia Gulf Transmission Co.,* Civil Action 05-4161 (E.D. La) (complaint filed Sept. 13, 2005) (alleging that oil companies' dredging of pipeline channels in Louisiana's coastal wetlands tortiously degraded storm surge mitigation capacity during Hurricane Katrina). Although the court dismissed the action due to the concern, expressed in the *Boomer* and *American Electric Power* cases discussed in chapter 5, that the case involved too many defendants spread over too amorphous an area, the court noted that "a more focused, less ambitious lawsuit between parties who are proximate in time and space, with a less attenuated connection between the defendant's conduct and the plaintiff's loss, would be the way to test their theory." *Barasich v. Columbia Gulf Transmission Co.,* 2006 WL 3333797, *18 (E.D. La. Sept. 28, 2006).

6. 2005 WL 1645974 (R.I. Super. Ct., July 5, 2005).
7. 2005 WL at *3.
8. 2005 WL at *5, 7.
9. 42 United States Code § 4332(2)(B).
10. Portions of Ruhl et al. (2003) are reproduced in this section with permission.

References

Abel, Thomas, and John R. Stepp. 2003. A new ecosystems ecology for anthropology. *Conservation Ecology* 7(3): 12, available at http://www.consecol.org/vol7/iss3/art12.

Abrams, Robert H. 2002. Interstate water allocation: A contemporary primer for eastern states. *University of Arkansas at Little Rock Law Review* 25(1): 155–73.

Acheson, James M. 1988. *The lobster gangs of Maine.* Hanover, NH: New England University Press.

Acheson, James M., and Jennifer Brewer. 2000. Capturing the commons: Social changes in the territorial system of the Maine lobster industry. Paper presented at Capturing the Commons: Crafting Sustainable Commons in the New Millennium, Eighth Conference of the International Association for the Study of Common Property, Bloomington, IN, available at Digital Library of the Commons, http://dlc.dlib.indiana.edu/archive/00000196/.

Acheson, James M., and Robert S. Steneck. 1997. Bust then boom in the Maine lobster fishery: Perspectives of fishers and biologists. *North American Journal of Fisheries Management* 17(4): 826–47.

Acheson, James M., and Laura Taylor. 2001. The anatomy of the Maine lobster comanagement law. *Society & Natural Resources* 14(5): 425–41.

Acheson, James M., James A. Wilson, and Robert S. Steneck. 1998. Managing chaotic fisheries. In *Linking social and ecological systems: Management practices and social mechanisms for building resilience,* ed. Fikret Berkes and Carl Folk, 390–413. Cambridge, UK: Cambridge University Press.

Ackerman, Frank, and Lisa Heinzerling. 2004. *Priceless: On knowing the price of everything and the value of nothing.* New York: The New Press.

Adams, Jane, Steven Kraft, J. B. Ruhl, Christopher Lant, Timothy Loftus, and Leslie Duram. 2005. Watershed planning: Pseudo-democracy and its alternatives—The case of the Cache River watershed, Illinois. *Agriculture and Human Values* 22(3): 327–38.

Adams, William H., Dan Brockington, Jane Dyson, and Bhaskar Vira. 2003. Managing tragedies: Understanding conflict over common pool resources. *Science* 302(5652): 1915–16.

Agrawal, Arun. 2002. Common resources and institutional sustainability. In *The drama of the commons,* ed. Elinor Ostrom, Thomas Dietz, Nives Dolsak, Paul C. Stern, Susan Sontich, and Elke U. Weber, 41–86. Washington, DC: National Academy Press.

Agricultural Stabilization and Conservation Service (ASCS). 1986. Conservation Reserve Program; Interim Rule. *Federal Register* 51(49): 780–87.

Alcamo, Joseph, Detlef van Vuuren, Claudia Ringler, Wolfgang Cramer, Toshihiko Masui, Jacqueline Alder, and Kerstin Schulze. 2006. Changes in nature's balance sheet: Model-based estimates of future worldwide ecosystem services. *Ecology and Society* 11(1): 19, available at http://www.ecologyandsociety.org/vol11/iss1/art19.

Alderton, Ben W. 2003–2004. Who drinks first? The devolution of western water rights in *Rio Grande Silvery Minnow v. Keys. Journal of Land Use & Environmental Law* 18(2): 271–86.

Ali, Paul A., and Kanako Yano. 2004. *Eco-finance: The legal design and regulation of market-based environmental instruments.* The Hague: Kluwer Law International.

Allan, J. David. 1995. *Stream ecology structure and functions of running waters.* Netherlands: Kluwer Academic Publishers.

Allan, Tony. 2001. *The Middle East water question: Hydropolitics and the global economy.* London: I.B. Tauris Publishers.

Allison, Helen E., and Richard J. Hobbs. 2004. Resilience, adaptive capacity, and the "lock-in trap" of the Western Australian agricultural region. *Ecology and Society* 9(1): 3, available at http://www.ecologyandsociety/vol9/iss1/art3.

American Farmland Trust. 1984. *Soil conservation in America: What do we have to lose?* Washington, DC: American Farmland Trust.

Anderson, Terry J. 2004. Viewing land conservation through Coase-colored glasses. *Natural Resources Journal* 44(2): 361–81.

Anderson, Terry J., and Donald R. Leal. 1992. Free market versus political environmentalism. *Harvard Journal of Law and Public Policy* 15(2): 297–310.

Andreen, William L. 2006. Developing a more holistic approach to water management in the United States. *Environmental Law Reporter News & Analysis* 36(4): 10277–89.

Anebo, Felix K. G. 2005. *The adoption of environmental policy innovation: Water pollution permit trading in the United States.* PhD diss., Dept. of Political Science, Southern Illinois University, Carbondale, IL.

Antle, John M., and Jetse J. Stoorvogel. 2006. Predicting the supply of ecosystem services from agriculture. *American Journal of Agricultural Economics* 88(5): 1174–80.

Antle, J. M., and L. M. Young. 2005. Policies and incentive mechanisms for the permanent adoption of agricultural carbon sequestration practices in industrialized and developing countries. In *Climate change and global food security,* ed. Rattan Lal, Norman Uphoff, Bobby A. Stewart, and David O. Hanson, 679–701. Boca Raton, FL: CRC Press.

Appel, Peter A. 2001. The power of Congress "without limitation": The property clause and federal regulation of private property. *Minnesota Law Review* 86(1): 1–130.

Arcadia Historical Museum. 2006. *Timeline for the Arcadia, Michigan area,* accessed at http://www.arcadiami.com/Timeline/timelinesettlers.htm on June 16 2006.

Arimura, Toshihide H. 2002. An empirical study of the SO_2 allowance market: Effects of PUC regulations. *Journal of Environmental Economics and Management* 44(2): 271–89.

Army Corps of Engineers, Environmental Protection Agency, National Resources Conservation Service, Fish and Wildlife Service, and National Marine Fisheries Service.

1995. Federal guidance for the establishment, use, and operation of mitigation banks. *Federal Register* 60(228): 58605–14.

Arndt, H. W. 1993. Sustainable development and the discount rate. *Economic Development and Cultural Change* 41(3): 651–61.

Arnold, Craig Anthony, ed. 2005. *Wet growth: Should water law control land use?* Washington, DC: Environmental Law Institute.

Athanas, Andrea, Joshua Bishop, Amy Cassara, Pamela Donauber, Chris Percival, Mohammad Rafiq, Janet Ranganathan, and Pernille Risgaard. 2006. *Business and ecosystems.* Washington, DC: Earthwatch Institute, World Resources Institute, World Business Council for Sustainable Development, and World Resources Institute.

Augustyniak, Christine M. 1993. Economic valuation of services provided by natural resources: Putting a price on the "priceless." *Baylor Law Review* 45(2): 389–403.

Aviram, Amitai. 2004. A paradox of spontaneous formation: The evolution of private legal systems. *Yale Law & Policy Review* 22(1): 1–68.

Bailey, Robert G. 1996. *Ecosystem geography.* New York: Springer-Verlag.

Baker, Katherine K. 1995. Consorting with forests: Rethinking our relationship to natural resources and how we should value their loss. *Ecology Law Quarterly* 22(4): 677–728.

Balmford, Andrew, Aarn Bruner, Philip Cooper, Robert Costanza, Stephen Farber, Rhys E. Green, Martin Jenkins, Paul Jeffriss, Valma Jessamy, John Madden, Kat Munroe, Norman Meyers, Shahid Naeem, Jouni Paavola, Matthew Rayment, Sergio Rosendo, Joan Roughgarden, Kate Trumper, R. Kerry Turner. 2002. Economic reasons for conserving wild nature. *Science* 297(5583): 950–53.

Banner, Stuart. 2002. Transitions between property regimes. *Journal of Legal Studies* 31(2): S359–S371.

Barbier, Edward B, and Geoffrey M. Heal. 2006. Valuing ecosystem services. *The Economists' Voice* 3(3): article 2, available at http://www.bepress.com/ev/vol3/iss3/art2.

Barling, Rowan D., and Ian D. Moore. 1994. Role of buffer strips in management of waterway pollution: A review. *Environmental Management* 18(4): 543–58.

Bascompte, Julian, Pedro Jordano, and Jens M. Olesen. 2006. Asymmetric coevolutionary networks facilitate biodiversity maintenance. *Science* 312(5772): 431–36.

Batjes, N. H., and W. G. Sombroek. 1997. Possibilities for carbon sequestration in tropical and subtropical soils. *Global Change Biology* 3(2): 161–73.

Bator, Francis M. 1958. Anatomy of market failure. *Quarterly Journal of Economics* 72(3): 351–79.

Baulch, Vivian M. n.d. The mystery of Pere Marquette's final resting place. *The Detroit News* accessed at http://info.detnews.com/history/story/index.cfm?id=158&category=people on June 15, 2006.

Baumol, William J., and Wallace E. Oates. 1988. *The theory of environmental policy.* 2nd ed. Cambridge, UK: Cambridge University Press.

BBC News. 2005. "Q&A: Common Agricultural Policy" accessed at http://news.bbc.co.uk/2/hi/europe/4407792.stm#whatis on Dec 22, 2005.

Bell, Abraham, and Gideaon Parchomovsky. 2003. Of property and antiproperty. *Michigan Law Review* 102(1): 1–70.

Bennett, David A., Ningchaun Xiao, and Marc P. Armstrong. 2004. Exploring the geographic consequences of public policies using evolutionary algorithms. *Annals of the Association of American Geographers* 94(4): 827–47.

Benson, Reed. 2006. "Adequate progress," or rivers left behind? Developments in Colorado and Wyoming instream flow laws since 2000. *Environmental Law* 36(4): 1283–1310.

Benson, Reed. 1997. Whose water is it? Private rights and public authority over reclamation project water. *Virginia Environmental Law Journal* 16(3): 363–427.

———. 2002. Giving suckers (and salmon) an even break: Klamath basin water and the Endangered Species Act. *Tulane Environmental Law Journal* 15(2): 197–238.

Bentham, Jeremy. 1882. *Theory of legislation.* 4th ed. London: Trubner.

Benzie County Patriot. 1969. Blaine land buying creates mystery. December 25.

Berry, Bill, ed. 2006. CRP: 1 million new acres, 13 million re-enrolled. *BufferNotes* June 2006, available at http://www.nacdnet.org/buffers/06June/index.html, accessed on July 17, 2006.

Binder, Denis. 2001. The spending clause as a positive source of environmental protection: A primer. *Chapman Law Review* 4(1): 147–62.

Bjorklund, Johanna, Karin E. Limburg, and Torbjörn Rydberg. 1999. Impact of production intensity on the ability of the agricultural landscape to generate ecosystem services: An example from Sweden. *Ecological Economics* 29(2): 269–91.

Blaikie, Piers 1986. *The political economy of soil erosion in developing countries.* Longman Science and Technology.

Blair, John M., Scott L. Collins, and Alan K. Knapp. 2000. Ecosystems as functional units in nature. *Natural Resources & Environment* 14(3): 150–55.

Blumm, Michael C., and Lucas Ritchie. 2005. Lucas's unlikely legacy: The rise of background principles as categorical takings defenses. *Harvard Environmental Law Review* 29(2): 321–68.

Bockstael, N., R. Costanza, I. Strand, W. Boynton, K. Bell, and L. Wainger. 1995. Ecological economic modeling and valuation of ecosystems. *Ecological Economics* 14(2–3): 143–59.

Bolund, P., and S. Hunhammar. 1999. Ecosystem services in urban areas. *Ecological Economics* 29(2): 293–301.

Bonham, Charles H. 2006. Perspectives from the field: A review of western instream flow issues and recommendations for a new water future. *Environmental Law* 36(4): 1205–35.

Bouckaert, Boudewejn. 2000. Original assignment of private property. In *Encyclopedia of law and economics,* vol. II, *Civil law and economics,* ed. Boudewejn Bouckaert and Gerritt DeGeest, 1–17. Cheltenham, UK: Edward Elgar.

Boulding, Kenneth E. 1966. The economics of the coming spaceship earth. In *Environmental quality in a growing economy,* ed. Henry Jarrett, 3–14. Baltimore, MD: Resources for the Future/Johns Hopkins University Press.

Boumans, Roelof, Robert Costanza, Joshua Farley, Matthew W. Wilson, Rosimeiry Portela, Jan Rotmans, Ferdinando Villa, and Monica Grasso. 2002. Modeling the dynamics of the integrated earth system and the value of global ecosystem services using the GUMBO model. *Ecological Economics* 41(2002): 529–60.

Boyd, James. 2004. What's nature worth? Using indicators to open the black box of ecological valuation. *Resources* 2004(154): 18–22.

Boyd, James, and Spencer Banzhaf. 2006. What are ecosystem services? The need for standardized environmental accounting units. Resources for the Future Discussion Paper 06-02. Washington, DC: Resources for the Future.

Boyd, James, and Lisa Wainger. 2002a. Landscape indicators of ecosystem service benefits. *American Journal of Agricultural Economics* 84(5): 1371–78.

———. 2002b. Measuring ecosystem service benefits for wetland mitigation. *National Wetlands Newsletter* 24(6): 1, 11–15.

Boyd, Jesse A. 2003. Hip deep: A survey of state instream flow law from the Rocky Mountains to the Pacific Ocean. *Natural Resources Journal* 43(4): 1151–1216.

Brock, William A., Karl-Goran Maler, and Charles Perrings. 2002. Resilience and sustainability: The economic analysis of nonlinear dynamic systems. In *Panarchy: Understanding transformations in human and natural systems,* ed. Lance H. Gunderson and C. S. Holling, 261–89. Washington, DC: Island Press.

Bromley, Daniel W. 1991. *Environment and economy: Property rights and public policy.* Oxford: Basil Blackwell.

Bromley, Daniel W. 1978. Property rules, liability rules, and environmental economics. *Journal of Economic Issues* 12: 43–60.

Brooker, Rob W. 2006. Plant–plant interactions and environmental change. *New Phytologist* 171(2): 271–84.

Brooks, Richard O., Ross Jones, and Ross A. Virginia. 2002. *Law and ecology: The rise of the ecosystem regime.* Burlington, VT: Ashgate.

Brouwer, Floor, and Philip Lowe, eds. 2000. *Cap regimes and the European countryside: Prospects for integration between agricultural, regional, and environmental policies.* CABI Publishing.

Brown, Phillip H., and Christopher L. Lant. 1999. The effect of wetland mitigation banking on the achievement of no-net-loss. *Environmental Management* 23(3): 333–45.

Buchanan, James M. 1965. An economic theory of clubs. *Economica* 32(Feb.): 1–14.

Buchanan, James M., and Gordon Tullock. 1962. *The calculus of consent: Logical foundations of constitutional democracy.* Ann Arbor: University of Michigan Press.

Buchanan, James M., and Yong J. Yoon. 2000. Symmetric tragedies: Commons and anticommons. *Journal of Law and Economics* 43(1): 1–13.

Buck, Susan J. 1998. *The global commons.* Washington, DC: Island Press.

Bulte, Erwin, Richard Damania, Lindsey Gillson, and Keith Lindsay. 2004. Space—The final frontier for economists and elephants. *Science* 306(5695): 420–21.

Burger, Joanna, and Michael Gochfeld. 1998. The tragedy of the commons 30 years on. *Environment* 40(10): 4–27.

Burtraw, Dallas. 2000. *Innovation under the tradable sulfur dioxide emission permits program in the U.S. electricity sector.* Washington, DC: Resources for the Future.

Butler, Colin D., and Willis Oluoch-Kosura. 2006. Linking future ecosystem services and future human well-being. *Ecology and Society* 11(1): 30, available at http://www.ecologyandsociety.org/vol11/iss1/art30.

Buzbee, William W. 2005. Contextual environmental federalism. *New York University Environmental Law Journal* 14(1): 108–29.

Cairns, John, Jr., and B. R. Niederlehner. 1994. Estimating effects of toxicants on ecosystem services. *Environmental Health Perspectives* 102(11): 936–39.

Cairns, Robert D., and Pierre Lasserre. 2006. Implementing carbon credits for forests based on green accounting. *Ecological Economics* 56(4): 610–21.

Calabresi, Guido, and A. Douglas Melamed. 1972. Property rules, liability rules, and inalienability: One view of the cathedral. *Harvard Law Review* 85(6): 1089–1128.

Callies, David L. 2000. Custom and public trust: Background principles of state property law? *Environmental Law Reporter* 30(1): 10003–23.

Callies, David L., and J. David Breemer. 2002. Selected legal and policy trends in takings law: Background principles, custom and public trust "exceptions" and the (mis)use of investment backed expectations. *Valparaiso University Law Review* 36(2): 339–79.

Callies, David L., Robert H. Freilich, and Thomas E. Roberts. 2004. *Cases and materials on land use.* Eagan, MI: West Group.

Carpenter, Stephen R., Elena M. Bennett, and Garry Peterson. 2006. Scenarios for ecosystem services: Overview. *Ecology and Society* 11(1): 29, available at http://www.ecologyandsociety.org/vol11/iss1/art29.

Carpenter, Stephen R., and William A. Brock. 2004. Spatial complexity, resilience, and policy diversity: Fishing on lake-rich landscapes. *Ecology and Society* 9(1): 8, available at http://www.ecologyandsociety.org/vol9/iss1/art8.

Castelle, A. J., A. W. Johnson, and C. Conolly. 1994. Wetlands and stream buffer size requirements—A review. *Journal of Environmental Quality* 23(5): 878–82.

Cattaneo, Andrea, Daniel Hellerstein, Cynthia Nickerson, and Christina Myers. 2006. *Balancing the multiple objectives of conservation programs.* ERS Rpt. No. 19. Washington, DC: Economic Research Service, U.S. Dept. of Agriculture.

Champ, Patricia A., Kevin J. Boyle, and Thomas C. Brown, eds. 2003. *A primer on non-market valuation.* Dordrecht, the Netherlands: Kluwer Academic Publishers.

Chapagain, A. K., and A. Y. Hoekstra, 2004. *Water footprints of nations,* vol. 1, *Main report.* Value of Water Research Series No. 16, UNESCO-IHP.

Chase, V., L. Demming, and F. Latawiec. 1995. *Buffers for wetlands and surface waters: A guidebook for New Hampshire municipalities.* Concord, NH: Audubon Society of New Hampshire.

Cheever, Federico. 1998. Four failed forest standards: What can we learn from the history of the National Forest Management Act's substantive timber management provisions? *Oregon Law Review* 77(2): 601–706.

Chen, Jim. 2004. Webs of life: Biodiversity conservation as a species of information policy. *Iowa Law Review* 89(2): 495–608.

Christensen, Norman L., Ann M. Bartuska, James H. Brown, Stephen Carpenter, Carla D'Antonio, Rober Francis, Jerry F. Franklin, James A. MacMahon, Reed F. Noss, David J. Parsons, Charles H. Peterson, Monica G. Turner, and Robert G. Woodmansee. 1996. The report of the Ecological Society of America Committee on the Scientific Basis for Ecosystem Management. *Ecological Applications* 6(3): 665–91.

Clark, Edwin H., Jennifer A. Haverkamp, and William Chapman. 1985. *Eroding soils: The off-farm impacts.* Washington, DC: The Conservation Foundation.

Clark, John G. 1995. Economic development vs. sustainable societies: Reflections on the players in a crucial contest. *Annual Review of Ecology and Systematics* 26: 225–48.

Coase, Ronald. 1937. The nature of the firm. *Economica* 4(16): 386–405.

———. 1960. The problem of social cost. *Journal of Law & Economics* 3(1): 1–44.

Colby, Bonnie G. 1989. Estimating the value of water in alternative uses. *Natural Resources Journal* 29(2): 511–27.

Cole, Daniel H. 2000. New forms of private property: Property rights in environmental goods. In *Encyclopedia of law and economics,* vol. II, *Civil law and economics,* ed. Boudewejn Bouckaert and Gerritt DeGeest, 274–314. Cheltenham, UK: Edward Elgar.

———. 2002. *Pollution and property: Comparing ownership institutions for environmental protection.* Cambridge, UK: Cambridge University Press.

Cole, Daniel H., and Peter Z. Grossman. 1999. When is command-and-control efficient? Institutions, technology, and the comparative efficiency of alternative regulatory regimes for environmental protection. *Wisconsin Law Review* 1999(5): 887–938.

———. 2002. The meaning of property rights: Law versus economics. *Land Economics* 78(3): 317–30.

Cole, Kenneth L., Forest Stearns, Glenn Guntenspergen, Margaret B. Davis, and Karen Walker. 2003. Historical landcover changes in the Great Lakes region. In *Land use history of North America*. U.S. Geological Survey. Accessed at http://biology.usgs.gov/luhna/chap6.html on March 15, 2006.

College of Agriculture and Life Sciences, University of Wisconsin–Madison. 2005. *The Wisconsin buffer initiative*. Madison: College of Agriculture and Life Sciences, Office of the Dean, University of Wisconsin–Madison.

Collier, K. J., A. B. Cooper, R. J. Davies, J. C. Rutherford, C. M. Smith, and R. B. Williamson. 1995a. *Managing riparian zones: A contribution to protecting New Zealand's rivers and streams*, vol. 1, *Concepts*. Wellington, NZ: Department of Conservation.

———. 1995b. *Managing riparian zones: A contribution to protecting New Zealand's rivers and streams*, vol. 2, *Guidelines*. Wellington, NZ: Department of Conservation.

Commission Regulation EC No 795/2004. Decoupling rules. *Official Journal of the European Union* 30(4): 2004.

Commission Regulation EC No 796/2004. Cross-compliance rules. *Official Journal of the European Union* 30(4): 2004.

Costanza, Robert. 1987. Social traps and environmental policy. *Bioscience* 37(6): 407–412.

———. 1996. Ecological economics: Reintegrating the study of humans and nature. *Ecological Applications* 6(4): 978–90.

———. 2000. Social goals and the valuation of ecosystem services. *Ecosystems* 3(1): 4–10.

———. 2001. Visions, values, valuation, and the need for an ecological economics. *Ecological Economics* 51(6): 459–68.

Costanza, Robert, ed. 1991. *Ecological economics: The science and management of sustainability*. New York: Columbia University Press.

Costanza, Robert, and Herman E. Daly. 1992. Natural capital and sustainable development. *Conservation Biology* 6(1): 37–46.

Costanza, Robert, and Steve Farber. 2002. Introduction to the special issue on the dynamics and value of ecosystem services: Integrating economic and ecological perspectives. *Ecological Economics* 41(3): 367–73.

Costanza, Robert, Ralph d'Arge, Rudolph deGroot, Stephan Farber, Monica Grasso, Bruce Hannon, Karin Limburg, Shahid Naeem, Rober V. O'Neill, Jose Paruelo, Robert G. Raskin, Paul Sutton, and Marjan van den Belt. 1997. The value of the world's ecosystem services and natural capital. *Nature* 387(6630): 253–60.

Cramton, Peter, and Suzi Kerr. 2002. Tradeable carbon permit auctions: How and why to auction not grandfather. *Energy Policy* 30(4): 333–45.

Cunjak, R. A. 1996. Winter habitat of selected stream fishes and potential impacts from land-use activity. *Canadian Journal of Fisheries and Aquatic Sciences* 53: S267–S282.

Daily, Gretchen C., ed. 1997. *Nature's services: Societal dependence on natural ecosystems*. Washington, DC: Island Press.

Daily, Gretchen C., and Shamik Dasgupta. 2001. Ecosystem services, concept of. In *Encyclopedia of Biodiversity*, vol. 2, ed. Simon Levin, 353–62. San Diego: Academic Press.

Daily, Gretchen C., and Katherine Ellison. 2002. *The new economy of nature: The quest to make conservation profitable.* Washington, DC: Island Press.

Daily, Gretchen C., Tore Soderqvist, Sara Aniyar, Kenneth Arrow, Partha Dasgupta, Paul R. Ehrlich, Carl Folke, AnnMarie Jansson, Bengt-Owe Jansson, Nils Kautsky, Simon Levin, Jane Lubchenco, Karl-Goram Maler, David Simpson, David Starrett, David Tilman, and Boran Walker. 2000. The value of nature and the nature of value. *Science* 289(5478): 395–96.

Daly, Herman, and Joshua Farley. 2004. *Ecological economics: Principles and applications.* Washington, DC: Island Press.

Dana, David. 1995. Natural preservation and the race to develop. *University of Pennsylvania Law Review* 143(3): 655–708.

Davidson, J. 2002. Protecting the still functioning ecosystem: The case of the prairie pothole wetlands. *Washington University Journal of Law & Public Policy* 9(2002): 123–62.

deBlij, H. J., and Peter O. Muller. 2006. *Geography: Realms, regions, and concepts.* 12th ed. John Wiley and Sons, Inc.

de Groot, Rudolf S. 1992. *Functions of nature.* Groningen: Wolters-Noordhroff.

de Groot, Rudolf S., Matthew A. Wilson, and Roelof M. J. Boumans. 2002. A typology for the classification, description, and valuation of ecosystem functions, goods, and services. *Ecological Economics* 41(3): 393–408.

Deason, Jonathan P., Theodore M. Schad, and George William Sherk. 2001. Water policy in the United States: A perspective. *Water Policy* 3(3): 175–92.

Deitz, Thomas, Elinor Ostrom, and Paul C. Stern. 2003. The struggle to govern the commons. *Science* 302(5652): 1907–12.

Dellapenna, Joseph W. 2002. The law of water allocation in the southeastern states at the opening of the twenty-first century. *University of Arkansas–Little Rock Law Review* 25(1): 9–88 (2002).

———. 2004. Adapting riparian rights to the twenty-first century. *West Virginia Law Review* 106(3): 539–93.

Demsetz, Harold D. 2002. Toward a theory of property rights, II: The competition between private and collective ownership. *Journal of Legal Studies* 31(2): S653–S672.

Department of Agriculture. 2000. National forest system land and resources management planning. *Federal Register* 65(218): 67514–81.

———. 2005. National forest system land and resources management planning. *Federal Register* 70(3): 1023–61.

Department of the Army and Environmental Protection Agency. 1990. Memorandum of Agreement between Department of the Army and the Environmental Protection Agency concerning the Clean Water Act Section 404(b)(1) Guidelines. *Federal Register* 55(48): 9210–13.

Department of Defense and Environmental Protection Agency. 2006. Compensatory mitigation for losses of aquatic resources; proposed rule. *Federal Register* 71(59): 15520–56.

Deutsch, Lisa, Carl Folke, and Kristian Skanberg. 2003. The critical natural capital of ecosystem performance as insurance for human well-being. *Ecological Economics* 44(2–3): 205–17.

Diamond, Jared. 1999. *Guns, Germs, and Steel: The fates of human societies.* New York: W.W. Norton and Company.

———. 2005. *Collapse: How societies choose to fail or succeed.* New York: Viking Books.

Donahue, Debra. 1999. *The western range revisited: Removing livestock from public lands to conserve native biodiversity.* Norman: University of Oklahoma Press.

Doremus, Holly. 2000. The rhetoric and reality of nature protection: Toward a new discourse. *Washington and Lee Law Review* 57(1): 11–73.

———. 2003a. A policy portfolio approach to biodiversity protection on private lands. *Environmental Science and Policy* 6: 217–32.

———. 2003b. Takings and transitions. *Journal of Land Use and Environmental Law* 19(1): 1–46.

Doremus, Holly, and A. Dan Tarlock. 2003. Fish, farms, and the clash of cultures in the Klamath basin. *Ecology Law Quarterly* 30(2): 279–350.

Driesen, David M. 2006. Is cost–benefit analysis neutral? *University of Colorado Law Review* 77(2): 335–404.

Dukeminier, Jesse, James E. Krier, Gregory S. Alexander, and Michael H. Schill. 2006. *Property.* 6th ed. New York: Aspen Publishers.

Duncan, Myrl L. 2002. Reconceiving the bundle of sticks: Land as a community-based resource. *Environmental Law* 32(4): 773–807.

Duram, Leslie A. 2005. *Good growing: Why organic farming works.* Lincoln: University of Nebraska Press.

Dziegielewski, Benedykt. 2005. Personal communication.

Eagle, Steven J. 2004. Environmental amenities, private property, and public policy. *Natural Resources Journal* 44(2): 425–44.

Economic Research Service. 1997. *Agricultural resources and environmental indicators, 1996–97.* Ag. Handbook No. 712. Washington, DC: Economic Research Service, U.S. Dept. of Agriculture.

Edwards, Steven F. 2003. Property rights to multi-attribute fishery resources. *Ecological Economics* 44(2–3): 309–23.

Ehrlich, Paul R., and Anne H. Ehrlich. 1992. The value of biodiversity. *Ambio* 21(3): 219–26.

Ekins, Paul. 2003. Identifying critical natural capital: Conclusions about critical natural capital. *Ecological Economics* 44(2–3): 277–92.

Ekins, Paul, Sandra Simon, Lisa Deutsch, Carl Folke, and Rudolf de Groot. 2003. A framework for the practical application of the concepts of critical natural capital and strong sustainability. *Ecological Economic* 44(2–3): 165–85.

Ellerman, A. Denny, Paul L. Joskow, Juan Pablo Montero, and Richard Schmalensee. 2003. Final report: Evaluation of phase II compliance with Title IV of the 1990 Clean Air Act Amendments. U.S. Environmental Protection Agency, National Center for Environmental Research.

Ellickson, Robert C. 1973. Alternatives to zoning: Covenants, nuisance rules, and fines as land use controls. *University of Chicago Law Review* 40(4): 681–781.

———. 1991. *Order without law: How neighbors settle disputes.* Cambridge, MA: Harvard University Press.

Elliott, E. Donald, Bruce A. Ackerman, and John C. Millian. 1985. Toward a theory of statutory evolution: The federalization of environmental law. *Journal of Law, Economics & Organizations* 1(2): 313–40.

Emel, Jacque L., and Elizabeth Brooks. 1988. Changes in form and function of property rights institutions under threatened resource scarcity. *Annals of the Association of American Geographers* 78(2): 241–52.

Engel, Kirsten H. 1997. State environmental standard setting: Is there a "race," and is it "to the bottom"? *Hastings Law Journal* 48(2): 271–398.

English, Burton C., James A. Maetzold, Brian R. Holding, and Earl O. Heady, eds. 1984. *Future agricultural technology and resource conservation.* Ames: Iowa State University Press.

Environmental Law Institute. 1993. *Wetland mitigation banking.* Washington, DC: Environmental Law Institute.

———. 2002. *Banks and fees: The status of off-site mitigation in the United States.* Washington, DC: Environmental Law Institute.

Epstein, Richard. 1995. *Simple rules for a complex world.* Cambridge, MA: Harvard University Press.

Esty, Daniel C. 1996. Revitalizing environmental federalism. *Michigan Law Review* 95(3): 570–653.

European Commission–Agriculture. 2005. CAP reform—A long-term perspective for sustainable agriculture. Accessed at http://europa.eu.int/comm/agriculture/capreform/index_en.htm on Dec. 14, 2005.

European Commission, Directorate General for Agriculture. 2003. Reform of the Common Agricultural Policy: A long-term perspective for sustainable agriculture: Impact analysis.

European Commission, Directorate General for Agriculture and Rural Development. 2005. Agri-Environment measures: Overview on general principles, types of measures, and application.

Falkenmark, Malin, and Johan Rockstrom. 2004. *Balancing water for humans and nature: The new approach in ecohydrology.* Sterling, VA: Earthscan.

Farber, Stephen C., Robert Costanza, and Matthew A. Wilson. 2002. Economic and ecological concepts for valuing ecosystem services. *Ecological Economics* 41(3): 375–92.

Farm Services Agency. 2003. *Conservation Reserve Program final programmatic environmental impact statement—Appendix D: Literature review and research recommendations for the Conservation Reserve Program.* Washington, DC: Farm Services Agency, U.S. Dept. of Agriculture.

———. 2006a. *Fact sheet: Conservation Reserve Program general sign-up 33 Environmental Benefits Index.* Accessed at http://www.fsa.usda.gov/in/crp33ebi06.pdf on July 17, 2006.

———. 2006b. *Conservation Reserve Program: Monthly summary—June 2006.* Accessed at http://www.fsa.usda.gov/dafp/cepd/stats/jun2006.pdf accessed on July 28, 2006.

Farnworth, Edward G., Thomas H. Tidrick, Carl F. Jordan, and Webb M. Smathers, Jr. 1981. The value of natural ecosystems: An economic and ecological framework. *Environmental Conservation* 8(4): 275–82.

Feather, P., D. Hellerstein, and L. Hansen. 1999. *Economic valuation of environmental benefits and the targeting of conservation programs: The case of the CRP.* Agricultural Economic Report No. 778. Washington, DC: Economic Research Service, U.S. Department of Agriculture.

Feller, Joseph M. 2001. Back to the present: The Supreme Court refuses to move public range law backward, but will the BLM move public range law forward? *Environmental Law Reporter News & Analysis* 31(1): 10021–39.

Feng, Hongli. 2005. The dynamics of carbon sequestration and alternative carbon accounting, with application to the upper Mississippi River Basin. *Ecological Economics* 54(1): 23–35.

Fennell, Lee Anne. 2004. Common interest tragedies. *Northwestern University Law Review* 98(3): 907–90.

Firey, Walter. 1960. *Man, mind and land: A theory of resource use.* Glencoe, IL: Free Press.

Fischer, Richard A., and J. Craig Fischenich. 2000. *Design recommendation for riparian corridors and vegetated buffer strips,* EMRRP Technical Notes Collection, Vicksburg, MS: U.S. Army Engineer Research and Development Center.

Fischman, Robert L. 2001. The EPA's NEPA duties and ecosystem services. *Stanford Environmental Law Journal* 20(2): 497–536.

———. 2006. Cooperative federalism and natural resources law. *New York University Environmental Law Journal* 14(1): 179–231.

Fitzgerald, Tim. 2005. Can ecosystem valuation create markets? *PERC Reports* 23(3): 7–8.

Fitzpatrick, Daniel. 2006. Evolution and chaos in property rights systems: The third world tragedy of contested access. *Yale Law Journal* 115(5): 996–1048.

Fitzsimmons, Allan. 1999. *Defending illusions: Federal protection of ecosystems.* Lanham, MD: Rowman & Littlefield Publishers.

Folke, Carl, C. S. Holling, and Charles Perrings. 1996. Biological diversity, ecosystems, and the human scale. *Ecological Applications* 6(4): 1018–24.

Forman, Richard T. T. and Michel Godron, 1986. *Landscape ecology.* New York: Wiley.

Frazier, Terry. 1998. Protecting ecological integrity within the balancing function of property law. *Environmental Law* 28(1): 53–112.

Freeman, Jody, and Charles D. Kolstad eds. 2007. *Moving to markets in environemtnal regulation: Lessons learned from 20 years of experience.* New York: Oxford University Press.

Freyfogle, Eric T. 2006. Goodbye to the public–private divide. *Environmental Law* 36(1): 7–24.

Friedman, Thomas, 2005. *The world is flat: A brief history of the twenty-first century.* New York: Farrar, Straus and Giroux.

Frischmann, Brett M. 2005. An economic theory of infrastructure and commons management. *Minnesota Law Review* 89(4): 917–1030.

Fuller, Lon. 1968. *Anatomy of the law.* New York: Praeger Press.

Fumero, John J. 2003. Florida water law and environmental water supply for Everglades restoration. *Journal of Land Use & Environmental Law* 18(2): 379–89.

Gardner, Royal C. 1993. Banking on entrepreneurs: Wetlands, mitigation banking, and takings. *Iowa Law Review* 81(3): 527–88.

———. 2005. Mitigation. In *Wetlands law and policy: Understanding Section 404,* ed. Kim Dana Connolly, Stephen M. Johnson, and Douglas R. Williams, 253–82. Chicago: American Bar Association Publishing.

Garrett, Gregory W. 2003. The economic value of the Apalachicola River and Bay. MS course paper, Florida State University Tallahassee.

Gatto, Marino, and Giulio A. De Leo. 2000. Pricing biodiversity and ecosystem services: The never-ending story. *BioScience* 50(4): 347–55.

Golley, Frank B. 1993. *The history of the ecosystem concept in ecology.* New Haven: Yale University Press.

Good, James. 1998. Effective coastal wetland management. *National Wetlands Newsletter* 20(4): 1–17.

Gottfried, Robert R. 1992. The value of a watershed as a series of linked multiproduct assets. *Ecological Economics* 5(2): 145–61.

Gottfried, Robert R., David Wear, and Robert Lee. 1996. Institutional solutions to market failure on a landscape scale. *Ecological Economics* 18(2): 133–40.

Goulder, Lawrence H., and Donald Kennedy. 1997. Valuing ecosystem services: Philosophical bases and empirical methods. In *Nature's services: Societal dependence on natural ecosystems,* ed. Gretchen C. Daily, 23–47. Washington, DC: Island Press.

Gowdy, John M., and Carl N. McDaniel. 1999. The physical destruction of Nauru: An example of weak sustainability. *Land Economics* 75(2): 333–38.

Grand Traverse Regional Land Conservancy (GTRLC). 2003. 2003 Recreation Grant Application: Michigan Department of Natural Resources. Traverse City, MI: Grand Traverse Regional Land Conservancy.

Green, Rhys E., Stephen J. Cornell, Jorn P. W. Scharlemann, and Andrew Balmford. 2005. Farming and the fate of nature. *Science* 307(5709): 550–55.

Gullo, T. 1994. *CBO Testimony on the Conservation Reserve Program.* Before Subcommittee on Environment, Credit, and Rural Development, House Agricultural Committee, 2 August 1994. Accessed at http://www.cbo.gov/ftpdocs/49xx/doc4908/doc76.pdf on July 27, 2006.

Guo, Zhongwei, Xiao Xiangming, and Dianmo Li. 2000. An assessment of ecosystem services: Water flow regulation and hydroelectric power production. *Ecological Applications* 10(3): 925–36.

Gustavson, Kent, Stephen C. Lonergan, and Jack Ruitenbeek. 2002. Measuring contributions to economic production—Use of an index of captured ecosystem value. *Ecological Economics* 41(3): 479–90.

Hall, Noah D. 2006. Toward a new horizontal federalism: Interstate water management in the Great Lakes region. *University of Colorado Law Review* 77(2): 405–56.

Hammond, F. M. 2002. *The effects of resort and residential development on black bears in Vermont. Final report.* Waterbury, VT: Fish and Wildlife Department, Vermont Agency of Natural Resources.

Hardin, Garrett. 1968. The tragedy of the commons. *Science* 162(3859): 1243–48.

Harlow, Jerry, T. 1994. *History of the Soil Conservation Service: National Resources Inventories.* Fort Worth, TX: Natural Resources Conservation Service.

Harrison, Ron. 2005. *Herring Lakes watershed management plan–2003 (Rev. No. 1 12/05).* Beulah, MI: Benzie Conservation District.

Hartung, R. E., and J. M. Kress. 1977. Woodlands of the Northeast: Erosion and sediment control guides. U.S. Soil Conservation Service and U.S. Forest Service, Upper Darby, Pennsylvania.

Hartvigsen, Gregg, Ann Kinzig, and Garry Peterson. 1998. Use and analysis of complex adaptive systems in ecosystem science: Overview of special section. *Ecosystems* 1(5): 427–30.

Hayes, David. 2005. Guest on *On-Point,* November 29, 2005. Produced by Energy and Environment TV.

Hayes, David J. 2003. Privatization and control of U.S. water supplies. *Natural Resources & Environment* 18(2): 19–24.

Heal, Geoffrey. 2000. *Nature and the marketplace: Capturing the value of ecosystem services.* Washington, DC: Island Press.

Heal, Geoffrey, Gretchen C. Daily, Paul R. Ehrlich, James Salzman, Carol Boggs, Jessica Hellman, Jennifer Hughes, Claire Kremen, and Taylor Ricketts. 2001. Protecting natural capital through ecosystem service districts. *Stanford Environmental Law Journal* 20(2): 333–64.

Heimlich, Ralph E., and Tim Osborn. 1994. Buying more environmental protection with limited dollars. In *When Conservation Reserve Program contracts expire: The policy options*. Ankeny, IA: Soil and Water Conservation Society.

Heller, Mark A. 1998. The tragedy of the anticommons: Property in the transition from Marx to markets. *Harvard Law Review* 111(3): 621–88.

Henderson, Henry L. 2002. Toward sustainable governance of the waters of the Great Lakes Basin. In *Improved decision-making for water resources: The key to sustainable development for metropolitan regions*, 29–49. Great Cities Institute/Illinois–Indiana Sea Grant College Program Urban Water Resources Conference Proceedings, Chicago, IL.

Hepple, Robert P., and Sally M. Benson. 2002. Implications of surface seepage on the effectiveness of geologic storage of carbon dioxide as a climate change mitigation strategy. In *Greenhouse gas control strategies* (Proceedings of the 6th International Conference on Greenhouse Gas Control Technologies), Kyoto, Japan, October 2002, vol. 1, ed. John Gale and Yoichi Kaya, 261–66. Oxford, UK: Pergamon.

Hirsch, Dennis. 2001. Second generation policy and the new economy. *Capital University Law Review* 29(1): 1–20.

———. 2004. Lean and green? Environmental law and policy for the flexible production economy. *Indiana Law Journal* 79(3): 611–66.

Holcombe, Justin K. 2005. Protecting ecosystems and natural resources by revising conceptions of ownership, rights, and valuation. *Journal of Land, Resources, and Environmental Law* 26(1): 83–110.

Holland, John. 1995. *Hidden order: How adaptation builds complexity.* Reading, MA: Addison-Wesley.

Holling, C. S. 1996. Surprise for science, resilience for ecosystems, and incentives for people. *Ecological Applications* 6(3): 733–35.

———. 2004. From complex regions to complex worlds. *Ecology and Society* 9(1): 11, available at http://www.ecologyandsociety.org/vol9/iss1/art11.

Holling, C. S., and Lance H. Gunderson. 2002. Resilience and adaptive cycles. In *Panarchy: Understanding transformations in human and natural systems*, ed. Lance H. Gunderson and C. S. Holling, 25–62. Washington, DC: Island Press.

Holling, C. S., Lance H. Gunderson, and Garry D. Peterson. 2002. Sustainability and panarchies. In *Panarchy: Understanding transformations in human and natural systems*, ed. Lance H. Gunderson and C. S. Holling, 63–102. Washington, DC: Island Press.

Holmes, K. John, and Robert M. Friedman. 2000. Design alternatives for a domestic carbon trading scheme in the United States. *Global Environmental Change* 10(4): 273–88.

Holmlund, Ceclia M., and Monica Hammer. 1999. Ecosystem services generated by fish populations. *Ecological Economics* 29(2): 253–68.

Houck, Oliver A. 1983. Land loss in coastal Louisiana: Causes, consequences, and remedies. *Tulane Law Review* 58(1): 3–96.

Houghton, R. A., J. L. Hackler, and K. T. Lawrence. 1999. The U.S. carbon budget: Contributions from land-use change. *Science* 285(5427): 574–78.

Howard, John H. 1929. *A history of Herring Lake.* Boston, MA: The Christopher Publishing House.

Hsu, Shi-Ling. 2003. A two-dimensional framework for analyzing property rights regimes. *U.C. Davis Law Review* 36(4): 813–93.

———. 2005. What is a tragedy of the commons: Overfishing and campaign spending problems. *Albany Law Review* 69(1): 75–138.

Hutson, Susan S., Nancy L. Barber, Joan F. Kenny, Kristin S. Linsey, Deborah S.

Lumia, and Molly A. Maupin. 2005. *Estimated use of water in the United States in 2000*. U.S. Geological Survey Circular 1268.

Hylton, Keith N. 2002. When should we prefer tort law to environmental regulation? *Washburn Law Journal* 41(3): 515–34.

Intergovernmental Panel on Climate Change (IPCC). 2007. Climate change 2007: The physical science basis summary for policymakers. UK: IPCC.

Jaber, Salahuddin M. 2006. *Monitoring spatial soil organic carbon variations using remote sensing and geographic information systems*. PhD diss. in Environmental Resources and Policy, Southern Illinois University, Carbondale, IL.

Jackson, Robert B., Esteban G. Jobbagy, Roni Avissar, Somnath B. Roy, Danian J. Barrett, Charles W. Cook, Kathleen A. Farley, David C. le Maitre, Bruce A. McCarl, and Brian C. Murray. 2005. Trading water for carbon with biological carbon sequestration. *Science* 310(5756): 1944–47.

Janssen, Marco A., 2002 ed. *Complexity and ecosystem management*. Cheltenham, UK: Edward Elgar Publishers.

Jervell, Anne Maxnes, and Desmond A. Jolly. 2003. *Beyond food: Toward a multifunctional agriculture*. Working Paper of the Norwegian Agricultural Economics Research Institute, Oslo.

Jennings, Ann, Roy Hoagland, and Eric Rudolph. 1999. Downsides to Virginia mitigation banking. *National Wetlands Newsletter* 21(1): 9–11.

Johnson, Barbara. 2005. *Conservation Reserve Program: Status and current issues*. CRS Report to Congress RS21613. Washington, DC: Congressional Research Service, Library of Congress.

Johnson, Stephen M. 2004. *Economics, equity, and the environment*. Washington, DC: Environmental Law Institute.

Kadlec, Robert H., and Robert L. Knight. 2004. *Treatment wetlands*. Boca Raton, FL: Lewis Publishers.

Kanner, Allan. 2005. The public trust doctrine, *parens patriae*, and the attorney general as the guardian of the state's natural resources. *Duke Environmental Law & Policy Forum* 16(1): 57–115.

Kaplow, Louis, and Steven Shavell. 1996. Property rules versus liability rules: An economic analysis. *Harvard Law Review* 109(4): 713–90.

Karkkainen, Bradley C. 2001. Information as environmental regulation: TRI and performance benchmarking, precursor to a new paradigm? *Georgetown Law Journal* 89(2): 257–370.

———. 2002. Toward a smarter NEPA: Monitoring and managing government's environmental performance. *Columbia Law Review* 102(4): 903–72.

Kates, Robert W., William C. Clark, Robert Corell, J. Michael Hall, Carlo C. Jarger, Ian Lowe, James J. McCarthy, Hans Joachim Schellnhuber, Bert Bolin, Nancy M. Dickson, Sylvie Fauchaux, Gilberto C. Gallopin, Arnulf Grübler, Brian Huntley, Jill Jäger, Narpat S. Jodha, Roger E. Kasperson, Akin Mabogunje, Pamela Matson, Harold Mooney, Berrien Moore III, Timothy O'Riordan, and Uno Svedin. 2001. Sustainability science. *Science* 292(5517): 641–642.

Katz, Michael L., and Harvey S. Rosen. 1998. *Microeconomics*. 3rd ed. Boston: Irwin, McGraw-Hill.

Kauffman, Stuart. 1995. *At home in the universe: The search for the laws of self-organization and complexity*. New York: Oxford University Press.

Kaufmann, J. K. 1992. Habitat use by wood turtles in Pennsylvania. *Journal of Herpetology* 26(3): 315–21.

Keller, Brant. 2005. What we always knew: Wetlands win hands down at pollution mitigation. *National Wetlands Newsletter* 27(5): 12–14.

Keller, Cherry M. E., Chandler S. Robbins, and Jeff S. Hatfield. 1993. Avian communities in riparian forests of different widths in Maryland and Delaware. *Wetlands* 13(2):137–44.

Kelso, M. M. 1977. Natural resource economics: The upsetting discipline. *American Journal of Agricultural Economics* 59: 814–23.

Keystone Center. 1996. *The Keystone national policy dialogue on ecosystem management.* Keystone, CO: Keystone Center.

King, Dennis M. 1997. Comparing ecosystem services and values. Paper prepared for the U.S. Department of Commerce.

King, Dennis M., and Lisa W. Herbert. 1997. The fungibility of wetlands. *National Wetlands Newsletter* 19(5): 10–13.

King, Dennis M., and Peter J. Kuch. 2003. Will nutrient credit trading ever work? An assessment of supply and demand problems and institutional obstacles. *Environmental Law Reporter: News and Analysis* 33(5): 10352–68.

King, Dennis M. 2004. Trade-based carbon sequestration accounting. *Environmental Management* 33(4): 559–571.

Klaassen, Ger, Andries Nentjes, and Mark Smith. 2005. Testing the theory of emissions trading: Experimental evidence on alternative mechanisms for global carbon trading. *Ecological Economics* 53(1): 47–58.

Klass, Alex. 2006. Adverse possession and conservation: Expanding traditional notions of use and possession. *University of Colorado Law Review* 77(2): 283–333.

Klein, Christine A. 2003. The environmental commerce clause. *Harvard Environmental Law Review* 27(1): 1–70.

Konarska, Keri M., Paul C. Sutton, and Michael Castellon. 2002. Evaluating scale dependence of ecosystem service valuation: A comparison of NOAA-AVHRR and Landsat TM datasets. *Ecological Economics* 41(3): 491–507.

Kraft, Steven E., and T. Toohill. 1984. Soil degradation and land-use changes: Agroecological data acquired through representative farm and linear programming analysis. *Journal of Soil and Water Conservation* 39(5): 334–38.

Kremen, Charles. 2005. Managing ecosystem services: What do we need to know about their ecology? *Ecology Letters* 8: 468–79.

Kremen, Charles, N. M. Williams, R. L. Bugg, J. P. Fay, and R. W. Thorp. 2004. The area requirements of an ecosystem service: Crop pollination by native bee communities in California. *Ecology Letters* 7: 1109–19.

Krier, James E. 1992. The tragedy of the commons, part two. *Harvard Journal of Law & Public Policy* 15(2): 325–47.

Krier, James E., and Stewart Schwab. 1995. Property rules and liability rules: The cathedral in another light. *New York University Law Review* 70(2): 440–83.

Krishna, Anirudh. 2002. *Active social capital.* New York: Columbia University Press.

Kyle, Thomas G. 1993. *Which way is the sky falling?* Minocqua, WI: Northword Press, Inc.

Kysar, Douglas A. 2001. Sustainability, distribution, and the macroeconomic analysis of law. *Boston College Law Review* 43(1): 1–71.

———. 2003. Law, environment, and vision. *Northwestern University Law Review* 97(2): 675–729.

Lahey, William L., and Cara M. Cheyette. 2002. The public trust doctrine in New England: An underused judicial tool. *Natural Resources & Environment* 17(2): 92–94, 124–25.

Laitos, Jan G., and Thomas A. Carr. 1999. The transformation on public lands. *Ecology Law Quarterly* 26(2): 140–242.

Lal, Rattan. 2001. Soils and the greenhouse effect. In *Soil carbon sequestration and the greenhouse effect*, ed. Rattan Lal, 1–8. USA: Soil Science Society of America, Inc.

Lal, R., J. M. Kimble, R. F. Follett, and C. V. Cole. 1998. *The potential of U.S. cropland to sequester carbon and mitigate the greenhouse effect*. Madison, WI: Ann Arbor Press.

Lampkin, Nicholas, Carolyn Foster, Susanne Padel, and Peter Midmore. 1999. *The policy and regulatory environment for organic farming in Europe*. Stuttgart: University of Hohenheim.

Lant, Christopher L., ed. 1999. *Human dimensions of watershed management*. American Water Resources Association Monograph.

Lant, Christopher. 2003. Watershed governance in the United States: The challenges ahead. *Water Resources Update* 126: 21–28.

Lant, Christopher L. 2006. Water resources sustainability: An ecological economics perspective. In *Water resources sustainability*, ed. Larry W. Mays, 55–72. New York: McGraw-Hill.

Lant, Christopher L., Steven E. Kraft, Jeffrey Beaulieu, David Bennett, Timothy Loftus, and John Nicklow. 2005. Using ecological–economic modeling to evaluate policies affecting agricultural watersheds. *Ecological Economics* 55(4): 467–84.

Lavigne, Peter. 2004. Watershed councils East and West: Advocacy, consensus and environmental progress. *UCLA Journal of Environmental Law and Policy* 22(2): 301–19.

Lazarus, Richard J. 1986. Changing conceptions of property and sovereignty in natural resources: Questioning the public trust doctrine. *Iowa Law Review* 71(3): 631–716.

———. 1993. Putting the correct "spin" on Lucas. *Stanford Law Review* 45(5): 1411–32.

———. 2004. *The making of environmental law.* Chicago: University of Chicago Press.

Lee, Robert G. 1992. Ecologically effective social organization as a requirement for sustaining watershed ecosystems. In *Watershed management: Balancing sustainability and environmental change,* ed. Robert J. Naiman, 73–90. New York: Springer-Verlag.

Legg, Wilfred. 2000. The environmental effect of reforming agricultural policies. *Cap regimes and the European countryside: Prospects for integration between agricultural, regional, and environmental policies,* ed. Floor Brouwer and Philip Lowe, 17–30. Wallingford, UK: CAB International.

Leshy, John. 2005. A conversation about takings and water rights. *Texas Law Review* 83(7): 1985–2026.

Levin, Simon A. 1998. Ecosystems and the biosphere as complex adaptive systems. *Ecosystems* 1(5):431–36.

———. 1999. *Fragile dominion: Complexity and the commons.* Cambridge, UK: Perseus.

Lewin, Jeff L. 1986. Compensated injunctions and the evolution of nuisance law. *Iowa Law Review* 71(3): 775–832.

Limburg, Karin E., Robert V. O'Neil, Robert Costanza, and Stephen Farber. 2002. Complex systems and valuation. *Ecological Economics* 41(3): 409–40.

Little, James R., and Bruce I. Knight. 2004. Memo: Conservation Buffer Newsletter Insert/Handout, 2 February 2004. Accessed at http://www.nrcs.usda.gov on July 9, 2006.

Loomis, John B. 1996. How large is the extent of the market for public goods: Evidence from a nationwide contingent valuation survey. *Applied Economics* 28(7): 779–782.

Loomis, John, Paula Kent, Liz Strange, Kurt Fausch, and Alan Covich. 2000. Measuring the total economic value of restoring ecosystem services in an impaired river basin: Results from a contingent value survey. *Ecological Economics* 33(1): 103–17.

Lowe, Philip, and David Baldock. 2000. Integration of environmental objectives into agricultural policy making. In *Cap regimes and the European countryside: Prospects for integration between agricultural, regional, and environmental policies,* ed. Floor Brouwer and Philip Lowe, 31–52. Wallingford, UK: CAB International Publishing.

Lowrance, Richard R., Robert L. Todd, Joseph Fail, Jr., Ole Hendrickson, Jr., Ralph Leonard, and Loris Asmussen. 1984. Riparian forests as nutrient filters in agricultural watersheds. *Bioscience* 34(6): 374–77.

Lum, Albert L. 2003. How goes the public trust doctrine: Is the common law shaping environmental policy? *Natural Resources & Environment* 18(2): 73–75.

Mackenzie, Fred T. 1998. *Our changing planet.* 2nd ed. Upper Saddle River, NJ: Prentice-Hall.

Malanson, George P. 1993. *Riparian landscapes.* Cambridge, UK: Cambridge University Press.

Malloy, Timothy F. 2002. Regulating by incentives: Myths, models, and micromarkets. *Texas Law Review* 80(3): 531–605.

Markandya, Anil, and David W. Pearce. 1991. Development, the environment, and the social discount rate. *World Bank Research Observer* 6(2): 137–52.

Marsh, George P. 1864. *Man and nature: Or, physical geography as modified by human action.* Cambridge, MA: Harvard University Press.

Marshall, Curtis H., Jr., Roger A. Pielke, Sr., and Louis T. Steyeart. 2003. Crop freezes and land use change in Florida. *Nature* 426(6962): 29–30.

Mathews, Ruth. 2006. Instream flow protection and restoration: Setting a new compass point. *Environmental Law* 36(4): 1311–29.

Mattson, Robert A. 2002. A resource-based framework for establishing freshwater inflow requirements for the Suwannee River estuary. *Estuaries* 25(6B): 1333–42.

Mayer, Leo. 1982. Farm Exports and Soil Conservation. *Proceedings of the Academy of Political Science* 34: 99–111.

McAllister, Stephen R., and Robert L. Glicksman. 2000. Federal environmental law in the "new" federalism era. *Environmental Law Reporter News & Analysis* 30(Dec.): 11122–43.

McCool, Daniel. 2005. A river commons: A new era in U.S. water policy. *Texas Law Review* 83(7): 1093–1928.

McElfish, James M., Jr. 1994. Property rights, property roots: Rediscovering the basis for legal protection of the environment. *Environmental Law Reporter News & Analysis* 24(May): 10231–49.

McLaughlin, Nancy A. 2005. Rethinking the perpetual nature of conservation easements. *Harvard Environmental Law Journal* 29(2): 421–521.

McMichael, A. J., C. D. Butler, and Carl Folke. 2003. New visions for addressing sustainability. *Science* 302(5652): 1919.

McNutt, Patrick. 2000. Public goods and club goods. In *Encyclopedia of law and economics,* vol. I, *The history and methodology of law and economics,* ed. Boudewejn Bouckaert and Gerritt DeGeest, 927–51. Cheltenham, UK: Edward Elgar.

Medema, Steven G., and Richard O. Zerbe. 2000. The Coase theorem. In *Encyclopedia of law and economics,* vol. I, *The history and methodology of law and economics,* ed. Boudewejn Bouckaert and Gerritt DeGeest, 836–92. Cheltenham, UK: Edward Elgar.

Meltz, Robert. 2005. Wetlands and regulatory takings. In *Wetlands law and policy: Understanding Section 404,* ed. Kim Dana Connolly, Stephen M. Johnson, and Douglas R. Williams, 427–44. Chicago: American Bar Association Publishing.

Miano, Steven T., and Michael E. Crane. 2003. Eastern water law: Historical perspectives and emerging trends. *Natural Resources & Environment* 18(2): 14–18.

Middleton, Beth. 1995. The role of flooding in seed dispersal: Restoration of cypress swamps along the Cache River, Illinois. University of Illinois at Urbana–Champaign, Water Resources Center, Research Report 220.

Millen, Ronny, and Christopher L. Burdett. 2005. Critical habitat in the balance: Science, economics, and other relevant factors. *Minnesota Journal of Law, Science & Technology* 7(1): 227–300.

Millennium Ecosystem Assessment. 2003. *Ecosystems and human well-being: A framework for assessment.* Washington, DC: Island Press.

———. 2004a. Conditions and Trends Assessment. Available at http://www.millenniumassessment.org/en/products.chapters.aspx#.

———. 2004b. Scenarios Assessment. Available at http://www.millenniumassessment.org/en/products.chapters.aspx#.

———. 2005. *Ecosystems and human well-being: Synthesis.* Washington, DC: Island Press.

Milne, Bruce T. 1998. Motivation and benefits of complex systems approaches in ecology. *Ecosystems* 1(5): 449–56.

Mitsch, William J. and James G. Gosselink. 2000. The value of wetlands: importance of scale and landscape setting. *Ecological Economics* 35(2000): 25–33.

Moberg, Fredrik, and Carl Folke. 1999. Ecological goods and services of coral reef ecosystems. *Ecological Economics* 29(2): 215–33.

Mogenson, Richard K. 2006. Mitigation banking: It's no myth. *National Wetlands Newsletter* 28(2): 15–16, 21.

Mooney, Harold A., and Paul R. Ehrlich. 1997. Ecosystem services: A fragmentary history. In *Nature's services: Societal dependence on natural ecosystems,* ed. Gretchen C. Daily, 11–19. Washington, DC: Island Press.

Moore, Grady C. 1999. Water wars: Interstate water allocation in the Southeast. *Natural Resources & Environment* 14(1): 5–10, 66–67.

Nagle, John Copeland, and J. B. Ruhl. 2006. *The law of biodiversity and ecosystem management.* 2nd ed. New York: Foundation Press.

Naiman, Robert J., and Kevin H. Rogers. 1997. Large animals and system-level characteristics in river corridors: Implication for river management. *BioScience* 47(8): 521–29.

National Research Council. 1986. *Soil conservation: Assessing the National Resources Inventory,* vol. 2. Washington, DC: National Academy Press.

———. 1994. *Nature's numbers: Expanding the national economic accounts to include the environment.* Washington, DC: National Academy Press.

———. 1999. *New strategies for America's watersheds.* Washington, DC: National Academy Press.

———. 2001. *Compensating for wetland losses under the Clean Water Act.* Washington, DC: National Academy Press.

————. 2004a. *Valuing ecosystem services.* Washington, DC: National Academy Press.

————. 2004b. *Endangered and threatened fishes in the Klamath River basin: Causes of decline and strategies for recovery.* Washington, DC: National Academy Press.

National Wildlife Federation. 2004. *The Bush administration's new strategy for undermining the Endangered Species Act.* Washington, DC: National Wildlife Federation.

Natural Resources Conservation Service. 1998. *Buffer strips: Common sense conservation.* USDA Natural Resources Conservation Service. Available at http://www.nhq.nrcs. usda.gov/CCS/Buffers.html

————. 2001. *1997 National Resources Inventory: A Guide for Users of 1997 NRI Data Files* (revised December 2000). Accessed at http://www.nrcs.usda.gov/technical/ NRI/1997/obtain_data.html on July 29, 2006.

————. 2004. NRCS This Week: Progress Toward Two-Million-Mile Buffer Goal. Accessed at http://www.nrcs.usda.gov/news/thisweek/2004/040310/febbuffers.html on July 15, 2006.

————. 2005. *Productive lands, healthy environment: Natural Resources Conservation Service Strategic Plan 2005–2010.* Washington, DC: Natural Resources Conservation Service, U.S. Dept. of Agriculture.

————. 2006. Buffer strips: Common sense conservation. NRCS Web access site for riparian buffer information and programs. Accessed at http://www.nrcs.usda.gov/ feature/buffers on July 9, 2006.

Neuman, Jan, Anne Squier, and Gail Achterman. 2006. Sometimes a great notion: Oregon's instream flow experiments. *Environmental Law* 36(4): 1125–55.

Nolan, John R. 2002. In praise of parochialism: The advent of local environmental law. *Harvard Environmental Law Review* 26(2): 365–416.

Office of Management and Budget. 2002. *Stimulating smarter regulation: 2001 report to Congress on the costs and benefits of regulations and unfunded mandates on state, local, and tribal entities.*

Oki, Taikan, and Shinjiro Kanae. 2006. Global hydrological cycles and world water resources. *Science* 313(5790): 1068–72.

Olschewski, Roland, Teja Tscharntke, Pablo C. Benitez, Stefan Schwarze, and Alexandra-Maria Klein. 2006. Economic evaluation of pollination services comparing coffee landscapes in Ecuador and Indonesia. *Ecology and Society* 11(1): 7, available at http://www.ecologyandsociety.org/vol11/iss1/art7.

Olson, Mancur. 1965. *The logic of collective action: Public goods and the theory of groups.* Cambridge, MA: Harvard University Press.

Olson, Richard K. 1992. Evaluating the role of created and natural wetlands in controlling nonpoint source pollution. *Ecological Engineering* 1(1): xi–xv.

Opschoor, J. B. 1998. The value of ecosystem services: Whose values? *Ecological Economics* 25(1): 41–43.

Organic-Europe 2005. Accessed at http://www.organic-europe.net/europe_eu/statistics.asp on Dec. 22, 2005.

Organization for Economic Cooperation and Development. 2001. *Multifunctionality: Toward an analytical framework.* Paris: Organization for Economic Cooperation and Development.

Orts, Eric W. 1995. Reflexive environmental law. *Northwestern University Law Review* 89(4): 1227–1340.

Osborn, C. Tim, Felix Llacuna, and Michael Linsenbigler. 1995. *The Conservation Reserve Program: Enrollment Statistics for signup Periods 1–12 and Fiscal Years*

1986–93. ERS Statistical Bul. No. 925. Washington, DC: Economic Research Service, U.S. Dept. of Agriculture.

Ostrom, Elinor. 1990. *Governing the commons: The evolution of institutions for collective action.* Cambridge, UK: Cambridge University Press.

———. 2000a. Private and common property rights. In *Encyclopedia of law and economics,* vol. II, *Civil law and economics,* ed. Boudewejn Bouckaert and Gerritt DeGeest, 274–314. Cheltenham, UK: Edward Elgar.

———. 2000b. Collective action and the evolution of social norms. *Journal of Economic Perspectives* 14(3): 137–58.

Ostrom, Elinor, Roy Gardner, and James Walker. 1994. *Rules, games, and common-pool resources.* Ann Arbor: University of Michigan Press.

Ostrom, Elinor, Joanna Burger, Christopher F. Field, Richard B. Norgaard, and David Policansky. 1999. Revisiting the commons: Local lessons, global challenges. *Science* 284(5412): 278–82.

Palmer, Margaret, Emily Bernhardt, Elizabeth Chornesky, Scoll Collins, Andrew Dobson, Clifford Duke, Barry Gold, Robert Jacobson, Sharon Kingsland, Rhonda Krantz, Michael Mappin, M. Louisa Martinez, Florenza Micheli, Jennifer Morse, Michael Pace, Mercedes Pascual, Stephen Palumbi, O. J. Reichman, Ashley Simons, Alan Townsend, and Monica Turner. 2004. Ecology for a crowded planet. *Science* 304(5675): 1251–52.

Parkhurst, Gregory M., and Jason F. Shogren. 2003. Evaluating incentive mechanisms for conserving habitat. *Natural Resources Journal* 43(4): 1093–1149.

Parobek, Cori S. 2003. Of farmers' takes and fishes' takings: Fifth Amendment compensation claims when the Endangered Species Act and western water rights collide. *Harvard Environmental Law Review* 27(1): 177–226.

Pate, Jennifer, and John Loomis. 1997. The effect of distance on willingness to pay values: A case study of wetlands and salmon in California. *Ecological Economics* 20(3): 199–207.

Pearce, David. 1998. Auditing the earth: The value of the world's ecosystem services and natural capital. *Environment* 40(2): 23–28.

Pearce, David, Anil Markandya, and Edward Barbier. 1989. *Blueprint for a green economy.* London: Earthscan Publications.

Percival, Robert, Christopher H. Schroeder, Alan S. Miller, and James P. Leape. 2003. *Environmental regulation.* 4th ed. New York: Aspen Publishers.

Peterjohn, William T., and David L. Correll. 1984. Nutrient dynamics in an agricultural watershed: Observation on the role of a riparian forest. *Ecology* 65(5): 1466–75.

Pickering, Kevin T., and Lewis A. Owen. 1994. *An introduction to global environmental issues.* London, UK: Routledge.

Pierce, Richard J., Jr., Sidney A. Shapiro, and Paul R. Verkuil. 1999. *Administrative law and process.* 3rd ed. New York: Foundation Press.

Pifher, Mark T. 2005. The Section 404(b)(1) guidelines and practicable alternatives analysis. In *Wetlands law and policy: Understanding Section 404,* ed. Kim Dana Connolly, Stephen M. Johnson, and Douglas R. Williams, 221–52. Chicago: American Bar Association Publishing.

Pimm, Stuart L. 1984. The complexity and stability of ecosystems. *Nature* 307: 321–26.

Platts, W. S. 1981. Sheep and streams. *Rangeland* 3(4): 158–60.

Portela, Rosimeiry, and Ida Rademacher. 2001. A dynamic model of patterns of deforestation and their effect on the ability of the Brazilian Amazonia to provide ecosystem services. *Ecological Modeling* 143(1–2): 115–46.

Posner, Richard A. 1979. Some uses and abuses of economics in law. *University of Chicago Law Review* 46(2): 281–315.

———. 1998. *Economic analysis of law.* Boston: Little, Brown.

Postel, Sandra, and Stephen Carpenter. 1997. Freshwater ecosystem services. In *Nature's Services,* ed. Gretchen Daily, 195–214. Washington, DC: Island Press.

Pretty, Jules. 2003. Social capital and the collective management of resources. *Science* 302(5652): 1912–14.

Pretty, J. N., C. Brett, D. Gee, R. E. Hine, C. F. Mason, J. I. L. Morison, H. Raven, M. D. Rayment, and G. Van der Blij. 2000. An assessment of the total external costs of UK agriculture. *Agricultural Systems* 65 (2000): 1134–36.

Proctor, Theresa B. 2004. Erosion of riparian rights along Florida's coast. *Journal of Land Use & Environmental Law* 20(1): 117–57.

Prosser, William L., W. Page Keeton, Dan B. Dobbs, Robert E. Keeton, and David G. Owen. 1984. *Prosser and Keeton's handbook of torts.* St. Paul, MN: West Publishing Co.

Purdy, Jedediah. 2006. The American transformation of waste doctrine: A pluralist interpretation. *Cornell Law Review* 91(3): 653–98.

Putnam, Robert. 2000. *Bowling alone.* New York: Simon & Schuster.

Randall, Alan. 1983. The problem of market failure. *Natural Resources Journal* 23(1): 133–48.

Rasband, James M. 2001. The rise of urban archipelagoes in the American West: A new reservation policy? *Environmental Law* 31(1): 1–93.

Raymond, Leigh. 2003. *Private rights in public resources: Equity and poperty allocation in market-based environmental policy* Washington, DC: Resources for the Future.

Repetto, Robert. 1992. Earth in the balance sheet. *Environment* 34(7): 12–20, 43–45.

Repetto, Robert, William Magrath, Michael Wells, Christine Beer, and Fabrizio Rossini. 1989. *Wasting assets: Natural resources in the national income accounts.* New York: World Resource Institute.

Revesz, Richard L. 2001. Federalism and environmental regulation: A public choice analysis. *Harvard Law Review* 115(2): 553–641.

Ribaudo, M. 1989. *Water quality benefits from the Conservation Reserve Program,* AER-606. Washington, DC: Economic Research Service, U.S. Dept. of Agriculture.

Ricketts, Taylor H. 2004. Tropical forest fragments enhance pollinator activity in nearby coffee crops. *Conservation Biology* 18(5): 1262–71.

Rifkin, Jeremy, 2005. *The European Dream: How Europe's vision of the future is quietly eclipsing the American Dream.* New York: Penguin.

Ritchie, Bruce. 2001a. Atlanta needs water to grow on. *Tallahassee Democrat,* Nov. 5.

———. 2001b. Less water produces less power. *Tallahassee Democrat,* Nov. 6.

———. 2001c. Does shipping have a future? *Tallahassee Democrat,* Nov. 7.

———. 2001d. Lake's low level has residents looking downstream. *Tallahassee Democrat,* Nov. 9.

Roberts, Callum M. 1995. Effects of fishing on the ecosystem structure of coral reefs. *Conservation Biology* 9(5): 988–95.

Robertson, Morgan M. 2006. Emerging ecosystem service markets: Trends in a decade of entrepreneurial wetland banking. *Frontiers in Ecology and the Environment* 4(6): 297–302.

Rodgers, William H., Jr. 1994. *Environmental law.* 2nd ed. St. Paul, MN: West Publishing.

Rodriguez, Jon Paul, T. Douglas Beard, Jr., Elena M. Bennett, Graeme S. Cumming, Steven J. Cork, John Agard, Andrew P. Dobson, and Garry D. Petersen. 2006. Trade-offs across space, time, and ecosystem services. *Ecology and Society* 11(1): 28, available at http://www.ecologyandsociety.org/vol11/iss1/art28.

Roodman, David Malin. 1998. *The natural wealth of nations: Harnessing the market for the environment.* New York: W. W. Norton.

Roos-Collins, Richard. 2005. A plan to restore the public trust uses of rivers and creeks. *Texas Law Review* 83(7): 1929–40.

Rose, Carol M. 1990a. Property as storytelling: Perspectives from game theory, narrative theory, feminist theory. *Yale Journal of Law & Humanities* 2(1): 37–57.

———. 1990b. Energy and efficiency in the realignment of common-law water rights. *Journal of Legal Studies* 19(2): 261–96.

———. 1998. The several futures of property: Of cyberspace and folk tales, emission trades and ecosystems. *Minnesota Law Review* 83(1): 129–82.

———. 2002. Common property, regulatory property, and environmental protection: Comparing community-based management to tradable environmental allowances. In *The drama of the commons,* ed. Elinor Ostrom, Thomas Dietz, Nives Dolsak, Paul C. Stern, Susan Stonich, and Elke U. Weber, 233–57. Washington, DC: National Academy Press.

Rossi, Jim. 1997. Participation run amok: The costs of mass participation for deliberative agency decision making. *Northwestern University Law Review* 92(1): 173–250.

Rossmann, Antonio. 2003. A new law and the "era of limits" on the Colorado. *Natural Resources & Environment* 18(2): 3–7.

Ruhl, J. B. 1995. Biodiversity conservation and the ever-expanding web of federal laws regulating nonfederal lands: Time for something completely different? *University of Colorado Law Review* 66(3): 559–673.

———. 1998. Valuing nature's services—The future of environmental law? *Natural Resources & Environment* 13(4): 359–60.

———. 1999. The (political) science of watershed management in the Ecosystem Age. *Journal of the American Water Resources Association* 35(3): 519–26.

———. 2000. Farms, their environmental harms, and environmental law. *Ecology Law Quarterly* 27(2): 263–349.

———. 2003. Equitable apportionment of ecosystem services: A new water law for a new water age. *Journal of Land Use & Environmental Law* 19(1): 47–57.

———. 2004. Endangered Species Act Innovations in the post-Babbittonian era—Are there any? *Duke Environmental Law & Policy Forum* 14(2): 419–39.

———. 2005a. Ecosystem services and the common law of the "fragile land system." *Natural Resources & Environment* 20(2): 3–9, 69.

———. 2005b. Cooperative federalism and the Endangered Species Act—Is there hope for something more? In *Strategies for environmental success,* ed. Michael A. Wolf, 325–37. Washington, DC: Environmental Law Institute.

Ruhl, J. B., and R. Juge Gregg. 2001. Integrating ecosystem services into environmental law: A case study of wetlands mitigation banking. *Stanford Environmental Law Journal* 20(2): 365–92.

Ruhl, J. B., and James Salzman. 2003. Mozart and the Red Queen: The problem of regulatory accretion in the administrative state. *Georgetown Law Journal* 91(4): 757–850.

———. 2006. The effects of wetland mitigation banking on people. *National Wetlands Newsletter* 28(2): 1, 9–14.

Ruhl, J. B., Christopher Lant, Timothy Loftus, Steven Kraft, Jane Adams, and Leslie Durham. 2003. Proposal for a model state watershed management act. *Environmental Law* 33(4): 929–47.

Ryan, Erin. 2001. Pubic trust and distrust: The theoretical implications of the public trust doctrine for natural resources management. *Environmental Law* 31(2): 477–96.

Sagoff, M. 1988. *The economy of the earth: Philosophy, law, and the environment.* New York: Cambridge University Press.

Salamon, Lester M. 2002. *The tools of government: A guide to the new governance.* Oxford, UK: Oxford University Press.

Salzman, James. 1997. Valuing ecosystem services. *Ecology Law Quarterly* 24(4): 887–903.

———. 2005. Creating markets for ecosystem services: Notes from the field. *New York University Law Review* 80(3): 870–961.

———. 2006. A field of green? The past and future of ecosystem services. *Journal of Land Use and Environmental Law* 21(2): 133–51.

Salzman, James, and J. B. Ruhl. 2000. Currencies and the commodification of environmental law. *Stanford Law Review* 53(3): 607–94.

Sauer, Carl O. 1925. The Morphology of landscape. *University of California Publications in Geography* 2:19–54.

Savage, Melissa, Bruce Sawhill, and Manor Askenazi. 2000. Community dynamics: What happens when we rerun the tape? *Journal of Theoretical Biology* 205(4): 515–26.

Sax, Joseph L. 1970. The public trust doctrine in natural resource law: Effective judicial intervention. *Michigan Law Review* 68(3): 471–567.

Schnepf, Max. 2006. Personal Communication from Max Schnepf, former national coordinator of the National Conservation Buffer Initiative, by e-mail communication, June 10, 2006.

Schroeder, Christopher H. 2000. Third way environmentalism. *University of Kansas Law Review* 48(4): 801–28.

———. 2005. Federalism's values in programs to protect the environment. In *Strategies for environmental success,* ed. Michael A. Wolf, 247–58. Washington, DC: Environmental Law Institute.

Scodari, Paul. 1997. *Measuring the benefits of federal wetland programs.* Washington, DC: Environmental Law Institute.

Scott, Geoffrey R. 1998. The expanding public trust doctrine: A warning to environmentalists and policy makers. *Fordham Environmental Law Journal* 10(1): 1–70.

Seidenfeld, Mark. 2000. Empowering stakeholders: Limits on collaboration as the basis for flexible regulation. *William and Mary Law Review* 41(2): 411–501.

Semlitsch, Raymond D. 1998. Biological delineation of terrestrial buffer zones for pond breeding amphibians. *Conservation Biology* 12(5): 1113–19.

———. 2000. Size does matter: The value of small isolated wetlands. *National Wetlands Newsletter* 22(1): 5–6.

Shapiro, Sidney A., and Robert L. Glicksman. 2000. Goals, instruments, and environmental policy choice. *Duke Environmental Law and Policy Forum* 10(2): 297–325.

Sibbing, Julie M. 2005. Mitigation banking: Will the myth ever die? *National Wetlands Newsletter* 27(6): 5–6, 8.

Simon, Herbert A. 1957. *Models of man: Social and rational.* New York: Wiley and Sons.

Simpson, R. David. 1998. Economic analysis and ecosystems: Some concepts and issues. *Ecological Applications* 8(2): 342–49.

Smith, Henry E. 2004. Property and property rules. *New York University Law Review* 79(5): 1719–98.

Smith, Katherine R. 2006. Public payments for environmental services from agriculture: Precedents and possibilities. *American Journal of Agricultural Economics* 88(5): 1167–73.

Society of Wetland Scientists. 2005. Wetland mitigation banking: Clarifying intent. *National Wetlands Newsletter* 27(5): 5–6.

Southwick, E. E., and L. Southwick, Jr. 1992. Estimating the economic value of honeybees (Hymenoptera apidae) as agricultural pollinators in the United States. *Journal of Economic Entomology* 85(3): 621–33.

Speth, James Gustave. 2004. *Red sky at morning: America and the crisis of the global environment.* New Haven, CT: Yale University Press.

Sprankling, John G. 1996. The antiwilderness bias in American property law. *University of Chicago Law Review* 63(2): 519–90.

Stanford Law Review. 1948–49. Who owns the clouds? *Stanford Law Review* 1(1): 43–63.

Stearns, F. W. 1997. History of the Lake States Forests: Natural and human impacts. In *Lake States regional forest resources assessment: Technical papers,* Gen. Tech. Rep. NC-189, ed. J. Vasievich, M. Webster, and H. Henry, 8–29. U.S. Department of Agriculture, Forest Service, North Central Forest Experiment Station, St. Paul, Minnesota, USA.

Steer, David, Todd Aseltyne, and Lauchlan Fraser. 2003. Life-cycle economic model of small treatment wetlands for domestic wastewater disposal. *Ecological Economics* 44(2–3): 359–69.

Stephenson, Dustin S. 2000. The tri-state compact: Falling waters and failing opportunities. *Journal of Land Use and Environmental Law* 16(1): 83–109.

Stewart, Richard B. 2001. A new generation of environmental regulation? *Capital University Law Review* 29(1): 21–182.

———. 2003. Administrative law in the twenty-first century. *New York University Law Review* 78(2): 437–60.

Stroup, Richard L., and John A. Baden. 1983. *Natural resources: Bureaucratic myths and environmental management.* San Francisco: Pacific Institute for Public Policy Research.

Stroup, Richard L., and Sandra L. Goodman. 1992. Property rights, environmental resources, and the future. *Harvard Journal of Law and Public Policy* 15(2): 427–41.

Subak, Susan. 2000. Agricultural soil carbon accumulation in North America: Considerations for climate policy. *Global Environmental Change* 10: 185–95.

Sugameli, Glenn. 1999. *Lucas v. South Carolina Coastal Council:* The categorical and other "exceptions" to liability for fifth amendment takings of private property far outweigh the "rule." *Environmental Law* 29(4): 939–88.

Swanson, Timothy, and Andreas Kontoleon. 2000. Nuisance. In *Encyclopedia of law and economics,* vol. II, *Civil law and economics,* ed. Boudewijn Bouckaert and Gerritt DeGeest, 380–402. Cheltenham, UK: Edward Elgar.

Swinton, Scott M., Frank Lupi, Philip Robertson, and Douglas A. Landis. 2006.

Ecosystem services from agriculture: Looking beyond the usual suspects. *American Journal of Agricultural Economics* 88(5): 1160–66.

Tansley, Arthur G. 1935. The use and abuse of vegetational concepts and terms. *Ecology* 16(3): 284–307.

Tarlock, A. Dan. 1985. The law of equitable apportionment revisited, updated, and restated. *University of Colorado Law Review* 56(3): 381–411.

———. 1991. New water transfer restrictions: The West turns to riparianism. *Water Resources Research* 27(6): 987–94.

———. 2000a. Prior appropriation: Rule, principle, or rhetoric? *North Dakota Law Review* 76(4): 881–910.

———. 2000b. Reconnecting property rights to watersheds. *William & Mary Environmental Law & Policy Review* 25(1): 69–112.

———. 2001. The future of prior appropriation in the new West. *Natural Resources Journal* 41(4): 769–86.

———. 2002. The potential role of local governments in watershed management. *Pace Environmental Law Review* 20(1): 149–76.

Thompson, Barton H. 1996. The search for regulatory alternatives. *Stanford Environmental Law Journal* 15(2): viii–xxi.

———. 2000. Markets for nature. *William & Mary Environmental Law & Policy Review* 25(2): 261–316.

Thomson, Rebecca W. 1995. "Ecosystem management"—Great idea, but what is it, will it work, and who will pay? *Natural Resources & Environment* 9(3): 42–45.

Thornley, Andrew. 2005. A tale of two river basins: The Southeast finds itself in a rare interstate water struggle. *University of Denver Water Law Review* 9(1): 97–119.

Tiessen, H., C. Feller, E.V.S.B. Sampaio, and P. Gavin. 1998. Carbon sequestration and turnover in semiarid savannas and dry forest. *Climatic Change* 40(1): 105–17.

Tietenberg, Tom. 2005. *Environmental and natural resources economics.* 7th ed. Boston: Addison Wesley Longman.

———. 2006. Tradable permits in principle and practice. *Penn State Environmental Law Review* 14(2): 251–81.

Tilman, David. 1999. The ecological consequences of changes in biodiversity: A search for general principles. *Ecology* 80(5): 1455–74.

Tilman, David, Kenneth G. Cassman, Pamela A. Matson, Rosamond Naylor, and Stephen Polasky. 2002. Agricultural sustainability and intensive production practices. *Nature* 418(6898): 671–77.

Toman, Michael. 1998. Why not to calculate the value of the world's ecosystem services and natural capital. *Ecological Economics* 25(1): 57–60.

Trowbridge, Gordon. 2001. Population shift strains shoreline—Traffic, crowds follow urban refugees to lakes. *Detroit News and Free Press,* April 22.

Turner, B. L., II. 2002. Contested identities: Human-environment geography and its disciplinary implications in a restructuring academy. *Annals of the Association of American Geographers* 92(1): 52–74.

Turner, R. Eugene, Ann M. Redmond, and Joy B. Zedler. 2001. Count it by acre or function—Mitigation adds up to net losses of wetlands. *National Wetlands Newsletter* 23(6): 5–6, 14–16.

United States Army Corps of Engineers. 1998. Draft environmental impact statement—main report: Water allocation for the Apalachicola–Chattahoochee–Flint (ACF) River Basin.

United States Department of Agriculture. 2000. Crop Production November 2000. Washington, DC: Agricultural Statistics Board, U.S. Department of Agriculture.

United States Environmental Protection Agency. 1998. Conservation buffers to be established nationwide—USDA and agribusinesses work together. *Nonpoint Source New Notes* no. 50. Accessed at http://notes.tetratech-ffx.com/ newsnotes.nsf/ 606a2768c7ff5f63852565ff0061ae0d/ab033bc2192464d18525667c0066b7d8? OpenDocument, last updated 22 April 02, on July 20, 2006.

———. 2004. Constructed treatment wetlands. Available at: http://www.epa.gov/ owow/wetlands/pdf/ConstructedW_pr.pdf accessed on December 5. 2006.

———. 2006. Ecological benefits assessment strategy plan, Washington, DC; U.S. Environmental Protection Agency, available online at http://www.epa.gov/economics.

United States Fish and Wildlife Service, 1990. *Cypress Creek National Wildlife Refuge: Final environmental assessment.* Twin Cities, MN: U.S. Fish and Wildlife Service.

———. 2003. Designation of critical habitat for the Blackburn's Sphinx moth. *Federal Register* 68(111): 34710–66.

United States Geological Survey. 2006. Hydrologic unit maps. Available at http://water.usgs.gov/GIS/huc.html.

United States Government Accountability Office. 2005. Wetlands protection: Corps of Engineers does not have an effective oversight approach to ensure that compensatory mitigation is occurring.

United States Government Accounting Office. 1994. *Ecosystem management: Additional actions to adequately test a promising approach.*

Van Wilgen, Brian W., Richard M. Cowling, and Chris J. Burgers. 1996. Valuation of ecosystem services. *BioScience* 46(3): 184–89.

Vandenberg, Michael P. 2005. Order without social norms: How personal norm activation can protect the environment. *Northwestern University Law Review* 99(3): 1101–66.

Veltman, Virginia C. 1995. Banking on the future of wetlands using federal law. *Northwestern University Law Review* 89(2): 654–89.

Verry, Elon S., James W. Hornbeck, and C. Andrew Dolloff, eds. 2000. *Riparian management in forests of continental eastern United States.* Boca Raton, FL: Lewis Publishers.

Villa, Ferdinando, Matthew A. Wilson, Rudolf de Groot, Stephen Farber, Robert Costanza, and Roelof M. J. Boumans. 2002. Designing an integrated knowledge base to support ecosystems services valuation. *Ecological Economics* 41(3): 445–56.

Virginia, Ross A., and Diana H. Wall. 2001. Ecosystem function, principles of. In *Encyclopedia of Biodiversity,* vol. 2, ed. Simon Levin, 345–52. San Diego: Academic Press.

Vitousek, Peter M., Harold A. Mooney, Jane Lubchenco, and Jerry M. Melillo. 1997. Human domination of Earth's ecosystems. *Science* 277(5325): 494–99.

Von Oppenfield, Rolf R. 2005. State roles in the implementation of the Section 404 program. In *Wetlands law and policy: Understanding Section 404,* ed. Kim Dana Connolly, Stephen M. Johnson, and Douglas R. Williams, 321–56. Chicago: American Bar Association Publishing.

Wainger, Lisa A., Dennis King, James Boyd, and James Salzman. 2001. Wetland value indicators for scoring mitigation trades. *Stanford Environmental Law Journal* 20(2): 413–78.

Walker, Brian, Lance Gunderson, Ann Kinzig, Carl Folke, Steve Carpenter, and Lisen Schultz. 2006. A handful of heuristics and some propositions for understanding resilience in social-ecological systems. *Ecology and Society* 11(1): 13. Available at http://www.ecologyandsociety.org/vol11/iss1/art13.

Wardle, David A., Richard D. Bardgett, John N. Klironomos, Heikki Setala, Wim H. van der Putten, and Diana H. Wall. 2004. Ecological linkages between aboveground and belowground biota. *Science* 304(5677): 1629–33.

Waters, Thomas. 1995. *Sediments in streams: Sources, biological effects, and control.* Bethesda, MD: American Fisheries Society.

Wegner, Seth. 1999. *A review of the scientific literature on riparian buffers width, extent and vegetation.* Athens, GA: Institute of Ecology, University of Georgia.

Wegner, Seth, and L. Fowler. 2000. *Protecting stream and river corridors: Creating effective local riparian buffer ordinances.* Public Policy Research Series. Athens, GA: Carl Vinson Institute of Government, University of Georgia.

Welsch, D. J. 1991. *Riparian forest buffers.* U.S. Department of Agriculture–Forest Service Publication Number NA-PR-07-91. Radnor, Pennsylvania.

Westman, Walter E. 1977. How much are nature's services worth? *Science* 197(4307): 960–64.

Wiebe, Keith, and Noel Gollehon, eds. 2006. *Agricultural resources and environmental indicators, 2006 edition.* Econ. Info. Bull. 16. Washington, DC: Economic Research Service, U.S. Dept. of Agriculture.

Wilkinson, Jessica, and Jared Thompson. 2006. *2005 status report on compensatory mitigation in the United States.* Washington, DC: Environmental Law Institute.

Williams, Douglas R., and Kim Dana Connolly. 2005. Federal wetlands regulation: An overview. In *Wetlands law and policy: Understanding Section 404,* ed. Kim Dana Connolly, Stephen M. Johnson, and Douglas R. Williams, 1–26. Chicago: American Bar Association Publishing.

Williams, Garnett P. 1986. Rivers meanders and channel size. *Journal of Hydrology* 88(1–2): 147–64.

Williams, Jeffrey R., Jeffrey M. Peterson, and Siân Monney. 2005. The value of carbon credits: Is there a final answer? *Journal of Soil and Water Conservation* 60(2): 36A–40A.

Williamson, Alex. 2005. Seeing the forest and the trees: The natural capital approach to Forest Service reform. *Tulane Law Review* 80(2): 683–711.

Williamson, Oliver E. 1979. Transaction-cost economics: The governance of contractual relationships. *Journal of Law and Economics* 22(2): 233–61.

Wilson, Edward O. 1992. *The diversity of life.* New York: W. W. Norton.

Wilson, James. 2001. Scientific uncertainty, complex systems, and the design of common-pool institutions. In *The drama of the commons,* ed. Elinor Ostrom, Thomas Dietz, Nives Dolsak, Paul C. Stern, Susan Stonich, and Elke U. Weber, 327–59. Washington, DC: National Academy Press.

Wilson, Jessica, and Jared Thompson. 2006. *Status report on compensatory mitigation in the United States.* Washington, DC: Environmental Law Institute.

Wilson, Matthew A., and Stephen R. Carpenter. 1999. Economic valuation of freshwater ecosystem services in the United States: 1971–1997. *Ecological Applications* 9(3): 772–83.

Wilson, Matthew A., Robert Costanza, Roelof Boumans, and Shuang Liu. 2005. Integrated assessment and valuation of ecosystem goods and services provided by coastal systems. In *The intertidal ecosystem: The value of Ireland's shores,* ed. James G. Wilson, 1–24. Dublin: Royal Irish Academy Press.

Woodall, S. L. 1985. Influence of land treatments on temperature variations of groundwater effluence to streams. PhD diss., University of Georgia, Athens, GA.

World Resources Institute. 2000. *World resources 2000–2001: People and ecosystems: The fraying web of life.* Washington, DC: World Resources Institute.

Wyman, Katrina M. 2005. From fur to fish: Reconsidering the evolution of private property. *New York University Law Review* 80(1): 117–240.

Yandle, Bruce, and Andrew P. Morriss. 2001. The technologies of property rights: Choice among alternative solutions to tragedies of the commons. *Ecology Law Quarterly* 28(1): 123–68.

Yang, Wanhong, Madhu Khanna, Richard Farnsworth, and Hayri Onal. 2005. Is geographical targeting cost-effective? The case of the Conservation Reserve Enhancement Program in Illinois. *Review of Agricultural Economics* 27(1): 70–88.

Young, I. M., and J. W. Crawford. 2004. Interactions and self-organization in the soil-microbe complex. *Science* 304(5677): 1634–37.

Young, R. A., C. A. Onstad, D. D. Bosch, and W. P. Anderson. 1989. AGNPS: A nonpoint-source pollution model for evaluating agriculture watersheds. *Journal of Soil and Water Conservation* 44(2): 168–73.

Zerbe, Richard O., Jr., and Howard E. McCurdy. 1999. The failure of market failure. *Journal of Policy Analysis and Management* 18(4): 558–78.

Zimmerman, Erich, 1951. *World Resources and Industries.* New York: Harper & Row.

Zinn, J. 1994. *Conservation Reserve Program: Policy issues for the 1995 Farm Bill.* CRS Report to Congress, 95-8 EVR. Washington, DC: Congressional Research Service, Library of Congress.

Zywicki, Todd J. 2002. Baptists? The political economy of environmental interest groups. *Case Western Reserve Law Review* 53(2): 315–51.

About the Authors

J. B. Ruhl is Matthews & Hawkins Professor of Property at the Florida State University College of Law and lives in Tallahassee, Florida, with his wife, Lisa LeMaster, and sons Grant and Grayson. He teaches and writes in the areas of ecosystem management and endangered species law. He holds a B.A. in Economics and a J.D. from the University of Virginia, an LL.M. in Environmental Law from George Washington University, and a Ph.D. in Geography from Southern Illinois University–Carbondale.

Steven E. Kraft is Professor and Chair of the Department of Agribusiness Economics and co-director of the Environmental Resources and Policy Ph.D. program at Southern Illinois University–Carbondale, where he teaches environmental resource economics and policy and farm management. He lives in Carbondale, Illinois, with his wife, Carol A. Burns, and their children, Tristan and Keya. He conducts research on soil and water conservation policy and watershed planning and is a Fellow of the Soil and Water Conservation Society. He holds a B.S. in Economics and International Relations from American University and an M.S. and a Ph.D. in Agricultural Economics from Cornell University.

Christopher L. Lant is Professor of Geography and Environmental Resources and co-director of the Environmental Resources and Policy Ph.D. program at Southern Illinois University–Carbondale, as well as executive director of the Universities Council on Water Resources. He teaches and writes in the area of water resources management and environmental policy. He lives in Carbondale with his wife, Sandra Charlson, and daughters Hannah and Helen. He holds an M.A. and a Ph.D. in Geography from the University of Iowa.

Index